RENEWALS 458-4574
DATE DUE

WITHDRAWN
UTSA Libraries

THE UNIFICATION AND DIVISION OF INDIA

B. B. MISRA

DELHI
OXFORD UNIVERSITY PRESS
BOMBAY CALCUTTA MADRAS
1990

Oxford University Press, Walton Street, Oxford OX2 6DP
New York Toronto
Delhi Bombay Calcutta Madras Karachi
Petaling Jaya Singapore Hong Kong Tokyo
Nairobi Dar es Salaam
Melbourne Auckland
and associates in
Berlin Ibadan

© Oxford University Press 1990

SBN 0 19 562615 X

Phototypeset by Spantech Publishers Pvt Ltd, New Delhi 110060
Printed by Rekha Printers (P) Ltd., New Delhi 110020
and published by S. K. Mookerjee, Oxford University Press
YMCA Library Building, Jai Singh Road, New Delhi 110001

CONTENTS

Introduction	v
The Instruments of Territorial Integration	1
The Delimitation of India's Frontiers	68
Territorial Reorganization for Security and Development	116
Institutional Agencies and Political Stability	183
Devisive Tendencies in a Plural Society	268
The Politics of Power and Party Rule 1937–47	316
NOTES	377
SELECT BIBLIOGRAPHY	397
INDEX	403

INTRODUCTION

THOUGH a unified geographical category equipped with national frontiers, the Indian subcontinent has hardly ever been a single, integrated political entity, except in modern times, during and following British rule. Sir John Strachey, who had a distinguished career in the Indian Civil Service, was not wrong when, in an address at Cambridge in 1884, he remarked that there 'never was an India'[1] of the kind that emerged under the British in the nineteenth century. His general observation about the absence of Indian nationalist sentiments was, of course, not unbiased; for although immediately centred around questions of employment in the public services, Indian nationalism had by that period begun to seriously organize a common forum for the presentation of demands of national importance. But Strachey's comment, that the kind of India forged by the British had earlier 'never' existed, could not be disputed. There are only a few, rare instances of all-India empires based on territorial expansion. But even then, there was no territorial integration signifying unity of political control and administrative cohesion.

The natural boundaries which the physical features of India provide, have constituted a source of strength in terms of geographical unity. At the same time, however, the compactness of the region's geographical conditions made it inward-looking and largely isolated from the rest of the world, despite a series of foreign invasions. People remained politically unmindful of what was happening outside the region, or even in its immediate frontier. All-India empires, which could otherwise engender an all-India territorial concept were, in addition, a rarity. And even if there was anything near an all-India empire, like that of Chandragupta Maurya in the fourth century B.C. or of Samudragupta a thousand years later, it proved to be short-lived, and an exception rather than the general rule. Besides, the political history of India has been affected by a series of foreign invasions, which seriously interfered with the process of political integration in the country. From very early times, people of various races, climes and cultures had thus been entering India and adding to the variety of its ethnographical groups, its religions and languages, its customs and usages. Despite India's geographical unity, these extraneous developments clouded, in the absence of an all-India polity, the prospects of its political unity.

Though weak and generally fairly undeveloped, communication within the country, an essential condition for the growth of an integrative process, was not altogether absent. Pilgrims travelled great distances to visit shrines distributed throughout the country. Adi Shankaracharya had, for example, established during his lifetime four historic temples at Sringeri in the south, Badrinath in the north, Dwarka in the west and Puri in the east. These were sanctified as holy places for Hindus and encouraged through travel a spirit of cultural cross-fertilization. Large periodical assemblages of people on sacred river banks also perhaps served similar objectives. But the motive force which impelled people to action in all these performances remained religion, not the integration of the country into a nation, a territorial concept. Even foreigners were accepted as rulers if they merely recognized the sanctity of religious places and did not interfere with the performance of religious rites. There was some resistance to foreigners, in the medieval period when religion became a focal point of attack by Islamic rulers. But in the absence of a territorially integrative force to provide political backing for religious freedom, this resistance was not very strong.

Territorial integration in modern times was the first step towards the political unity that India attained at a national level. But this integration was accompanied by a certain degree of homogeneity in the making of society, a homogeneity, which, in terms of India's historical development and its social organization, was found completely absent even as late as the closing decades of the nineteenth century. There was, no doubt, a semblance of unity among educated Indian elites in their seeking a share in power. But it did not have deep roots in a society which, in the words of Lord Dufferin, consisted of a 'congeries' of mutually exclusive elements, professing various religions, practising various rites, speaking different languages, and separated from one another by discordant prejudices, by conflicting source of usage, and antagonistic material interests, the most patent characteristic of the Indian cosmos being its division between the Hindus, with 'their elaborate caste distinctions', their ploytheistic beliefs, and 'their habits of submission to successive conquerors', and the Muslims with their monotheism, 'social equality' and their 'iconoclastic fanaticism'. These conditions were highly prejudicial to the emergence of India as a nation, although by 1858 the British had brought the whole country under a unified control, a control which imparted political reality to its geographical unity. The situation was far worse in pre-British days, when even the minimum condition of a single all-India political authority did not exist.

A study of India's political history lends itself to the inevitable conclusion that, within the broad limits of its natural frontiers, there existed throughout several times a serious conflict of interest between a Centre reflecting the cultural dominance of the ruling potentates, often aliens or invaders, and its various regions, each bounded by certain geographical barriers and distinguished by its own characteristic cultural norms and linguistic forms. While the former attempted to introduce institutional uniformity as an instrument of administrative control over its provinces, the latter remained in search of an opportunity to declare independence and reject any central right to interference with regional affairs. The existence of mutually exclusive groups among regional interests was, of course, always present. But opportunities permitting, they often coalesced to oppose any attempt at centralization. The failure of Muhammad Tughlak to establish his capital in the Deccan is a good illustration of the manner in which both Muslim and Hindu chiefs combined to frustrate the designs of Delhi to subdue the southern kingdoms. The failure of Aurangzeb to bring these kingdoms under his thumb is another example of a Muslim-Maratha combination frustrating Delhi; in fact, Delhi could not stabilize its rule even in such provinces as Bengal and Gujarat, Malwa and Khandesh. The entire system of communication was, of course, too weak to permit any effective control from the centre. But even the subsequent development of communications did not reduce the adherence to regional interests. The perennial clash between the central government, which generally kept on changing its complexion over the centuries, and the provincial authorities, which traditionally commanded regional loyalties, constituted a regular threat to India's political unity and stability.

The weakness of the central authority of the Mughals even during the height of their glory was highlighted especially by the servants of the East India Company who, in their early days in Bengal under the Emperor Shah Jahan, operated under the delusion that the Mughal Empire was so compactly organized that the written word of the Emperor would be regarded as law by all his subordinates. However, they soon discovered that the Emperor's written orders had no value unless the Company was at 'peace with the Governor'.[2] There was no fixity of norms in the conduct of public business, no limitation on the exercise of discretion. Even after the Company's servants had provided themselves with a fortress in Bengal and the means of offering retaliation, the concessions accorded to them in 1717 proved altogether nugatory; for whatever the Mughal Emperor might choose to order, his Governor at Murshidabad would not allow the Company to realize.[3] It was not till 1765 that the military supremacy of the English in Bengal enabled Clive to win, with the collaboration of its Nawab, the irrevocable grant of Diwani from the Mughal Emperor, Shah Alam.

The general trend of Mughal governors was, in fact, to identify themselves with regional interests, by an alliance especially with such dominant elements of ruling groups as might contribute to their independence of Delhi, a development which was encouraged by the absence of fixity in the tenure of governorships and the competing claims of the royal heirs who sought the help of provincial potentates to ensure success in their own projected contests for the throne.

The character of Mughal government, especially in a distant province such as Bengal, was predominantly that of military occupation. While district officers had the means of providing for their own remuneration by a grant of land revenue in return for military service, the troops employed even during the best days of the Mughals were left in arrears of pay and dependent on the exactions they could wring from the terror-stricken cultivators. In 1757, Clive referred to the practice and custom in the country of holding back a quarter of the army's pay;[4] for it was believed that a soldier with considerable outstandings of pay would be less likely to desert than one who had received everything. The garrison troops maintained for security purposes were, however, given small allotments of land in lieu of cash payment.[5] The Mughal infantry was in fact more a rabble of half-armed men than a body of trained and disciplined soldiers. It consisted chiefly of levies brought to the battlegrounds by zamindars or tribal leaders to meet the requirements of specific situations,[6] a traditional and virtually private source of militia supplied to the Mughal army independently of whether it was imperial or provincial. This want of central control over military organization and the supply of armed resources was a serious weakness of Mughal policy. It strengthened the forces of regionalism. The Subahdar, who under the Mughals was not infrequently invested with the powers of both financial and executive administration, used his position to establish a semi-autonomy; and independent governorships emerged despite a division of the Subahdar's functions between a Diwan, holding charge of revenue, and a Naib or Nawab Nazim, invested with the administration of law and order.

However, this general predisposition towards regionalism conduced to moderation in the functioning of Mughal administration. The whole system was doubtless based on a military despotism, but it was not oppressive to the masses. The function of the Mughal government was limited. It did not introduce any change in the character and institutions of rural society beyond simplifying the mode of land revenue collection by an attempt to exclude intermediaries between the state and cultivators. Following the example of **Sher Shah**, it limited its own power over the people by fixing its demand which was moderate and not subject to being varied at will. This gave cultivators security against the rapacity of collectors. The people knew what they had to pay and the government knew what it had to receive. Unlike the change-oriented principle of *Contract*, which British rule later introduced in the administration of the Indian subcontinent, the Mughal system was in fact no more than an instrument of *status quo ante*.

During the best days of the Mughals both the central and regional forces were evenly balanced. The decline of the Empire, which had already begun during Aurangzeb's lifetime, proceeded in three main directions, namely, the independence of Mughal governors, the emergence of resistance movements by non-Muslim country powers, and, finally, the political involvement of Europeans, an alternative group of aliens who superseded the first two groups and established their own empire, though of a kind that had never existed before. This new empire not only integrated the country territorially, but made its geographical entity a political reality.

THE process by which India gained political unity under British rule had no precedent in the history of the country. Qualitatively, the process was unique in terms of principle and organization, in approach and mode of operation. The element of conquest was no doubt present, but the conquest was of a different kind. Unlike under the Mughals or their predecessors, the country never remained subject to military occupation for any length of time, for a rule bound civil administration always dominated.

Unlike the land-locked Muslim invaders from the north-west frontier and Central Asia, who initially set up a central government at Delhi or Lahore before proceeding to reduce the provinces, Britain was a sea-power known for its resilience and mobility, a power which moved cautiously in India from areas of regional influence, and consolidated each of its territorial gains before taking steps in the direction of suzerainty.

The aim of British policy was to concentrate on control over provincial administration without seeming to deny imperial authority. They were aware of the basic weakness in the political structure of the Mughals, but chose to desist from hasty action for financial reasons and constraints imposed by their Home authorities and the fear of political involvement with country powers which could prove financially ruinous for a commercial body like the Company.

Indeed, the rise of British power in its first phase during the eighteenth century was a result not so much of wars, conquests and annexations, as of peaceful penetration through the prevailing system of territorial assignment which conferred rights to collect land revenue and administer justice, law and order that went with it. In other words, they rose to power as a 'zamindar' in the contemporary connotation of the term, not as a conqueror. Doubtless, the authorities of the Company in India were at times induced to get politically embroiled and take advantage of internecine wars among Indian potentates. But, unlike the French, the Directors in London advised restraint from both political and financial considerations, a policy which Pitt's India Act confirmed with an emphasis which even Cornwallis found difficult to have modified.

The march of the East India Company towards political supremacy in India began with a change of policy in favour of wars, annexations and subsidiary alliances. This change is mainly attributed to the Governor-Generalship of Wellesley (1798–1806) who justified it on the grounds of the consequences which had flowed from a general adherence to the policy laid down by the India Act of 1784, and Act which declared under its Article 34 that 'to pursue schemes of conquest and extension of dominion in India are . . . measures repugnant to the wish, honour, and policy of this nation'.

The circumstances which followed the passage of the India Act, did not permit the restraint it had sought to impose. In the midst of warring country powers which were in a state of perpetual unrest and lawlessness, the British could not but fight for the protection of the territories already acquired. They had before the turn of the century come, in fact to hold a position where the alternative was either expansion or destruction. Since they could not afford to lose what they had already gained, expansion remained the only choice, a choice which clearly involved aggression.

However, the main consideration which dictated the necessity of aggression was British concern for the exclusion of the French from Indian politics. For increasing lawlessness and unrest apart, the policy of non-intervention served as an incentive to the revival of French influence. It was to put a check to this trend that, despite his belief in the policy of quiescence, Cornwallis had perforce to wage war against Tipu and annex his territories as a safeguard against future attack. But the safeguard did not work: a defensive alliance was no doubt concluded with the Nizam and the Marathas in 1792, and it was hoped that this Triple Alliance would ensure peace. In the time of Sir John Shore, a Civilian who succeeded Cornwallis, one of the Allies (the Nizam) was attacked by the other, the Marathas. As an honest and law-abiding civil servant Shore decided that, in keeping with the terms of the India Act of 1784, the Company was not required to intervene. The result was that the Nizam capitulated and was reduced to the status of a tributary of the Marathas. His ministers were appointed by them; his army was disbanded, and only a small corps was left for the support of his authority. This was viewed as a complete collapse of British prestige, and indignant at what he considered an act of betrayal by the Company, the Nizam welcomed French officers to organize an army for him. These officers were given possession of a considerable portion of his territory from which they collected revenues for payment to their troops. The Nizam's dominion was increasingly coming under French influence and there was no likelihood of his helping the British in the expected war with Tipu and his French allies.

The Marathas who were looking forward to an all-India empire within their own reach and regarded the Company as their main obstacle, also took the aid of French officers, for they realized that in the existing state of military science and technology a resort to European methods of organization was essential for success. Anxious for revenge, Tipu too was in search of an opportunity to strike and was already in open negotiation with the French. The Company's fear of the French arose additionally from the fact that the French Republic had about this time entered upon its career of conquest, and was anxious to recover the position in India which the Bourbon monarchy had lost: Napoleon was already in Egypt. A policy of non-intervention by the British had in fact brought about a situation which could only be remedied by its complete reversal.

Another salient feature of the policy which Wellesley adopted on his arrival in May 1798 was the introudction of what is known as the subsidiary alliance system with the country powers that recognized British paramountcy. Alliances in the past had doubtless been formed with Indian states. But these were formed generally after the dependent state had been conquered. The principal stipulation was uniformly protection by the British in consideration of a subsidy to be paid by the state which was, however, declared or considered independent in the management of all its internal concerns. Thus, in 1798 Saddat Ali, the reigning Prince of Oudh, contracted an alliance with the Company at a time when he was insecure in his kingdom; his army was disaffected, the security of his own person was constantly threatened, and when an invasion was expected from Zeman Shah, the King of Kabul. The object of the alliance was to meet, by introducing into the Nawab's territories 3,000 additional British troops, a particular contingency on payment of an annual subsidy of Rs 76 lakhs of Oudh sicca rupees. But the subsidy was neither regularly paid nor was the Nawab's independence ensured without the Company's army, a situation which opened the door for regular British interference.

The defensive alliance which Cornwallis concluded with the Nizam after Tipu's defeat and the Treaty at Seringapatam in 1792 had likewise its own limitations. It did not precisely determine the relationship in which the Nizam would stand with the British or with the Marathas, another partner of the Triple Alliance of 1792. Moreover, the British troops stationed in the Nizam's territories were neither to defend him against the Marathas nor to overawe and bring his tributaries under his subjection, tributaries whose insubordination and rebellious conduct seriously impeded the collection of revenue in parts of the Nizam's dominion located on the frontiers of Mysore. Between 1792 and 1798 the government of the Nizam had come under Maratha pressure, which brought about the capitulation of Kurdla in 1795 and defeated completely the purpose of the Triple Alliance.

To restore the confidence of the Nizam and enable him to discharge the obligations of the defensive alliance of Cornwallis, Wellesley's government started secret negotiations with him at Hyderabad. It was thus suggested that the French officers and their troops should be dismissed from the service of the Nizam, and an offer was made

Introduction

to supply British troops immediately to enable the Nizam to perform his treaty obligations against Tipu. It was agreed that the British troops would remain at Hyderabad as substitutes for the troops under the command of the French officers. A treaty was accordingly concluded with utmost dispatch and secrecy on 1 September 1798. A body of British troops entered the Nizam's territories and had moved rapidly towards Hyderabad before the French became aware of the entire proceedings. This body of troops, having joined those already at Hyderabad, performed their assigned task without any bloodshed. The Nizam was thus restored to the power of performing his defensive engagements with the Company.

Though designed to serve as a defensive alliance against Tipu, the treaty of 1798 was also a temporary arrangement to meet a contingency. The Peshwa, too, was sounded to join the English on Wellesley's promise to give him a share in the conquests. Though temporary, the alliance did achieve the object intended. Tipu was defeated and killed during his defence of Seringapatam in May 1799. An offer was made to the Marathas to accept a couple of districts lying to the north-west of Mysore, but they declined. The Nizam was given the districts lying to the north-east near his own kingdom. The English retained Canara on the west, Wynaad in the south-east, Coimbatore and Daraporam in the south, and a couple of tracts in the east, including the town and island of Seringapatam. The rest of the kingdom was given to a boy of the old Hindu reigning dynasty, who accepted the terms of the subsidiary alliance which made the Company supreme in both internal and external matters. A British force was stationed in his territory against payment of an annual subsidy and at the discretion of the Governor-General who could, if necessary, take over the entire internal administration of the state.

It was the conquest of Mysore and the consequent elimination of the French that put the East India Company on the road to paramountcy. The principle of the subsidiary alliance which guaranteed the protection of a dependent state against internal unrest and external aggression flowed from this paramountcy. Under the treaty of 1 September 1798 the Nizam was not entitled to the assistance of the British troops stationed in his territories, either to defend himself against the Marathas or to suppress lawlessness and insubordination on the part of his tributaries. It was under the assumed right to paramountcy arising from the conquest of Mysore that the British government 'was obliged to determine the precise nature of the relation in which it would stand, as well to the Nizam's as to the Maratha government, and that in which those powers should stand to each other'.[7] The subsidiary alliance which the British made with the Nizam in 1800 was in principle based on these terms. It extended the basis of the treaty of 1798 and made it generally defensive against all powers, instead of being confined to Tipu. In fact, the treaty of 1800 placed the Nizam under the protection of the British government, a protection which also covered security against such rebellious elements as tended to disrupt the internal peace and tranquillity of the country, especially in the districts which earlier formed part of Mysore.

The subsidiary engagements of 1800 were a permanent defensive arrangement and introduced another change with a view to obviating the embarrassment of interference with regular internal concerns. By the treaty of 1798 the British government was, for instance, bound to furnish the Nizam a subsidiary force in return for payment of a stipulated sum every month. This subsidy usually fell into arrears and the arrears over a period of time gave rise to complaints and remonstrances from the British government, a development which tended to put a serious strain on the good understanding which was sought to be promoted between the two contracting powers. The subsidiary alliance of 1800 removed all causes of misunderstanding by a grant of territorial security in lieu of the subsidy so stipulated. The territory granted to the Company had earlier formed part of Tipu's Kingdom and been ceded to the Nizam by the treaties of Seringapatam concluded in 1792 and 1799.

The principle of 'territorial security' which applied to the subsidiary alliance with the Nizam, was imposed upon the Nawab of Oudh by a treaty of 1801. This treaty brought half the territory of that state under direct British rule and reduced the remainder to a more complete dependence on the British government. The districts ceded to the Company were made to yield a revenue which not only met the old subsidy of Rs 76 lakhs a year as stipulated under the treaty of 1798, but also provided for an increase in the number of troops stationed in the Nawab's territory.

Annexation was yet another instrument which Wellesley used for territorial expansion to reinforce British claims to political supremacy. The annexation of the Carnatic in 1801 was in this regard specially important from both political and administrative considerations, for it was indicative of the nature of British policy in relation to its tributary states in general.

Under the terms of the treaty of annexation the whole civil and military government of the Carnatic was transferred forever to the East India Company. The Nawab (Azimuddaula) and his heirs were to retain their title and dignity, and to receive as pension one-fifth of the net revenues of the territory. An additional security was given to the British government to the extent of £800,000 per annum added to their financial resources. The annexation of the Carnatic was thus another important step in the direction of the Company being viewed as a sovereign power. This annexation could have effected without a fetish being made of treaty obligations to justify the overthrow of the Nawab's regime by a direct assumption of responsibility. The continued play of corruption could alone have justified the adoption of such a measure. But the respect shown for a due observance of legal norms underlying the terms of treaties was not only a general characteristic of British concern for the rule of law, but also an integral part of British policy to achieve its political objective peacefully, if possible; without any sudden jolt to public opinion.

The rulers of Tanjore and Surat were likewise compelled to quit and remain satisfied with guaranteed pensions. In the case of Tanjore, a Maratha principality, the occasion for Wellesley's interference arose from a disputed succession which finally led to the conclusion of a subsidiary treaty in October 1799; under this the entire civil and military adminstration of the state was transferred to the Company in return for a pension of £40,000 a year. Surat went the same way though for a different reason, which was the failure of the Nawab to defray the arrears of the expenses involved in the maintenance of a garrison in the state on behalf of the Mughal Emperor since 1759. When the old Nawab died in January 1799, his brother was obliged to surrender the administration of the territory to the Company in March 1800.

As for the Rajas and zamindars of Malabar, ever since the peace of 1792 with Tipu Sultan, they had remained hostile to the British. With the exception of the Raja of Coorg, who had been granted independence, the English did not have any firm control over Malabar which had been given to them by Tipu under the Treaty of Seringapatam (1792). The Malabar chiefs kept a considerable proportion of the Bombay army occupied in constant operations, and only with Tipu's death and defeat in 1799 did the tide turn in favour of the British. The extension of the Company's territories to the entire stretch of the Malabar coast, the transfer to it of all of the Nizam's acquisitions from Mysore, and the annexation of the Carnatic and

Tanjore established beyond doubt the *Pax Britannica* from the Malabar to the Coromandel coasts.

The Marathas were doubtless a match for the Mughals, but they, too, succumbed to the British. Balaji Baji Rao had been on reasonably good terms with the English who had helped him in putting down the rebel Angria in 1756. But the British withheld their support when the Marathas proceeded two years later to conquer Janjira from the Siddi; for the British feared that, should the Marathas take possession of Janjira, Bombay might become their next target. This refusal to help Balaji Baji Rao against the Siddi in 1758 doubtless caused umbrage, but to soothe the Peshwa in the course of the Seven Years War, the English took care to send an envoy to Poona in September 1759. By 1767 the next Peshwa Madhav Rao, had raised the prestige of Maratha power once again and to a height which caused concern to the British. The Mughal Emperor, Shah Alam, surrendered himself to the Marathas in 1770, but by an arrangement made with the Nawab of Oudh soon after, the English deprived the Emperor of the districts of Allahabad and Kora which were handed over to the Nawab in return for fifty lakhs and an additional sum to maintain a garrison against the Marathas. The Treaty of Benares (1773) confirmed this arrangement which acquired added significance on the death of Madhav Rao in November 1772 and the murder of his successor, Narayan Rao, in August 1773. The murder which was committed at the instance of Raghunath Rao, who was recognized as Peshwa for a few months, but opposed by a strong party at Poona led by Nana Fadnis and Mahadji Sindia. They set up a council of regency, when a posthumous son was born in 1774 to the late Peshwa's wife, Ganga Bai, and recognized the infant as the next Peshwa who was named Madhav Rao Narayan. Driven out of his home provinces, Raghunath Rao proceeded to seek the help of the English in Bombay who concluded with him the Treaty of Surat on 7 March 1775. The British agreed to help Raghunath Rao with a force of 2,500 at his own cost, while he, on his own part, undertook to cede to the English both Salsette and Bassein, in addition to a share in the revenues of Broach and Surat districts. It was further stipulated that he would not enter into any alliance with the enemies of the Company, or by-pass the British in peace negotiation with the Poona government. The first Anglo-Maratha War, which arose in 1775 from the extension of British support to a rebel against the central authority of Poona, continued for a number of years until peace was restored by the Treaty of Salbai, concluded in May 1782, but ratified by Nana Fadnis only in February 1783.

The Treaty of Purandhar, which was concluded in March 1776, annulled the earlier Treaty of Surat and the British renounced on certain conditions the cause of Raghoba. But foreign politics, arising from the American War of Independence came in the way of stable Anglo-Maratha relations. While the home authorities of the East India Company lent their full support to the Bombay government who gave shelter to Raghunath Rao in direct violation of the Treaty of Purandhar, Nana Fadnis received Chevalier de St. Lubin, a French agent, who reached Poona in 1777 and proposed an alliance with the Marathas, promising French help against the English in return for the grant of a French port in western India.

It was this French dimension in Indian politics that gave a fresh lease of life to Maratha power under Nana Fadnis and Mahadji Sindia who, in spite of internal bickerings and defections, preserved the overall unity of the Maratha confederates, a unity which was further reinforced by the British policy of non-intervention which Pitt's India Act (1784) emphasized as an essential guide to diplomatic conduct.

In the course of a decade which followed the Treaty of Salbai, Mahadji Sindia endeavoured to carry out the old Maratha policy of controlling the affairs of the Mughal Emperor at Delhi and to re-establishing Maratha supremacy over the Rajputs and other tributary states, a supremacy which had come to be shaken during the war with the English. And both the tasks were sought to be accomplished by openly giving up the old guerila tactics and raising a new army on the European model under the direction of Benoit de Boigne, a leading French general. This decision flowed from his experience of the first Anglo-Maratha war itself. For he had closely watched 'the victorious march of General Goddard from the river Jumna to Burhanpur and to Surat, splitting the whole of north India as it were into halves like a piece of bamboo'. Mahadji was more especially impressed by 'the havoc which the British guns had made during the campaign of Talegaum and the ease with which the British were quietly strengthening their position on the west coast by the capture of Bassein and Thana'.[8] He realized that, without modernizing his army on European lines, the Marathas could not hope to compete with the British as the inheritors of the Mughal empire; for the real question which had by then emerged was whether the supreme power was to be British or Maratha. There was no third alternative.

But an army, however efficient, is of little or no avail in the absence of a sound political system. It was here that the Marathas were found sadly wanting. The Maratha empire was, for example, considered at times to be a union of a number of chiefs possessing their own territory and power, though acknowledging the Peshwa as their nominal head. At other times, the Peshwa was regarded as the real head of a government of which Sindia and other chiefs were only powerful officers. There was, in fact, no fixity of constitutional norms. There were cases where the Peshwa made treaties independently of the Maratha chiefs, as with the Treaty of Seringapatam (1792) and the Treaty of Triple Alliance in 1792, which was concluded despite the objection of Mahadji Sindia and Tukoji Holkar. The Maratha chiefs, on the other hand, made war and peace in their own names without using the Peshwa's name or referring to his authority. The Peshwa was obeyed by the feudatories of the southern Maratha country, but the great hereditary chieftains who held sway in the non-Maratha regions of the empire obeyed him only when it suited them. Like the provincial governors of the Mughals, they were their own masters in war and peace, making treaties or breaking them as they pleased.

The English, on the other hand, were different. After initial abuses, they had established a reign of law and an efficient system of government and administration functioning under a well-defined constitution and unified control from their Home authorities. Even a sympathetic Maratha historian does not hesitate to marvel at 'the slow but steady progress of the British nation in the work of empire-building in India'. He admits that, unlike the British who made full use of their corporate machinery of decision-making, the Marathas were guided not so much by larger interests as by considerations of personal or factional animus and their own 'immediate concerns of the moment'.[9] Balaji Baji Rao, for instance, 'took British help to crush the Maratha navy headed by the Angria, his own naval commander', and 'utterly neglected to support the Bhonsle's claims in Bengal, when Siraj-ud-daula was being hard pressed by the British before the battle of Plassey'.[10] Nana Fadnis, who was otherwise endowed with great diplomatic skill, dispensed with the council of regency, a broad-based advisory body formed in 1774 to guide the affairs of the Maratha state during the infancy of the posthumous son born to Ganga Bai. He was doubtless effective so long as he worked in concert with Mahadji Sindia, but they often pursued their own independent courses of action which adversely affected Maratha fortunes. Administratively too, as Sardesai points out, there was generally 'no unity of command, no distribution of work and power, no clear-cut assignment of duites, no method, no rule,[11] a weakness which was temporarily kept in check by powerful personalities, but given to disintegration in the absence of loyalty to a common goal and corporate leadership.

For a couple of decades after Salbai the British pursued a policy of non-intervention in relation to the Marathas. One of the reasons was of course their declared adherence to the directive laid down by the India Act (1784). But the immediate object of an overt pro-Maratha stance was to secure their cooperation in the projected Mysore War against Tipu Sultan, the avowed ally of the French in India. The Triple Alliance which Lord Cornwallis formed in 1792 with the Marathas and the Nizam as partners was designed to ward off this potential danger.

With the death of the Peshwa, Madhav Narayan Rao in October 1795, a contest ensued for succession to the office of the Peshwa. After a series of debates and much wrangling, Raghoba's son, Baji Rao II, was brought out of confinement and established as the new Peshwa in December 1796, with Nana Fadnis, the ancient enemy of his father, as chief minister. The real power and influence of the government, however, came into the hands of Daulat Rao Sindia who, by right of succession to his relative, Mahadji Sindia, had all the Maratha territories in Hindustan, the main instrument of support being an army disciplined and commanded by French officers, in addition to the backing of the Maratha government at Poona.

This was the general state of Maratha affairs in 1798 when Wellesley took over as Governor-General in India.

While Wellesley was fully prepared after his conquest of Mysore to enforce the terms of his subsidiary alliances, the Marathas were involved in mutual conflicts and intrigues, more especially between Sindia and Holkar. Tukoji Holkar, who had been next in rank and power to Sindia, died in 1797, and while the succession to his power was being disputed by various claimants, attacked Malhar Rao Holkar, one of the claimants, and put him to death. The struggle within Maratha ranks eventually led to the Peshwa, Baji Rao II, seeking British protection and signing the Treaty of Bassein on 31 December 1802, a treaty which put an end to the Maratha empire and clinched the British claim to Indian sovereignty. By the Treaty of Bassein the Peshwa agreed 'to receive, and the Company to furnish, a permanent subsidiary force of not less than six thousand regular native infantry, with the usual proportion of field pieces, and European artillery men attached'. This subsidiary force was to be stationed in perpetuity in the Peshwa's territories, and he agreed to allot a part of his territories for the payment of the troops to be so detached. The territories yielding revenues worth twenty-six lakhs were specified in a schedule attached to the Treaty. Baji Rao II also agreed not to entertain in his service any European hostile to the English. He was in fact not to enter into any treaty or correspondence with any foreign power except with the prior knowledge and consent of the Company's government. Even differences between him and the Nizam or any other ally were to be subject to arbitration by the Company. It was emphasized that the friends and enemies of the contracting parties were without fail to be the friends and enemies of both.[12] Consistent with these terms of the Treaty a British force under Arthur Wellesley conducted Baji Rao II to his capital and restored him to his Peshwaship on 13 May 1803, for Holkar had already withdrawn from Poona for the north and left the city for the British army to occupy.

The Treaty of Bassein, however, wounded the feelings of other Maratha leaders who viewed it as a complete surrender of their national independence. Raghuji Bhonsle of Berar came forward on this occasion and took steps to patch up a peace between Daulat Rao Sindia and Yashwant Rao Holkar, the three together forming a confederacy against the British government. Sindia and Holkar agreed to come together at the instance of Bhonsle, for both had lost the power and influence they had earlier wielded with the Peshwa's government, which they could not under the new arrangement now hope to recover. Their immediate basis of unity was power for themselves, not the larger goal of national independence. Thus, while Sindia and Bhonsle proceeded to mobilize their troops, Holkar retired to Malwa to watch the turn of events. The Gaikwar of Baroda remained neutral throughout, and when Holkar finally chose to take the field, it was already too late.

The British, on the other hand, started from a position of strength and advantage. They had already conquered Mysore and entered into a subsidiary alliance with its Raja and the Nizam; they had annexed the Carnatic and made treaties with the Gaikwar and Oudh, they had subdued the chieftains of Malabar and, above all, brought the Peshwa himself under their control. They were fully prepared for the hostilities which commenced early in August 1803.

Introduction

Within five months of the outbreak of hostilities Bhonsle and Sindia were both obliged to make two separate treaties with the English. By the Treaty of Deogaon concluded on 17 December 1803, Bhonsle was forced to cede to the Company the province of Cuttack, including Balasore, which established direct communication between Fort William in Bengal and Fort St George in Madras. The territories ceded in the Deccan comprised all the country lying to the west of the Warda river, including Burhanpur and Asirgarh. Bhonsle also agreed that he would not employ any European or American hostile to the English, or even a British subject, without the consent of the British government, which was vested with powers to settle all such disputes as arose between the Raja and the Nizam or the Peshwa. With his consent to maintain a British Resident, Mountstuart Elphinstone was stationed at Nagpur. Sindia, on the other hand, concluded with the English the Treaty of Surji-Arjangaon on 30 December 1803. He ceded to the English all his territories lying on both sides of the Jumna as well as his forts and territories to the north of the Rajput states of Jaipur, Jodhpur and Gohad. On the western side the English were given Ahmadnagar, Broach and Sindia's territories lying to the west of the Ajanta Hills. He was in addition obliged to relinquish all the influence which he had exercised over the person and power of the Mughal king, and to admit the independence of all the petty states in the north which had during the war-transferred their allegiance by treaty to the Company. Like Bhonsle, Sindia, too, had to agree that he would not employ Europeans or British subjects without the permission of the British government which, in fact, destroyed entirely the corps which was commanded and officered by Frenchmen. Sir John Malcolm was appointed Resident at the court of Sindia who, by another treaty concluded on 27 February 1804, entered into a subsidiary alliance under which a defence force was to be stationed near the frontier of his territory.

By the treaties so concluded with Bhonsle and Sindia, the Nizam acquired complete exemption from demands on his territories by the confederates. In addition, he secured a well-defined boundary along the dominions of Sindia to the south of the Ajanta hills, while from Bhonsle he acquired the province of Berar, with the Warda river and the hills on the north as another frontier. The fort and territory of Ahmadnagar, which earlier belonged to Sindia, were made over to the Peshwa, who ceased to have anything to do with any of the confederates. These were all valuable and important possessions which Wellesley's treaties secured for the Company. General Lake in the meantime had captured Delhi and Agra, and routed the northern army of Sindia, a victory which compelled him to sue for peace.

Peace was finally concluded with Sindia in November 1805. Gwalior and Gohad were restored to him with a pledge that the Company would not enter into treaties with the Rajput states to bring them under its own protection. Lake, in the meantime, hunted Holkar up to Amritsar where the latter sought the help of the Sikhs against the English. On his failure to get that he was obliged to sign a treaty in January 1806. Holkar gave up his claims to Tonk, Bundelkhand and all the other states lying to the north of the Chambal, but all his other lost territories were given back to him. By Declaratory Articles, even Tonk and Rampura were to be surrendered to him. The Rajput states were, in addition, left to take care of themselves in the absence of British protection which was withdrawn. They once again became subject to Maratha inroads.

The period of quiescence which followed Wellesley's recall in 1805, gave the English the opportunity to organize the governance of the vast new provinces acquired under him. But the policy directives favouring British neutrality regarding the Indian states that were not in alliance with the British also encouraged the forces of anarchy and lawlessness. In the wake of uncertainties created by the steady fall of Maratha power without a firm and immediate infrastructure emerging in its place, the Pindaris were soon threatening the peace. A class of freebooters who were known to be under the protection of the great Maratha princes, the Pindari raids involved the Company in hostilities not only in Central India, their main field of operation, but also in such territories as belonged to the Nizam or other British allies in Orissa and the Northern Circars.

The government of Lord Moira (later created Marquess of Hastings) marked a radical departure from the quiescence of his predecessor. Soon after his arrival in India, he noted the 'unfortunate' plight of Jaipur, Jodhpur, Udaipur and other states that were 'mercilessly wasted' by Sindia and Holkar as well as by Amir Khan, a Pathan leader, and the Pindaris. Lord Moria proceeded to take steps to suppress the lawless elements. The Company's Directors too came round in the meantime to support measures against the the Pindaris, but they were also anxious to avoid doing anything that might lead to war with the Marathas. The depredations carried on by the Pindaris in the territories of the Nizam and Orissa in December 1816 induced the Governor-General and his colleagues in the Council to decide unanimously 'that the extirpation of the Pindaris must be undertaken, notwithstanding the orders of the Court of Directors against adopting any measures against those predatory associations, which might embroil us with Sindia. Lord Moira argued in fact that no step could be taken to suppress the Pindaris without involving the English in hostilities with Sindia and Holkar, who treated the Pindari's as their 'dependents'.[13] The Pindaris were thus the first cause of the last war with the Marathas, which saw the end of their empire.

The second important cause was a serious attempt by Peshwa Baji Rao II to re-establish a confederacy of Maratha chiefs through secret negotiations carried on towards the close of 1816 with Bhonsle, Sindia, Holkar and the Gaekwar. Amir Khan, the Pathan leader of Central India, and the Pindaris were also within the ambit of a league against the British government. While Daulat Rao Sindia officially told the British that he would help suppress the Pindaris, he was simultaneously urging Holkar, Amir Khan and Ranjit Singh, to join him in opposition to the British.

Despite treaties to the contrary, the Marathas, particularly actively engaged in secret preparations for war against the English. The Peshwa sacked and burnt the English Residency at Poona on 6 November 1817, just a day after Sindia had signed a subsidiary treaty, and attacked with a considerable force a small British army of about 2,800 men under Colonel Burr. But the Peshwa was soon completely defeated at Khirki by Elphinstone. Appa Saheb of Nagpur, however, followed suit and attacked the British Residency there on 27 November 1817. His ambitions too were extinguished in the coming months. An attempt by Holkar to revive the Peshwas resistance to the British ended when he was forced to sign the Treaty of Mandasor on 6 January 1818. The Peshwa continued to fight two more battles after his defeat at Khirki, but he was defeated and forced to surrender to Sir John Malcolm on 3 June 1818.

The fall of the Marathas in 1818 not only contributed to the territorial expansion of the Company's government in India, but transferred to it the sovereignty of the subcontinent for all practical purposes. India's political unity, which proceeded from this British victory, was by far the most important result of this resolution of the Anglo-Maratha match for paramountcy. The hereditary and dynastic rule of the Peshwa was done away with. The Peshwaship itself was abolished. The dominions of Baji Rao II were placed under British control and he was removed to Bithur near Kanpur, where he was allowed to spend his last days on an annual pension of Rs 8 lakhs. Trimbakji was imprisoned for life and a small kingdom of Satara was created for Pratap Simhe, a lineal descendant of Shivaji, a state which came under the British after Dalhousie's doctrine of lapse was applied to it.

By the Treaty of Mandasor, on the other hand, Holkar was obliged to give up all claims on the Rajput states and surrender to Company's government extensive districts on both sides of the Narbada. The territories left to him were reduced to a narrow compass; for while a large portion of his country was transferred to the state of Kota, Tonk was ceded to the Pathan leader, Amir Khan, who had been won over by the English. Holkar, in addition, had to accept a subsidiary force within his territory and submit all his foreign relations for approval to the British government. A permanent British Resident was stationed at Indore, his capital. The Treaty of Mandasor also created the independent entity of Jaora, which was given to Gafur Khan, the son-in-law of Amir Khan who was recognized as the nawab of Tonk. The nawab of Bhopal, too, entered into a defensive and subordinate alliance with the Company. Holkar was thus reduced to a figurehead for all practical purposes. The fate of the Bhonsle kingdom at Nagpur was even worse. After his defeat, Appa Saheb fled to the Punjab and then to Jodhpur where he died. All the districts lying to the north of the Narbada were annexed to British territories, with Jabalpur as the headquarters of a British military station.

Sindia alone did not incur any forfeiture. While watching the course of the Peshwa's movement without actively supporting the English, he did not swerve from the terms of his subsidiary treaty, nor did his troops ever impede the marches of the retreating Pindaris who were being chased by the Company's troops through his territories. The extirpation of the Pindaris, whom Sindia had once fostered and protected was by itself considered sufficient punishment for Sindia. His agreement to renounce all his claims on the Rajput princes and to acknowledge the right of the British to enter into treaty relations with them or to form a league of those states under British protection earned Sindia some kindness from the British.

Thus, Lord Moira's most outstanding political achievements was the establishment of a pacific system in Central India which was of immense strategic importance for both defensive and offensive purposes. The friendship and alliances with the Rajput states formed an integral part of this system. Kota was the first to enter into this alliance on 26 December 1817. From January to December 1818, the states which joined the alliance included Udaipur, Bundi, Kishangarh, Bikaner, Jaipur, three sub-branches of the Udaipur house, and Jaisalmer. Sirohi came last and joined the league in 1823. The minor states of Malwa and Bundelkhand also came to acknowledge British paramountcy, which by now covered most parts of the subcontinent. The work of adminstrative reconstruction and consolidation was completed by a band of able officers, such as Elphinstone in Bombay, Munro in Madras, Malcolm in Central India, Metcalf in Delhi, and Tod and Ochterlony in Rajasthan.

THE establishment of the virtual paramountcy of the East India Company in 1818, however, did not put an end to wars, for India's frontiers still remained to be ascertained and duly delimited. These frontiers lay on the Burma border in the east and the north-east of Assam, along the foothills of the Himalayan states in the north and the Punjab, and, finally, the north-west and Sind.

The Burmese had seized Tenasserim from Siam in 1766, brought Arakan with its capital Mrohaung under their direct central authority in 1784 and conquered Manipur in 1813.

The Burmese advance on India's eastern and north-eastern frontiers proceeded from the still undefined character of the Company's territorial boundaries. And since its government remained politically preoccupied with the settlement of affairs in other parts of India, it made a seies of attempts to resolve differences through diplomatic channels, not through armed conflicts. While the Marquess of Hastings was engaged in the suppression of the Pindaris in India, the King of Ava demanded control of Chittagong, Dacca, Murshidabad and Cassimbazar, which once paid tribute to the rulers of Arakan. Finding little or no resistance on the Assamese frontier, the Burmese commander proceeded to conquer Assam in 1821–2 and captured in 1823 the Company's Shahpuri island near Chittagong in preparation for an attack on Bengal. In these circumstances Lord Amherst, the Governor-General, declared war in 1824. Instead of proceeding to Ava through the vast expanse of forests and mountain ranges, the British troops recruited mostly from Madras made an attack on Rangoon by sea and planned to move up the Irrawaddy through

Pegu and Prome before an assault on the landlocked capital of Ava. Rangoon was captured in May 1824 and Prome a year later. The British troops then marched to Yandaboo, situated within sixty miles of the Burmese capital. Burma concluded a treaty on 26 February 1826 on the terms dictated by Archibald Campbell, the British general. The King of Ava agreed to pay a crore of rupees as war indemnity; surrendered the provinces of Arakan and Tenasserim; abstain from interference in Assam, Cachar and Jaintia; recognize Manipur as an independent state under British suzerainty; conclude a commerical treaty on principles of reciprocity, and consent to the admission of a British Resident at Ava and a Burmese envoy at Calcutta.

With the military conquest of Assam by the English, the direction of all civil matters connected with the province was entrusted to David Scott, the Governor-General's Agent in the North-East Frontier. The Officer in Command of the troops in Upper Assam was later associated with Scott in a Commission for general administration, when the conquest was complete. Upper Assam was formally placed under Captain Neufville in subordination to David Scott. In making this arrangement the Calcutta government was guided by immediate considerations of security, for the royal house of Assam lacked the capacity to govern the province in a manner acceptable to the people of the region. It appears that, although averse initially to taking absolute possession of the province, security considerations in the eastern districts compelled the retention of strong military control over the north-east frontier. But once this was provided, the Government restored native management in that part of the Assam Valley which was not required for the maintenance of British troops. Accordingly, Upper Assam, with the exception of the tract about Sadiya and Muttuck, was made over in 1852 to Purander Singh who was regarded as being the most eligible representative of the royal family. He became a protected prince, guaranteed against invasion on condition that he paid an annual tribute of Rs 50,000. This arrangement, however, did not succeed for Purander failed to deliver the goods. His territories were thus placed under the direct management of British officers in 1858. Assam became a non-regulation province of India. The conduct of topographical and route surveys, the opening of road communications, and the establishment of a regular military outpost, all followed in the wake of the conquest and occurred simultaneously with reforms in revenue administration. Such measures contributed to the integration of the province with the rest of the subcontinent, an integration which was later sought to be reinforced by a sound policy to conciliate the hill tribes of the north-east frontier.

The Treaty of Yandaboo had doubtless deprived the Burmese of the greater part of their sea-coast and brought Assam, Cachar and Manipur under British protection. But Later, the new King of Burma Tharrawaddy (1837–1845), refused to confirm the Treaty of Yandaboo. British interests were affected in other ways also. While Burmese treatment of the British Resident at the Court of Ava remained unsatisfactory, British merchants felt they were being oppressed by the governor of Rangoon. In response, Lord Dalhousie, the Governor-General, sent a frigate under Commodore Lambert to Pagan (1845–52), who had succeeded Tharrawaddy, his father. A demand was made to remove the governor of Rangoon and meet the losses of British merchants. Keen to avoid war, the Burmese king ordered the removal of the governor and deputed senior officers to settle the whole matter peacefully. But no settlement occurred, wherupon Commodore Lambert declared the port of Rangoon subject to a complete blockade and seized a ship of the Burmese king. The Burmese opened fire on the British frigate and the Commodore returned the fire. The Burmese government was then served with an ultimatum to pay by 1 April 1852 a compensation and indemnity of the order of a £100,000 an ultimatum which remained unanswered till the expiry of the date. British forces, which had been alerted for the eventuality, stormed Rangoon in April 1852. Martaban fell quickly and the Irrawaddy delta was taken possession of in quick succession. Dalhousie himself reached Rangoon in September. Prome was taken in October, and Pegu in November. On the refusal of the Burmese king to recognize these conquests, Dalhorsie annexed Pegu and the whole of Lower Burma by a proclamation issued on 20 December 1852.

The annexation of Pegu and Lower Burma extended India's eastern frontier to the banks of the Salween river, with British control fully established along the whole of the eastern coast of the Bay of Bengal. Immediate arrangements were made for the administration of the newly acquired British province of Lower Burma which extended to the north for about fifty miles beyond Prome. The conquest of Upper Burma, which was linked to India's north-eastern frontier and the Chinese territory of Yunnan, was effected later in 1885.

The Himalaya mountains, which extend for nearly two thousand miles from Afghanistan to Burma, serve as a great divide between Central Asia and South Asia between the British territories of Bengal, Bihar and the North-Western Provinces and Nepal and Bhutan there lies a wide and extensive tract of foothills called the *duars* or *tarai* which earlier formed a natural division between the British dominions and the hill states. As their frontiers remained unsurveyed and undefined, there always existed the potential danger of conflict between the two sides, more especially because the latter, though situated on the southern slopes of the Himalaya, had long continued to be under the influence of Tibet and China. The British secured a stable northern frontier not only to bring under their sovereignty the entire tract of plains along the foothills stretching between the Sutlej in the west and the Tista in the east, but also increasingly drew into the British Indian sphere of influence all the hill states by completely alienating them from Tibet and China. It was perhaps to achieve this two-fold objective that Tibet was sought to be brought under British influence as a bulwark for both commerical and political purposes.

Of the hill states, Bhutan was the first to come to terms with the British. An Anglo-Bhutanese clash ocurred in 1772, when the Deb Raja of Bhutan raided British territory in Cooch Bihar in the foothills of north Bengal, beyond the district of Rangpur. In the absence of any resistance, the Raja acquired the territory, but soon a detachment of infantry was sent to dispossess the invaders and drive them back to their own frontier. Weary of the conflict and anxious about the safety of his own dominions, the Raja sought the mediation of the Teshoo Lama, the regent of Tibet and guardian of the infant Dalai Lama. Known for his benevolence and as a person who united in his office both political authority and religious sanction, the Teshoo Lama also 'acknowledged the sovereignty of the Emperor of China, who had a delegate with a small military force' residing at Lhasa, the capital of the Dalai Lama, 'but who had not yet much interposed in the interior government of either division of the province'. The Teshoo Lama thus occupied a position which entitled him to the privileges of a plenipotentiary. And since he was immediately 'interested for the safety of Bootan, which was a dependency of Tibet', he felt 'moved' by the 'prayers' of the Raja and 'sent a deputation to Calcutta'. In a letter to Warren Hastings in March 1774, the Teshoo Lama offered to mediate. His letter also served as a basis for establishing trade relations with the trans-Himalayan state of Tibet and, eventually, as an agency for promoting British interests in China also.

The tenor of the Teshoo Lama's communication recognized Bhutan as being 'submissive' and friendly. The Council at Fort William, likewise expressed itself in favour of peace with Bhutan. But more than that, Warren Hastings embraced the opportunity which he thought the latter afforded 'of extending the British connexion to a quarter of the world, with which we had hitherto no intercourse, and of opening new sources of commerce'. He was not yet clear about the 'physical or political accommodations or difficulties' that might promote or obstruct British interest in the northern countries, but he wanted 'to explore an unknown region' of Tibet and its contiguity 'to the western frontier of China'.

Impressed by the need for a 'forward policy', Warren Hastings proceeded immediately to take steps to secure his borders, not only by striking when danger threatened, but more especially by taking a long view to promote ordinary neighbourly intercourse with those on the other side of the Himalaya. Consistent with both commercial and political objectives, Hastings wrote back to the Teshoo Lama with a proposal for a treaty of friendship and cooperation in the peaceful pursuit of commerce. George Bogle was deputed to inquire into the nature of the roads between the borders of Bengal and Lhasa, including the borders of the area lying in between; the state of communications between Lhasa and the neighbouring countries, with relevant information on their governments and revenue, manners and customs. With these instructions Bogle set out from Calcutta for Tibet in the middle of May 1774 and, travelling through Bhutan, arrived on 12 October at Desheripgay near Shigatse, where Teshoo Lama was at the time staying. He was received with great hospitality and kindness. Bogle did not himself go to Lhasa as commissioned, for the Chinese regent was very averse to such a plan. The British emissary even suggested the expediency of some form of an alliance which might be used to restrain the Gurkhas of Nepal from attacks on Tibet and its feudatories. But since Tibet remained formally subject to the Chinese government, the Teshoo Lama, though favourably inclined, could do nothing to meet Bogle's proposals. Opposition to British plans came not only from China, but also from the King of Nepal who, in a letter to the Teshoo Lama, expressed his desire to have a few trading factories established on the Tibetan border and advised the Tibetans to have no connexion at all with the English.[15] Bogle returned to Bengal empty-handed in April 1775.

Even so, Warren Hastings did not lose hope. He maintained regular contact with Bhutan and the Teshoo Lama. The death of both the Lama and of Bogle in quick succession temporarily delayed further development, but in February 1782 news reached Calcutta that, in accordance with the Tibetan concept of re-incarnation, the Teshoo Lama had re-appeared in the person of an infant. Hastings therefore appointed Captain Samuel Turner at the head of another mission to Tibet at the start of 1783. Lieutenant Samuel Davis and Robert Saunders were included in the commission, the former as Draftsman and Surveyor, and the latter as Surgeon.

Turner's political and economic analysis of Tibet was of great significance to the British government in Bengal, and even to India as a whole. As far as trade was concerned, Turner's mission was not without its successes. He obtained from the Lama's regent the 'promise of encouragement to all merchants, natives of India, that may be sent to traffic in Tibet, on behalf of the government of Bengal'. He made it clear in his report to the Governor-General that no impediment 'now remains in the way of merchants, to prevent their carrying their concerns into Tartary. Your authority alone is requisite to secure them the protection of the Regent of Teshoo Loomboo, who has promised to grant free admission into Tibet to all such merchants, natives of India, as shall come recommended by you; to yield them every assistance requisite for the transport of their goods from the frontier of Bootan; and to assign them a place of residence for vending their commodities, either within the monastery, or, should it be considered as more eligible, in the town itself.'

The friendly relations so established with Tibet through the Regent of Teshoo Lama, were thwarted for a while by the Gorkha rulers of Nepal who endangered the security of merchants and interrupted through an invasion of Tibet in 1791 the peaceful conduct of commerce via Nepal, which had earlier been the only known channel of communication between Tibet and Bengal. By the first decade of the nineteenth century, the Gorkha rulers of Nepal had actually pushed their territorial conquests up to Bhutan in the east and Punjab in the west. With the British acquisition of the ceded districts from Oudh up to Gorakhpur, the northern frontier of the Company's dominions became coterminous with the territories of the Gorkhas who, because of illdefined frontiers on both sides, claimed a long stretch of plains in the foothills from Purnea in the east to Gorakhpur, Basti and Bareilly in the west. The border districts naturally became subject to incessant inroads under the Pande and Thapa ministers of King Girvanayuddha Vikrama (1798–1816), who also conquered Garhwal in the west and a part of western Sikkim in the east.

The Gorkha inroads into the southern Tarai districts led both Sir George Barlow and Lord Minto to remonstrate, though without any effect. But a Gorkha attack on Butwal, a part of the modern district of Basti in U.P., led to armed conflict. Lord Hastings, who planned the campaign, ordered an attack on the enemy simultaneously from four different points along the entire frontier from the Sutlej in the Punjab to the Kosi in Bihar. Geographical difficulties, however, stood in the way and led to serious British reverses. The advances of the Company's troops towards the Nepal capital was delayed. But British losses were more than retrieved when Almora in Kumaon was captured in April 1815. Amar Singha Thapa, the Gorkha leader, surrendered the fort of Malaon on 15 May and sued for peace. The Anglo-Nepalese Treaty of Peace was signed at Sugauli on 2 December 1815 and ratified on 4 March 1816 after the British had marched within fifty miles of the Nepal capital. It fixed the Kali river as the western limit of Nepal and the Mechi river as its eastern limit. Nepal gave up all claims to places in the Tarai along its southern frontiers, ceded to the English the districts of Kumaon and Garhwal in the west, withdrew from Sikkim and recognized the right of the Company to arbitrate in its territorial disputes. Nepal had also to accept a British Resident at Kathmandu and agreed to enter into diplomatic relations with the Company. The Residency so established tended gradually to arrogate to itself the role of paramountcy and served as an agency to watch Chinese activity in Tibet. British India's north-west frontier, which acquired a deep thrust into the mountains, provided for important hill-stations and summer capitals like Simla, Mussoorie, Almora, Ranikhet, Landour and Naini Tal. All these facilitated communications with Central Asian regions. The Raja of Sikkim was, on the other hand, given that part of western Sikkim which Nepal was under the Treaty of Sugauli obliged to cede to the Company. By a treaty concluded with the Raja in 1817 Sikkim became a barrier between eastern Nepal and western Bhutan.

However, the Treaty of Sugauli, did not mark the end of Nepalese hostility to Britain. In 1829 the Company and Nepal both signed an agreement laying down procedures for the alignment of their respective boundaries; for Bhimsen Thapa, who had been prime minister of Nepal since 1804, remained unreconciled to the humiliation which that Treaty signified. However, he was deposed by King Rajendra Vikram Shah in 1831. Matabar Singh Thapa, a nephew of Bhimsen, was favourably disposed towards the English but palace intrigues led to his assassination in 1845. Jang Bahadur, the son of a sister of Matabar Singh, then took over as prime minister and assumed the title of Rana. There began with him a new epoch of friendly relations betwen Nepal and the Company's government in India. He lent his full support to the British in the course of the Mutiny in 1857.

Like Bhutan and Sikkim, Nepal too was thus drawn into the sphere of British India and alienated from the north, with the crest of the Himalayan range standing in between. This alienation proceeded partly from the weakness of the Manchus in the post-1846 period, a weakness which left Tibet practically independent and free to act as buffers between the hill-states on the southern slopes of the Himalaya and the Central Asian countries. But the British did not attempt to destroy the independence of Nepal as a separate entity. On the contrary, they returned to Nepal, the western part of the Tarai, as a friendly gesture to Rana Jang Bahadur who had helped the British in 1857, and did not annex either Sikkim or Bhutan.

PUNJAB AND THE NORTH-WEST FRONTIER

THE territories which formed part of this frontier before their annexation to British India consisted of central Punjab with Lahore as its capial, the cis-Sutlej states lying to the east of the Sutlej river, Multan and Sind in the south and south-west, Kashmir in the north, and the trans-Indus lands extending to Peshawar and the Khyber Pass in the north-west. Barring the cis-Sutlej states which came under British protection in 1809, these territories had been conquered and consolidated as a sovereign Sikh state by Ranjit Singh. Beyond its trans-Indus territory, which included the fertile region of the Punjab in the north and the desert area of Sind in the south, on the borders of Afghanistan rose the great mountain-wall, the natural frontier of India in this direction, a frontier which Ranjit Singh had with skill, assiduity and endurance maintained against stiff Afghan opposition. The annexation of this sovereign Sikh kingdom of Punjab, which followed within a decade of the death of its founder in 1839, made political India for the first time correspond with geographical India.

The annexation of the Punjab was the result of two Anglo-Sikh wars fought between 1845 and 1849. It is needless here to go into the details of these wars, but it is relevant to note that, while a series of weak rulers after the death of Ranjit Singh led to the ascendancy of the Khalsa army without any restraint from a central authority, during 1844–5 the English were arranging to acquire boats to cross the Sutlej in the north and preparing to use their newly-conquered territory of Sind as a base for an attack on Multan in the south. The Sikh cause was in fact almost deemed lost before the war broke out; for while Dalip Singh, a minor, was acknowledged as king with his mother, Rani Jhindan, as Regent, the Lahore Darbar, the centre of political leadership, was planning to get rid of the ungovernable army by inducing it to invade British territory across the Sutlej, so that it would either be destroyed or exhaust itself in the military operations. The Sikh army thus fought against the English without any unity of command and was routed.

The Governor-General with his victorious army crossed the Sutlej on 13 February 1846 and occupied Lahore a week later. The Sikhs had no alternative but to accept any arrangement offered to them by the victor. Under the treaty Lord Hardinge dictated to the Sikhs on 9 March 1846 at Lahore, the Sikhs were required to cede to the British all territories lying to the left of the Sutlej, together with the whole of the Jullundur Doab between the Sutlej and the Beas. A war indemnity of a crore and a half rupees had to be paid, in addition partly in cash and partly by the cession of the hill districts between the Beas and the Indus, including Kashmir and Hazara. The size of the Sikh army was reduced and territorial delimitation as well as control over foreign relations made subject to British approval. Dalip Singh was recognized as the Maharaja and Rani Jhindan as his Regent. Lord Hardinge assured non-interference in internal administration, but a small British force was stationed at Lahore for the protection of Dalip Singh who was to be guided by Lal Singh, his minister appointed with the approval of the Governor-General.

By a separate treaty concluded at Amritsar on 16 March 1846 with Gulab Singh, a sardar of the Lahore Durbar, Kashmir was sold to him for one million sterling. Article V of the treaty with Gulab Singh declared that disputes with neighbouring states were to be referred to the arbitration of the British government and its decision was to be accepted.

The original treaty of Lahore was in the meantime revised on 16 December 1846 to bring the Punjab under more effective British control. The Lahore administration was consequently transferred to a council of regency consisting of eight sardars who were to act under the direction of the British Resident, Sir Henry Lawrence. A British force was, in addition, maintained at Lahore on payment of Rs 22 lakhs. The new arrangement so made was to continue for such period as the Governor-General and the Lahore Durbar considered necessary.

Lord Hardinge left for England in January 1848, and so did the Resident, Sir Henry Lawrence, who was for a short while succeeded by Sir John Lawrence, his brother, before Sir Frederick Currie took over in April 1848. However, Punjab did not settle down. The removal of Rani Jhindan from Lahore on a charge of conspiracy against the British Resident doubtless caused disaffection. But the development which led to the outbreak of the second Anglo-Sikh war in September 1848 was the murder of two British officers at Multan, a deed which was believed to have been committed at the instance of Diwan Mulraj, its governor. The involvement of Rani Jhindan as well as of Afghans supporting Mulraj and his collaborators added fuel to the fire, which assumed the character of a Sikh national upsurge. Determined to meet the challenge, Lord Dalhousie declared war on 10 October 1848. The English won at Chilianwala as well as at Multan. But they achieved their decisive victory at Gujrat, near the Chenab, on 21 February 1849. By March 1849, the Sikh chiefs and soldiers laid down their arms, while the Afghans were driven back to the Khyber Pass and Kabul.

Through a proclamation issued on 30 March 1849, Lord Dalhousie annexed the Punjab. Dalip Singh was removed on an annual pension of Rs 5 lakhs and sent to England with his mother, Rani Jhindan, who died in London. The Sikhs soon showed a loyalty to the British cause and faithfully served the Government of India during the second Anglo-Burmese war and the Mutiny of 1857.

Chapter I

THE INSTRUMENTS OF TERRITORIAL INTEGRATON

The establishment of an all-India state under British rule could not by itself result in political unity for any length of time, for the area needed to be reinforced and integrated territorially. It is proposed in this chapter to indicate some of the essential instruments or agencies of territorial integration which not only provided a coherent political system that imparted stability, but also led to the emergence of institutional arrangements capable of promoting internal change to satisfy the requirements of particular situations. It was these instruments of integration that imparted to the new all-India sovereignty under British rule its distinctive character.

Besides institutional provisions, the process of integration doubtless involved an emotional dimension also, which partook of the character of nationalism in its broad territorial sense. This chapter, however, will mainly review institutions and policies, the primary sources from which nationalism later flowed.

Of the several instruments of territorial integration, the political system, the framework of law and the civil administration were the most significant. The evolution of the Indian army and the system of communications which developed with the growth of science and technology, acquired added significance and acted as reinforcements from about the second half of the nineteenth century onwards, when the spread of Western education and the Indian press created conditions for a social base to sustain government and administration, conditions which were conducive to both continuity and change from within the existing institional frame itself.

The Political System

British India

The political system which developed in India under British rule was essentially civil in character, a character which flowed initially over a couple of centuries from the emphasis of the East India Company on trade and commerce, not on war and conquest. The system which so developed under its government and administration was organized mainly at three levels, namely, the controlling authorities in London, the supreme or central government and subordinate presidencies in India, and the Indian states held in varying degrees of subjection to the paramount power. The pattern of parliamentary control which the India Act introduced in 1784 covered the entire field of political operations and continued with some changes till 1858, when the Company itself was abolished and its government transferred to the Crown. The original trading character of the Company had none the less imparted to its Indian administration certain features which distinguished it from the administration of the country powers, which were essentially military and despotic in nature. The Company discouraged military aggrandisement and generally discountenanced even military establishments beyond what was considered absolutely necessary for the defence of its territories and the protection of its commerce. At times, the inadequacy of its military strength tended to prejudice the preservation of law and order. Even so, it resented the use of its armed forces in the discharge of civil duties and remained generally averse to recruitment to the civil service from its military cadres. In the very early part of its rule, the Company had in fact ordered the disbandment of what was known as the *sebandi* corps or military police which the Mughals and other country powers used in the collection of revenue.

These early characteristic features of the Company's government in India, which influenced subsequent developments, were immediately related to its rule-based political apparatus emerging from its erstwhile commercial constitution. Its supreme or central government which was constituted at Fort William in Bengal under the Regulating Act of 1773, was, for instance, preceded by a system under which each of its three Presidencies was governed by a President or Governor and Council who enjoyed equal rank and authority. The Council consisted of ten to sixteen senior servants appointed by the Company under the powers delegated to it by the charters of the British Crown and Parliament. 'All power was lodged in the President and Council jointly and nothing could be transacted except by a majority of votes.'[1]

The function of the President and Council during the Company's early years was purely commercial. But commerce could not be said to be entirely free from political involvements. Indians, for example, also built houses within the bounds of the Company's factories in order to seek its protection against Mughal officers, a development which involved jurisdictional issues and at times occasioned clashes.

Aungier, the Governor of Bombay, even pursued in 1669 a definite policy of encouraging foreign merchants to settle in Bombay by 'giving them protection for five years from liability to be arrested or sued in Bombay for previous debts contracted elsewhere'. For, in the absence of this protection, it was realized, 'they were much molested by arrests and suits' brought against them. In 1677 this privilege was sought to be extended 'so as to exempt all inhabitants of Bombay from liability to be sued by foreigners on bonds or other obligations'.[2] The weavers of the province were similarly secured against the oppressive conduct of the Nawab's government. It is clear then that the Company's factories and settlements were even during this early period emerging as extraterritorial units and that the consultative character of the President and Council in each Presidency provided a security and protection hardly conceived under despotism or military rule. The exercise of any territorial function, however, formed no part of their responsibility, except that a Select Committee later came into being under Clive for the conduct of secret political transactions.

But when later in Bengal and elsewhere the Company came to acquire considerable territories there was hardly any new provision for a constitution beyond what could be traced in the old charters, which had been framed for the conduct of business in trading settlements. The President and Council who carried on that business needed a principle of government adequate to its substance, and a coercive authority to enforce that principle. They could manage the sales of the Company's exports and provide for its annual investment, but lacked the powers to preserve its territorial wealth against private violence and embezzlement.[3]

What were the options open to the men in authority? Clive, who had acquired a first-hand knowledge and experience of the Indian situation, emphasized the need to subject the Company's territorial acquisitions to parliamentary control, for he felt that the conquest of the country was within easy reach and that the establishment of parliamentary sovereignty was merely a question of time. His confidence flowed not from any accession to the Company's military strength from England, but from his knowledge of the utter incapacity of Indian rulers and the mercenary character of their soldiery, which came to his notice in the overthrow of Siraj-ud-daula. In a letter to Lawrence Sulivan on 30 December 1758, he thus pointed out:[4]

The opportunities afforded me by the late Revolution have given me a just knowledge of the subject I am writing upon; experience, not conjecture, or the report of others, has made me well-acquainted with the genius of the people and nature of the country, and I can assert with some degree of confidence, that this rich and flourishing kingdom may be totally subdued by so small a force as two thousand Europeans.... The Moors as well as Gentoos are indolent ... the soldiers, if they deserve the name, have not the least attachment to their prince, they only can expect service from him who pays them best, but it is a matter of great indifference to them whom they serve.

In a long letter of 7 January 1759 to William Pitt, then Prime Minister, Clive expressed the same view with much greater confidence, but Pitt did not regard parliamentary sovereignty as a practical proposition. He viewed it as 'chimerical' in the absence of competent men to sustain the proposed take-over, more especially when Clive himself was scheduled to leave for Europe.[5]

The social pressure which exerted itself against a take-over by Parliament was, however, that of the 'Indian' interests in England, which commanded power-over the direction of Indian affairs and influence in Parliament. They were represented partly by those Englishmen who, while in India, engaged in private trade and returned to England as 'Nabobs'. They constituted a new class who invested their commercial gains and ill-gotten wealth in the acquisition of landed property. They included in their rank such Englishmen as were creditors of the Carnatic Nawab in Madras and carried on private business on their own without payment of transit duties in the post-Plassey period in Bengal. The 'Indian' interests in England were, in fact, generally opposed to any radical change in the constitution and status of the Company. The Mughal grant of the diwani of Bengal, Bihar and Orissa was in these circumstances a compromise to lay aside the basic question of legislative authority for a change in the political system. It was designed to justify the Company's occupation of Bengal on its right to function as a revenue agency of the Mughal emperor and to use that agency as a legitimate opportunity for the introduction of necessary reforms in administration, without a change at the political level.

Reforms were doubtless introduced after the grant of the diwani in 1765.[6] But corruption continued. Warren Hastings, who became the governor of Bengal on 13 April 1772, made it clear to the Court of Directors in his letter of 11 November 1773 that the conduct of individuals could not be reformed without a constitution of government based on political principles and invested with coercive powers to enforce obedience to law: a government which, instead of 'deriving support from the unremitted labour and personal exertion of individuals in power' might depend 'on the vital influence which flows through the channels of a regular constitution'.[7] His emphasis was on the law of a constitution acting as an integrating force, not on the strength of an individual ruler.

It took some time before Hastings' principle of government was fully put into practice as a guide to political conduct and administrative behaviour. The limitations of Parliament arose from its difficulties in deciding on the nature and extent of the sovereignty which the Company exercised in actual practice. The grant of diwani which formed the basis of its authority, signified in law an office, not property. How then could Parliament attach to its sovereignty a property which, if the Proprietors so wished, might be declared as held in benefice? The interest of the Company demanded the continuation of this state of obscurity concerning its political authority. It was thus considered, as part of policy, to leave the political dimension of the Company undefined 'in order that the English might treat the Princes in whose name they governed as realities or nonentities, just as might be more convenient'.[8] It was also argued that the other European powers would become jealous if the English Company determined to stand out too openly. Had the Company's virtual sovereignty in the diwani provinces been declared so by Parliament it would have undermined the very foundation of Mughal authority, which it was considered expedient to maintain formally.

Despite the limitations imposed by the realities of the Indian situation, the Regulating Act of 1773, which follow the accession of Warren Hastings to the government of the Company in Bengal, recognized the right of Parliament to regulate its civil, military and financial affairs. It constituted for the first time a supreme government, with a Governor-General and four Councillors with a view to reducing the authority of Fort St George and Bombay in matters of war and peace, which now became subject to prior approval by the central government, except in cases of imminent necessity where the postponement of hostilities or treaties was considered dangerous. This subordination to the supreme government in diplomatic relations was a significant step towards integration. In addition to their controlling authority in a limited, though important, sphere of activity, the Governor-General and Council were, however, made primarily responsible for the local administration of the Presidency of Fort William in Bengal, a responsibility which they were to exercise in the same way as the President and Council or Select Committee before the enactment of the new measure. The Regulating Act was, in fact, statutory acknowledgement of a *fait accompli*, which avoided the inconvenience or even risk involved in the precise definition of authority for legislation.

The dignity of the new government was sought to be raised in the estimate of the Company's servants. The first Governor-General and his four Councillors were thus named in the Act itself, which laid down that they were not to be removed from office except by the King on the recommendation of the Court of Directors. Its hand was further strengthened by statutory provisions for the punishment of bribery and corruption on the part either of the Company's servants or those appointed by the Crown. The Act authorized the Governor-General and Council to make and issue rules, ordinances and regulations for the order and civil government of the Company's settlement at Fort William and other factories and places subordinate to it. It was specially emphasized that the measures to be so enacted for good government were not to be repugnant to the liberal laws of England.

The Governor-General, the head of the executive government, remained none the less subject to majority rule, a continuing legacy of the erstwhile presidential system. He had only a casting vote when votes were equally divided. This clogged the machinery of government and tended to obstruct its business during the first phase of Hastings' government, when three of the four councillors, namely Clavering, Monson and Francis, sided with the opposition to the Governor-General, who secured a majority in Council by a casting vote only on Clavering's death in November 1776. The Amending Act of 1786 invested the Governor-General with power to override the decision of his Council in executive matters and act without its concurrence in extraordinary cases involving in his judgement the interests of the Company or the safety and tranquillity of British possessions in India. The measure so enacted was thus in full conformity with the principle of government which Warren Hastings had earlier suggested to the Court of Directors in his letter of 11 November 1772.

Under the India Act of 1784 the territorial administration of the imperial dominion was to be placed with a representative body of Parliament, while ccmmerce was to be left in the hands of the Company. The government in India, which carried on both, was to be run in the name of the Company, but subject, in political and revenue matters, to the proposed Parliamentary body, a controlling authority, which was to leave the Company's right to patronage unaffected in all cases, excepting over the appointment of the Governor-General, the Governors of Fort St George and Bombay, and the Commanders-in-Chief of the three Presidencies. Both the appointment and recall of these principal servants of the Company were to be vested in the Court of Directors, subject to the approbation of the Crown, which meant the ministry concerned in actual practice.

Pitt's India Act of 1784 constituted a body of six commissioners for the affairs of India which was popularly known as the Board of Control. It was presided over by a Secretary of State, one of the six commissioners, and as a corporate body was authorized and empowered 'to superintend, direct and control all acts, operations and concerns' which related to 'the civil or military government or revenue of the British territorial possessions in the East Indies'. It was allowed full access to all papers, orders and proceedings of the Company's Directors who were required to deliver to the Board of Control not only copies or extracts of what transpired between them and their governments in India, but also their own orders, minutes or proceedings relating to matters of civil or military government as well as territorial revenues and diplomatic relations. 'The Board might approve, disapprove or modify the dispatches proposed to be sent by the Directors, might require the Directors to send out the dispatches as modified and, in case of neglect or delay, might require their own orders to be sent out without waiting for the concurrence of the Directors'.[9] The Board was in addition authorized by the Act to transmit through a secret committee of three Directors their own secret orders to India on the subject of war, peace or diplomatic negotiation with any of the country powers. Even the Court of Proprietors, the general body of the Company, could no longer revoke or modify a decision taken by the Court of Directors with the approval of the Board of Control. The Company retained its exclusive control of commerce and the right to patronage in accordance with the plan already accepted. The approbation of both the Court and the Crown was involved only in the case of the principal servants of the Company, which was, however, required to go by the statutory provisions that subjected promotion to the principle of seniority and reserved appointment to the executive council exclusively to members of the Company's covenanted service.

The India Act thus established an effective political system of Parliamentary control over the Company's civil and military government in India, a system which remained operative with slight alterations till 1858. As regards the Indian government, the Act made it subject to a system of dual control under which the Company mostly retained its power to initiate proposals, subject to the revising and directing authority of the Board, an authority which, in effect, produced unity in the direction of Indian political affairs without its being directly involved on the spot.

The council form of central government at Fort William was retained, but the strength of the Council was reduced from four to three. Moreover, under the Act the Governor-General could secure prior consideration of his proposals in the Council by postponing those of his councillors by forty-eight hours.[10] The Commander-in-Chief, on the other hand, was to have 'voice and precedence in Council next after the Governor-General'.[11]

As an integrating political instrument, the powers of the supreme government at Fort William over the subordinate Presidencies were enlarged and made precise and effective in all matters that related to transactions with the country powers, or to war, or peace, or the application of the revenues or forces of such Presidencies in times of war. The entire diplomatic relations of the Company in India as well as the finances necessary to support them came to be specifically entrusted to the supreme government in Bengal, and no subordinate Presidency was to be allowed to conduct negotiations with country powers except with the permission and direction of the Governor-General in Council,[12] whose ratification was made indispensable for any treaty or peace concluded at the instance of a subordinate Presidency. The orders of the supreme government had to be obeyed 'in all cases whatever'. The only exception was where a subordinate Presidency government received positive orders or instructions to the contrary from the Court of Directors or from their secret committee. The Presidency concerned was in that case required to send to the Governor-General in Council true copies of the order received from London. Distance was doubtless a serious handicap in those early days, but it accounted for the political system being a government of record which tended to reduce the evils of discretion and constituted a positive and firm basis for cool and responsible decisions.

As the local agency of the home authorities, the Governor-General in Council was invested with considerable powers of discretion, exercisable with restraint and responsibility. The India Act specifically declared that the supreme government at Fort William should have 'sufficient power for all the purposes of emergency, and all the occasions which immense distance might give rise to'.[13] But at the same time, considerations of political unity demanded that the sovereign (though yet undeclared) authority in London must alone be empowered to make final decisions on such political issues as war,

The Instruments of Territorial Integration

peace and diplomatic relations. And as the late wars with Mysore and the Marathas declared at the instance of local authorities prior to 1783 had endangered the security of British possessions in India, the India Act laid down a policy guideline that imposed constraints on the declaration of war with country powers without the orders of the home authorities. It was made clear that 'it shall not be lawful for the Governor-General and Council' to implement 'schemes of conquest and extension of dominion in India' without 'the express command of the Court of Directors, or of the Secret Committee of the Court of Directors, in any case, except where hostilities have already been commenced, or preparations actually made for the commencement of hostilities, against the British nation in India or against some of the princes or states dependent thereon'.[14] The provision so enacted not only signified in organizational terms an emphasis on the unity of command in the conduct of political and diplomatic affairs, but, even otherwise, contributed to the consolidation of gains through financial discipline and administrative reform in the public interest.

Consistent with the increasing concentration of authority in London, the executive head of its supreme government in India too was empowered to overrule his Council in order to effectively use his discretion in a state of emergency. This was done at the instance of Lord Cornwallis who made it a condition for his acceptance of the Governor-Generalship. The Amending Act of 1786, which invested the Governor-General with power to override his Council in extraordinary cases, however, applied only to matters of executive administration. It did not extend to legislative functions or the exercise of any judicial authority vested in it as a court of appeal. Unlike the Mughal Emperor, who was the 'head of the government, the commander of the state forces, the fountain of justice, the chief legislator and the final authority in the country',[15] the Company's administration was even in its early stages based on constitutional principles; the head of its supreme government was not to exercise any function vested collectively in the Council as a corporate body. Even the exercise of his veto was limited to extraordinary situations and there, too, its justification was subject to being duly recorded in proceedings. These restrictions on the exercise of power accounted for the growth of responsibility and conduced to the mildness of government, an important integrating factor.

The growing shift towards parliamentary control and centralization in policy matters was a response to the demands of *laissez faire*. It struck at the root of landed aristocracy in society and government, and of monopoly in trade and commerce. It called for an extension of the social base for economic and political operations, giving rise to industrial capitalism in business and democracy in politics. The philosophy of Liberalism flowed from it as an ideological efflorescence and found support with such intellectuals as Jeremy Bentham and the famous Mills, who led the Utilitarians. The opening of the Company's trade in 1813 and its complete abolition a couple of decades later were a logical consequence of the growth of 'English capitalists'[16] and their free industrial enterprise which under the Government of India Act (1833) the Company itself was obliged to encourage in return for an extension of its rule for another twenty years.

The dispatch of 10 December 1834, which the Court of Directors sent to the Government of India along with the Act, emphasized in clear terms that all the 'laws and regulations' enacted in India as well as all the 'executive proceedings in relation to the admission and settlement of Europeans, like that law of the Imperial Legislature out of which they grow, must, generally speaking and on the whole, be framed on a principle not of restriction but of encouragement'.[17] The Act of 1833 'unsealed' the doors of entry into India for British subjects of European birth and asked the Government of India to see that none of its laws or executive actions restricted their flow. Hitherto they had either been adventurers or remained in India on mere sufferance. They now acquired under the Act a right to live in the country and even to hold land in increasing numbers.[18] This was a new development which necessarily called for an all-India policy and a legal system to secure the protection of Indians against Europeans in the interior at an all-India level.

The years which followed the opening of the Company's trade were marked in England by unprecedented progress. Canning's appointment as the leader of the House of Commons under Lord Liverpool marked the first official recognition of that principle of Liberalism which had been growing in England as an expression of capitalism. The criminal laws of the country, which earlier punished even slight offences with death, were being slowly modified. The rigid laws which kept the working classes bound as serfs to their employers, were being repealed, and their demands for combined action to secure better wages and conditions of work were being recognized. The extension of franchise likewise formed part of the movement for reforms, and the first Reform Bill of 1832, which introduced extended representation, marked the beginning of a series of Liberal measures which covered education, child welfare, criminal laws, the press, Catholic emancipation and factory legislation.

During this period conditions were ripe in India too for British rulers to apply themselves to the task of beneficent government as well as the formulation of what might be called an Indian policy which encompassed all-India interests and not merely those of a region or province. The realization that Britain was now responsible for the government of all India led to the urgency of formulating new political aims to guide the conduct of government in relation to the Indian people.

The development of a new Liberal concept of the functions of government proceeded in India contemporaneously with that in England in the 1820's, although it began expressing itself most markedly in the legislative and administrative action taken during the Governor-Generalship of Lord William Bentinck (1828–35). It also owed much to the celebrated group of outstanding scholar-statesmen who rose to eminence in the service of the Company and shaped its Indian policy in the light of British Liberalism and their own experiences of Indian society. For instance, Sir Thomas Munro, the Governor of Madras, was known for his ardent support of the raiyatwari settlement of land revenue; Mountstuart Elphinstone, the Governor of Bombay, had previously reorganized the lands conquered from the Marathas on a principle recognizing peasant proprietorship; and Sir John Malcolm, Elphinstone's successor to the Governorship of Bombay, had established order in the administration of Central India, the stronghold of the Marathas.

The formulation of the new Liberal policy at an all-India level was not easy. It involved a serious clash between India's traditional society and the forces of change born of an advanced capitalist system which recognized the rule of law irrespective of sex or social position. A dichotomy naturally emerged between the supporters of tradition and the advocates of change. While the former reacted strongly against English ideas and methods, the latter, more especially the rising middle classes of Bengal, had made their choice in favour of change. In an address to Lord Amherst, the Governor-General, Raja Rammohan Roy, as early as 11 December 1823 had questioned the usefulness of public money being spent on the revival and encouragement of Oriental Studies, and emphasized the need for 'planting in Asia the Arts and Sciences of modern Europe'.[19]

But Calcutta, the headquarters of the Company's supreme government, which enjoyed the full benefits of British trade and industry, was not all of India. Nor did Munro, Malcolm, Elphinstone and Metcalfe ever lend their support to anything like 'rash change and innovation'.[20] They in fact strongly disapproved of ideas of 'denationalizing India' by an imposition of western views of 'religion and social policy'. In his instructions to officers acting under his orders in Central India, Malcolm made it clear that British power in India 'rests on the general opinion of the natives of the comparative good faith, wisdom, and strength of their own rulers', an impression which 'will be improved by the consideration we show to their habits, institutions and religion—by the moderation, temper, and kindness with which we conduct ourselves towards them'. He added that the reputation of the British government would increase all the more when 'contrasted with the misrule and violence to which a great part of the population of India have for more than a century been exposed'.[21] Malcolm thus advised his officers to use restraint in the exercise of their power and see that the religious susceptibilities of the Indian people were respected.

This declared policy of respect for Indian traditions, however, did not signify any abdication of the duty of the government to introduce into India the results of Western science and literature, or to assume an attitude of indifference towards such practices as sati and infanticide. Thus, the aim of the Indian policy which emerged at this period was not to make Indians discontented but, at the same time, not to sacrifice Western, liberal values and morality which the British government in India represented. This policy meant showing respect for traditional beliefs, 'without the appearance of weakness' and defending western ideas and institutions 'without the appearance of brutality'.[22]

The development of this all-India view in British policy was by itself integrative by virtue of its attempts to accommodate modernity within tradition and to provide for law and order in place of misrule and violence, to which most provinces had remained exposed for more than a century. The political system which this policy called into being under the Government of India Act (1833) was designed to reinforce territorial integration with a highly centralized machinery of government.

The immediate occasion for centralization was the need for economy in the administration of British territories so that the Government of India could pay the Company an annuity of £630,000, the dividend then received by its Proprietors who, subject to the payment of this annuity over a period of forty years, surrendered by an agreement with the

Crown all the Company's real and personal property in India; they held it now in trust for the Crown. The Company was allowed to retain as security for the regular payment of its annuity the British territories and their administration in India for a period of twenty years. If deprived of the Government of India at the close of that period, it could demand the repayment of its capital stock, or, alternatively, the continuance of the annuity for forty years more. All financial control was therefore centralized with the Government of India: the Act of 1833 laid down that 'no Governor or Governor in Council shall have the power of creating any new office or granting any salary, gratuity, or allowance, without the previous sanction of the Governor-General of India in Council'.[23]

But a more fundamental reason for a highly centralized and cohesive constitution was the social and economic pressure reflected in the abolition of the Company's trade under the 1833 Act, which opened Indian markets to Britain's expanding industries, resulting in an increasing influx of Europeans and British-born subjects who were encouraged to settle in the interior of India. Since British-born subjects were not amenable to laws other than those enacted by the legislature of their own country, their settlement in the country posed a constitutional problem. It became necessary not only to bring the immigrants within the bounds of Indian laws to protect Indians against their ruling influence, but also to reconstitute the legislative authority of the government in India on principles not repugnant to those in England. This task called for legislative centralization, another instrument of territorial integration and a welfare-motivated social policy which needed the direction and firmness of a central authority.

In other respects, too, the Act provided for an equally clear and politically unifying all-India perspective. Ever since the India Act of 1784, there was, of course, a steady rise in the authority of the Governor-General in Council.[24] The system of direct private correspondence which developed from the time of Cornwallis between the Governor-General and the President of the Board of Control increased the powers and influence of both at the cost of the Court of Directors. The avowal of British sovereignty on the principles suggested by Lord Grenville in 1813, the end of the fiction of Mughal sovereignty and the removal of the image of the Mughal Emperor from the Company's coinage for the first time in 1833 were all measures which raised the dignity of the executive head of the British government in India. Statutory provisions had likewise been made since 1784 to strengthen the authority of the supreme government in Bengal, but conquests had raised serious financial and administrative problems. The constitutional arrangement of 1784 could not cope with the extent of new territorial acquisitions. The Civil Finance Committee appointed by Bentinck in 1828 reported that, despite the statutory provisions of control, the authority of the supreme government over subordinate Presidencies remained nominal in practice. The basic defect of the existing political apparatus lay in 'the want of a supreme government divested of local charge and disjoined from the government of a separate Presidency.'[25]

The Act of 1833 recognized the recently established supremacy of British power and the established unity of India by a change in the full title of the Governor-General. Hitherto, he was 'the Governor-General of Fort William in Bengal' and recognized only as possessing control over the 'British territorial possessions in India'. He was now styled 'the Governor-General of India' and his supreme council at Fort William as the 'Indian Council'. For purposes of all-India unity provision was made to enable the Indian Council to meet in any part of India, a feature which, however, remained inoperative until 1865. Another provision of equal importance in terms of India's unity was the power given to the Governor-General in Council to pass from time to time a law authorizing the Governor-General to exercise in the course of tours to any part of India all the powers vested in the Council collectively, excepting those for making laws and regulations. In such cases he was required to nominate a President from among his councillors to officiate during his absence. Apart from central control over public finance, the authority of the Government of India over subordinate Presidencies was further strengthened by the latter being required to submit to the former true copies of all their orders and proceedings. Information proved, in fact, the best lever of control for integration.

Starting from the local or provincial centres of their separate Presidencies without any obvious attempt to interfere with the central authority of the Mughals, the English not only emerged as the only all-India alternative to Mughal sovereignty, but set up a central political system which could lend itself to change without prejudice to its basic infrastructural unity and strength, a system which had never so far existed in the history of the country. The Government of India so constituted under the Act of 1833 continued till 1858, when the Crown took over from the Company.

The Indian States

The political constitution provided for British India did not apply to the 600-odd Indian states which survived the establishment of the British dominion. According to the Report of the Butler Committee (1828–9) these states covered about two-fifths of the area and one-fifth of the population of India. Politically, there were in fact 'two Indias' existing under British paramountcy—British India, governed directly by the British according to the statutes of Parliament and enactments of the Indian legislature, and the Indian states, still mostly under the personal rule of their feudatory chiefs or their traditional princely families. The Government of India Act (1858), which put an end to the administration of the Company, did not give the Crown any new powers which it had not previously possessed. It merely changed the machinery through which the Crown exercised its powers.

Before the take-over by the Crown in 1858, the position of about seven-eighths of the 600 states represented a traditional phenomenon in Indian politics. 'Every Indian conqueror', as Dodwell points out, 'had found himself embarrassed by the difficulties of administering the great extent of India, and had always left more or less undisturbed a great number of local chiefs who thus fell into dependence without ever undergoing the rigours of conquest. Their position had always depended on the attitude and might of the dominant power; and what they had been under the Moghul emperor they continued to be under the East India Company.'[26]

With the remaining eighth of the states, the Company's relations had doubtless been defined by a series of treaties which, while stipulating its control over external relations, differed greatly in terms of the right to interfere in the internal affairs of a state. While the Company had wide powers of interference in the case of Baroda, Mysore and Oudh, the Rajput princes were allowed to remain absolute rulers within their own territories. They had no international status, but under their treaty stipulations they retained wide sovereign powers which could be exercised as they pleased in their internal affairs. The case of the Nizam of Hyderabad stood on a different footing. Originally, his relations with the Company were at least on equal terms, but this was reduced after his defeat by the Marathas in 1795 and his subsequent acceptance of the terms of a subsidiary alliance under Wellesley. The Company's government justified interference even for reasons of failure to meet financial stipulations or where a treaty happened to be silent on the point of internal management. In some cases the Company had specifically agreed to protect a prince not only against external attack, but also against internal rebellion, a provision which involved a right to intervene despite an express treaty stipulation against interference. The Company's relations with the Indian states prior to 1858 were, in fact, altogether vague and indefinite. For, in addition to the rights vested in the Company by warring states for security against predatory and other attacks 'there had arisen under no sanction but that of [its] superior power on the one side and reluctant acquiescence on the other a body of precedents relating to successions and to interference in the internal administration of the states. Together these constituted the Company's paramountcy, undefined, undefinable but always tending to expand under the strong pressure of political circumstances. The process ... was a constitutional, not a diplomatic development. The princes who in the eighteenth century had been *de facto* sovereigns but *de jure* dependents, had become *de facto* dependents though possessing treaties many of which recognized them as *de jure* sovereigns.[27]

There are several studies on the subject of the Indian states and their relations with the paramount power, and it is not intended here to deal with these relations in any depth. It is necessary, however, to throw some light on the 'two Indias' phenomenon and its relevance to political integration, the recognition of which symbolized the statesmanship needed to hold both together within the framework of India's geographical indivisibility. For, as the Montford Report later recognized, it was through the 'two Indias' concept that British policy towards the states passed from the original plan of non-intervention in matters beyond its own 'ring-fence', to the policy of 'subordinate isolation' initiated by Lord Hastings; which, in its turn, finally led to what might be described as a policy of union and cooperation on the part of the Indian states with the paramount power, signifying the growth of their interest in many matters 'common to the land to which they and the British provinces alike' belonged as part of 'one geographical whole'.[28] The Chamber of Princes established in 1921 provided a common institutional forum that strengthened the forces of integration.

It was on the basis of 'two Indias' under one paramountcy that successions by adoption in the princely states continued to be respected till 1841, although annexation based on lapse or escheat on the failure of legitimate offspring, or the existence of any near relation, had already started occurring with the accession of Lord Auckland to the office of Governor-General. Based on the *House of Commons' Return on Successions by Adoption* (1850) Ludlow has mentioned several cases of succession,[29] including the Bhopal case (1819–20), which show that the British at that stage did not consider annexing a princely state by lapse. The Bhopal case showed clearly that, apart from Hindus, the adoption even by a Muhammadan princess of a son or successor to her husband dying without a male issue was sanctioned as a matter of course. In the Kotah case, which occurred in 1829, the Government of India held that the ruler 'must be considered to possess the right, in common with all other Hindoos, of making an adoption in conformity with the rules of the Shaster'.[30] In the first Gwalior case, adoption was not only sanctioned, but pressed upon the prince during his lifetime. In the first Indore succession case (1833–4) no objection was made to adoption by the widow of Holkar, although a collateral heir of full age made good his claim against the infant heir by adoption. These cases indicate the statesmanship which the paramount power exhibited in its attempt to bind the two-Indias together and its respect for legitimation founded on history and religion, law and usage, custom and consensus.

However, about this time economic force of free enterprise began to press for statutory recognition as an aid to Britain's industrial capital and demanded that British-born European subjects be given the right to enter and settle in India to promote markets for the consumption of British manufacture. The Government of India Act (1833) provided for the desired settlement of Europeans in the interior of India, and the Court's dispatch of 10 December 1834 ordered the Government to encourage it. This new situation influenced the authorities in India to take suitable steps to promote colonization and became a reason for the acquisition of territories. The settlers naturally found specially favourable conditions in the hills, about which *The Times* of 9 November 1858 made an interesting and relevant comment:

Some years ago, it was our policy to avoid the hills, and to shun any accession of territory among them. We therefore scrupulously respected the Hindoo right of adoption, and in defect of heirs we sought out distant relatives, and placed them on the *guddees* of their little principalities. Now our policy is altered. We desire the acquisition of territory in the hills. There is a day-dream of colonization and tea-planting in the minds of some of our people, and we wish to define our frontier; therefore the right of adoption is denied. We rigidly scrutinize the claims and the legitimacy of heirs, and inquire into the purity of pregnant *ranees*, and the natives, whose memories are long, look on and wonder.[31]

Despite their differences on other counts, Bentinck and members of his Council were unanimous in assuming that, 'if India was to be Westernized and modernized, the creative forces to produce change and development in the civil service and in India generally had to be European, and that European colonization in India was therefore immediately desirable on some scale'.[32] But while advocating 'an active policy of European colonization' as the only effective agency of 'reform and modernization', Bentinck had his own reservations. Guided in his conviction by Liberal social policy, Bentinck held firmly that, if British rule over Indians was to have a justification other than force and financial profit, European colonization for India's social and economic development must be 'by a small, sober class of British with capital and skill, mainly artisans, teachers and craftsmen', who should possess 'a character of solidity'.[33] This condition, however, was not acceptable to the Company's Directors who wanted to have their rule extended in their own interest for another twenty years, in return for statutory recognition of the rights of European settlers. The Charter Act (1833) conceded the right of European colonization, requiring the Government of India to implement it on the lines indicated in the Court's dispatch of 10 December 1834. The policy of free colonization engendered a spirit of annexation which soon found expression in the dispatches of British Agents to princely states. For example, Anslie, the Agent for Bundelkhand, doubted the sanctity of adoption in the Dutteah succession case, although he recognized that in the past adoption had been given precedence over the right of the British government to annex by escheat. Simon Fraser, a subsequent Agent for Bundelkhand, who was murdered later at Delhi in 1857, wrote in the same strain, and held that, where there was a treaty with a prince, his heirs and successors, the character of the engagement was altered if, by usurpation or otherwise, a person other than the rightful heir succeeded. But the old practice of succession by adoption none the less remained operative in the states of Bundelkhand, thanks to the intervention of Metcalfe, the Lieutenant-Governor of the North-Western Provinces and the support he received from Auckland, the Governor-General.

However, a new era of annexation began with the Colaba Case (1841–4), where the widows of Raghoji Angria were not allowed to adopt 'illegitimate' heirs to the principality. The Home Government agreed with the Government of India, whose policy now was not to abandon what it considered to be a 'just and honourable accession of territory or revenue.'

The Instruments of Territorial Integration 19

The Colaba Case was soon followed in 1842 by that of Mandavi, where lapse was enforced against the claims of adoption by widows and those of collaterals. It is true that adoptions might still be sanctioned, and that the Governor-General did reject suggestions to enforce escheat in the case of larger states. But the suggestions for annexation, instead of coming from the Residents, began now to originate in the headquarters of the Government of India itself. These suggestions remained unexecuted only when a subordinate officer resisted the temptation to please his superiors. Colonel Sutherland, an Agent for Rajputana, was, for example, a person who consistently pursued a course which maintained the integrity of the Rajput states—except perhaps in the case of Kotah in 1844 where, on the advice of the Resident at Indore, the succession in that state was sought to be limited to male heirs of the ruling prince. However, in a number of cases involving Rajput states, the principle of adoption by a widow was duly recognized. Sutherland even insisted on a second adoption being accepted if the first child died either before or after attaining maturity.

However, it was not easy for subordinate officers in the political department to preserve their independence in the face of the supreme government itself being committed to annexation Auckland, Ellenborough and Hardinge were keen to reduce the scope of adoption either by insisting on a direct lineal and legitimate heir of the deceased ruler or by treating sovereign princes as feudatories, and not entitled to any right of adoption. In the Indore case, for instance, Hardinge made it clear that the chiefship of the state must descend to the male heirs of the prince who were 'lawfully begotten in due succession from generation to generation'.[34] The person whom the Resident had recommended for nomination on the ground of popularity and general support was outright rejected. The principality of the Holkar was sought to be transmuted into an entailed state by stamping adoption as an act of grace on the part of the paramount power. Succession was in fact stripped of all considerations of either hereditary right, or right to adoption, or even consensus recognized earlier as a factor in the preservation of law and order, peace and tranquillity.

This line of policy indicated by Auckland and his two immediate successors was systematically pursued by Lord Dalhousie as an act of faith in colonization, which had been steadily increasing since 1837. The number of European settlers did not rise rapidly, but even so, a parliamentary report of 1852 indicated that the total number resident in India but not in the service of the Company had risen to a little more than 10,000. The systematic use by Lord Dalhousie of the doctrine of lapse could not but produce a result which forfeited the goodwill of Indians. Dalhousie assumed the Governor-Generalship in January 1848 and proceeded to wield the doctrine of lapse with terrific effect: Satara was annexed in 1848, Nagpur in 1853, Jhansi in 1854 and Oudh in 1856.

The respect for 'two Indias' as an integrating force was, however, restored by the Queen's Proclamation[35] of 2 December 1858. 'We hereby announce to the native princes of India', it said, 'that all treaties and engagements made with them by or under the authority of the Honourable East India Company are by us accepted, and will be scrupulously maintained; and we look for the like observance on their part.' The Proclamation further made it clear that the British government would not any more pursue a policy of annexation: 'We desire no extension of our present territorial possessions, and, while we will permit no aggressions upon our dominions or our rights to be flouted with impunity, we shall sanction no encroachment on those of others. We shall and we desire that they, as well as our own subjects, shall enjoy that prosperity and that social advancement which can only be secured by internal peace and good government.' The Proclamation thus signified clear recognition in principle of two Indias bound together politically under the Crown 'by the same obligations of duty which bind us to all our other subjects'.

Law and Legislative Authority

A general system of judicial establishments and police, as well as a legal system and legislative authority applicable in common to all classes and conditions of people in the Company's territories constituted a cohesive force, providing in course of time institutional mechanisms to link even legislators with their constituents for purposes of political stability.

Position Prior to 1834

The institutional frame which was sought to be evolved in the first instance in Bengal, emerged from an interaction between the Mughal system called *adalats* and the Supreme Court of judicature established at Fort William in Calcutta under the Regulating Act (1773). It authorized George III to erect or establish such a court by charter or letters patent, with a chief justice and three judges, being barristers in England or Ireland of not less than five years' standing, who were to be from time to time named by the King, his heirs and successors.

Under the Mughal system the Emperor himself was the supreme judicial authority functioning as the highest court of appeal and acting at times as a court of the first. The *qazi* was the judicial officer appointed at both superior and subordinate levels to administer civil and criminal justice according to Islamic law. But since the bulk of the population happened to be non-Muslim, a *daroga-i-adalat* was appointed for the determination of civil causes, subject to final approval by the *diwan-i-ala*, a minister at the centre who had his counterpart in the provinces also. The administration of criminal justice was, however, governed in all cases by the canon law.

The governor of a province was responsible for the maintenance of law and order, which covered the administration of criminal justice and police. He could punish offenders 'by reprimands, threats, imprisonment, stripes or amputation of limb', and he was expected to use 'the utmost deliberation before severing the bond of the principle of life.'[36]

With the decline of the Mughal empire, chaos resulted in the administration of law and order, which came to be arrogated by a numerous class of 'assignees' or farmers of land revenue passing under the common parlance of zamindars. After the grant of *Diwani* in 1765 no reforms worth the name were introduced by the Company till 1772. It then stood forth as diwan directly through the agency of its own servants, and Warren Hastings constituted under a European collector in each district a civil court, called the *diwani adalat*, and a criminal court, called the *faujdari adalat*, with their appellate counterparts established in Calcutta under a member of Council and known as the *sadar diwani adalat* and *sadar faujdari adalat*. But when collectors were recalled in 1774 at the instance of the Directors, districts were grouped together into six divisions,[37] each being placed under a chief and four or five members, who for the mostly happened to be the erstwhile collectors themselves. The chief and the members so constituted were together known as the Provincial Council, except in the case of Calcutta, which was supervised by a committee of revenue consisting of two members of the Council and three servants below the rank of Councillors. In addition to their revenue duties, each Provincial Council had under it a superintendent of the *sadar diwani adalat* attached for the administration of civil justice. The Councils themselves could also function as superior *diwani adalats*, hearing appeals from the decisions of deputies or Indian *naibs* in the districts. The administration of criminal justice, however, continued to be supervised by Mahomed Reza Khan who, though removed for a time, was appointed again to act as *naib nazim* and oversee the administration of criminal justice in accordance with the provisions of Muhammadan law.

Warren Hastings had in his judicial reforms proceeded on the assumption that if the courts of adalat worked under proper supervision, they would produce the same degree of efficiency as under the Mughals. His policy was therefore to restore them to their original design. But he was no Mughal Emperor or *nazim*. The political framework within which he was called upon to function was altogether different. Whatever the several other defects which Hastings sought to remedy under his judicial plan of August 1772,[38] the adalats in the past had not been courts of record or able to put an end to the practice of the same case being tried a number of times by the same authority for personal gain. In the administration of criminal justice, too, he proposed several amendments to remove the iniquities[39] of Islamic law governing the punishment of crimes. But the Muhammadan officers of justice, who were entrusted with that responsibility, regarded the task as blasphemous. The result was that punishments continued to be determined by the Quranic law, which left a wide field of discretion exercisable by qazis in such matters as did not relate to apostasy, a crime punishable with death. The Muhammadan criminal law did not recognize the principle of the king's peace and, in the absence of any law governing criminal procedure, the exercise of discretion degenerated into corruption. Corruption increased considerably in a situation where the servants of the Company wielded power without the question of its source of authority being in law defined or even ascertained.

In *The Judicial Administration of the East India Company in Bengal*[40] several cases of iniquity committed in the administration of criminal justice have been discussed, cases which European *supervisors* reported on their appointment to the several districts in 1769. They were surprised to see that the penalty inflicted was not related to the nature of the crime committed but the circumstances of the accused. While theft and murder were compounded for four or five rupees, fornication and witchcraft were punished with four or five thousand.

The establishment of the King's court in Calcutta, as in Madras and Bombay, marked a new phase in the evolution of law and legis-

The Instruments of Territorial Integration

lative authority in India. The freedom of the judiciary from executive interference was the basic concept of the new system, a concept foreign to the earlier authoritarian constitution of government and society in the country. It introduced trial by jury in criminal cases and maintained regular records of judicial proceedings.

Irregularities naturally occurred in the administration of criminal justice even in the Mayor's court established at Fort William in Bengal.[41] But in spite of its lmitations, it marked a significant qualitative change from the native legal system. Although the judges of the King's court were no more than ordinary covenanted servants of the Company, the introduction of attorneys in the conduct of judicial proceedings tended to impart to adjudication a degree of independence which clashed with the wishes even of the government under which they served. It was this independence which in course of time created confidence in the judiciary and law as an integrating force.

The Supreme Court of judicature established under the Regulating Act was designed to remove the limitations of the erstwhile King's court. No covenanted servant of the Company was to be its judge. It consisted of the chief justice and three other judges, all professionally qualified; for the Act itself laid down their minimum qualifications for appointment. The Supreme Court, which had full power and authority to exercise a civil, criminal, admiralty and ecclesiastical jurisdiction, was itself responsible for the establishment of such rules of practice and procedure as it deemed necessary for the administration of justice in accordance with the terms of the charter under which it was constituted. It was at all times to be a court of record, a court of oyer and terminer and gaol delivery within the limits of its jurisdiction.

Conflict between the Supreme Court and the supreme council at Fort William in Bengal proceeded from an extension of the court's jurisdiction to meet the requirements of the Company's territorial expansion. In addition to the town of Calcutta and the factories subordinate to it, the powers and authorities granted to the Supreme Court under its charter extended to all British subjects residing in the provinces of Bengal, Bihar and Orissa under the protection of the East India Company. The charter provided that the jurisdiction of the Supreme Court in civil and criminal cases would extend not only to English subjects in Bengal, Bihar and Orissa, but also to any person employed directly or indirectly in the service of the Company. In other words, in both the categories of personal jurisdiction, the court was invested with full power and authority to hear and determine complaints for all crimes, misdemeanours or oppression in the three provinces. But as the Company had acquired these provinces on grant from the Mughal Emperor without any reference to British sovereignty being involved, the charter-based power and authority exercisable by the Supreme Court could not have the full backing of law in case the Company chose to argue that it held these provinces not as its real property but as an officer (*diwan*) of the Mughal Emperor authorized to organize adalats.

The Supreme Court and the adalats thus derived their authority and jurisdiction from two distinct sources, the former from the English king under the Regulating Act, and the latter from the Mughal grant of the diwani. Troubles did not arise initially because the jurisdiction of the executive-nominated judges of the Mayor's court was restricted to the settlement of Calcutta and its subordinate factories, and confined to British subjects of European origin. But conflicts between the Supreme Court and the adalats were bound to arise, as they did, when the Regulating Act provided for the appointment of completely independent judges with powers to exercise a jurisdiction partially concurrent with the jurisdiction of the Company's *adalats* without unifying the sources of sovereignty from which each derived its power and authority.

The Governor-General thus proposed reconstitution of the *sadar diwani adalat*; since the task demanded knowledge and training in the theory and practice of law which members of the Council at Fort William did not possess, Warren Hastings himself took the initiative and, on 29 September 1780, proposed the appointment of Sir Elijah Impey as judge of the *sadar diwani adalat*, so that he might prepare a series of regulations for the *adalats*.[42] The regulations so prepared by Sir Elijah received the assent of the Governor-General and Council on 3 November. These were subsequently incorporated with additions and amendments in a code which the government adopted on 5 July 1781. It imparted to the *sadar* and provincial *adalats* a legal form and content which bore the stamp of Sir Elijah's professional quality and attainment. The diwani *adalats* or civil courts were, for example, divided at two levels, mufassil *diwani adalats* in the districts and *sadar diwani adalats* at the headquarters of government. And the landed aristocracy was in law divested of the privilege it enjoyed earlier. Every judge of the mufassil *diwani adalat* was now vested with powers to summon any zamindar or farmer of revenue to appear in person or through his vakil to answer an action lying in the court but not relating to public revenues.

Another important change which Sir Elijah's code of 5 July 1781 brought about was uniformity of judicial practice, a powerful intergrating element. Every mufassil *diwani adalat* was, for instance, authorized to make standing rules of judicial practice. All such rules and orders required the approval of the *sadar diwani adalat* and the assent of the Governor-General and Council. But once they were so approved and confirmed, they became applicable not only to the court of origin, but to every other mufassil *diwani adalat*. These rules were circulated for information and guidance and duly maintained in serial order. Like the Supreme Court of judicature, a court of *adalat* was also being induced to become a court of record.

The *sadar diwani adalat* normally acted as a court of appeal. But in case a lower court neglected to entertain a cause, the *sadar diwani adalat* could intercede and step in as a court of original jurisdiction, or ask the mufassil *diwani adalat* concerned to entertain and decide it. Similarly, the *sadar* court was vested with full powers and authority to frame rules of judicial practice for general guidance, and to revise, alter or disapprove those already made by subordinate courts. Judges were now required to take an oath of loyalty and uprightness of conduct in the discharge of their duties. No judge was to issue oral orders in the administration of justice. The whole emphasis of Sir Elijah Impey was, in fact, to see that the gap between the King's court and the Company's courts was reduced in respect of rules governing constitutional and procedural matters.

As regards criminal justice, the law of England applied to Europeans and remained limited to the town of Calcutta and its subordinate factories. The Mughal rulers had, in fact, extended this privilege to all British subjects who enjoyed freedom in the exercise of their own law within the limits of their factories, a freedom which applied to all European merchants as well to the Company's servants in general. The Indian settlers within the bounds of these factories were, however, subject to Muhammadan law and subject to the jurisdiction of a qazi. But in practice they also received the protection of British laws, for the qazis were often induced to remain indifferent and connived in avoiding the jurisdictional conflicts. The operation of the English law thus remained for a time restricted to the pockets of the Company's trade and manufacture.

The judges of the Supreme Court of judicature were under the Regulating Act empowered to frame their own rules of business. These rules were to be in consonance with those found in the courts in England. The Act therefore provided specifically that all offences cognizable by the Supreme Court must be tried by a jury[43] of British subjects resident in Calcutta. A trial by jury in a criminal case was indeed a cumbrous process, designed, from considerations of civil liberty, to make conviction difficult. It consisted of a lengthy operation of 'calling and swearing the jury, of a speech by the counsel for the plaintiff, the examination, cross-examination and re-examination of his witnesses: a speech by the counsel for the defendant followed by the examination, cross-examination and re-examination of his witnesses, and summing up of their evidence by him: the reply or speech by the plaintiff's counsel: the summing up of the whole case by the judge of the jury: and, lastly, the jury's verdict'.

As these legal procedures curbed the arbitrary proceedings of many Europeans, too, they aroused the combined opposition of both European and indigenous interests. The *Amending Act of 1781* therefore recognized on religious grounds the existing barriers of caste and the differences of status based on it. It also removed the jurisdiction of the Supreme Court from the Company's revenue agents, who exercised arbitrary powers of punishment in the collection of their demands—a concession to the various categories of agencies employed by the Company.

In their instructions to Lord Cornwallis the Directors, for political reasons, expressed themselves strongly against innovations borrowed from abroad in the administration of justice in India. They asked him to accommodate the views and interests of the Indian people and respect their customs and usages rather than rely on 'any abstract theories drawn from other countries, or applicable to a different state of things'. Even in suits between Indians within the jurisdiction of the Supreme Court, it was suggested, care should be taken to consider the personal law of the parties concerned, whether Hindu or Muhammadan. It is not that the Directors did not realize the need for change. Changes and modifications, as they said, were necessary from time to time. And these changes were actually reflected in the regulations enacted systematically by the Governor-General in Council for Bengal in 1793, and by the governments of Madras and Bombay who were granted similar powers in 1800 and 1823 respectively. Respect was thus restored for popular beliefs and usages which, though a negation of legal uniformity, was considered politically expedient.

After the Amending Act of 1781, the judicial system of Bengal witnessed the division of jurisdiction between the King's and the Company's courts. Sir Charles Grey, chief justice of Bengal, for instance, noticed in 1822 the 'utter want of connection between the

Supreme Court and the provincial courts and the two sorts of legal process which were employed in them'. And when supreme courts were established in 1824 in Madras and Bombay under the Indian Bishops and Courts Act of 1823, the experience of Bengal was repeated. Sir Erskine Perry, chief justice of Bombay, later referred to 'the strange anomaly in the jurisprudential condition of British India which consists in the three capital cities having systems of law different from those of the countries of which they are the capitals'.[44]

Apart from the jurisprudential anomaly, there was the inconvenience and delay flowing from the exclusive jurisdiction which the Supreme Courts of the presidency towns alone exercised over Europeans residing outside. The Charter Act of 1813 had, of course, done something to mitigate this inconvenience. It provided that British subjects residing, trading or holding immovable property more than ten miles beyond the presidency towns would be considered subject to local civil courts without a loss of their right to the Supreme Court; and justices of the peace, who were covenanted servants until 1832, were appointed to deal with debts due by them to the extent of not more than Rs 50 and cases of trespass and assault against them for which a fine of Rs 500 would be sufficient punishment. All serious cases, however, still remained subject to the Supreme Courts in Bengal and Madras and the recorder's court in Bombay, which was succeeded by the Supreme Court established under the Act of 1823.

A resort to local custom and usage as a guiding principle of legislation for Indians, and dependence on English law for Europeans, led in practice to confusion. Though considered politically convenient this proved to be jurisprudentially unsound. Sir Courtenay Ilbert summarized the situation that prevailed before the enactment of 1833.

At that date there were five different bodies of statute law in force in the Empire. First, there was the whole body of statute law existing, so far as it was applicable, which was introduced by the charter of George I, and which applied, at least, in the presidency towns. Secondly, all English Acts subsequent to that date, which were expressly extended to any part of India. Thirdly, the regulations of the Governor-General's Council which commence with the Revised Code of 1793, containing forty-eight regulations, all passed on the same day (which embraced the results of twelve years' antecedent legislation) and were continued down to the year 1834. They only had force in the territories of Bengal. Fourthly, the regulations of the Madras Council which spread over the period of thirty-two years from 1802 to 1834 and [were] in force in the Presidency of Fort St George. Fifthly, the regulations of the Bombay code which began with the revised code of Mr Mountstuart Elphinstone in 1827, comprising the results of twenty-eight years' previous legislation, and which were also continued until 1834, having force and validity in the Presidency of Bombay.[45]

The judicial system which recognized jurisdictional divisions between the King's and the Company's courts and showed such variations, could not meet the requirements of a new situation that was emerging from three separate, though related, premises, namely, the establishment of British paramountcy, the growth of an Indian policy based on Liberalism, and mounting pressure for colonization seeking the right to immigration into India, which the Charter Act of 1833 conceded to European settlers. All these elements not only called for a new centralized political system, but also demanded uniformity in the preparation of a common code of law and procedure which, for both national and integrative purposes, began in 1834. And as the task of forming such a code for both Europeans and Indians involved the creation in India of a centralized legislative authority, separate and distinct from the executive government, the year 1834 also marked a beginning in the evolution of an Indian legislature.

Common Code of Law and Procedure 1834–62

The Charter Act of 1833 provided for appointment by the Governor-General in Council of a Law Commission on to inquire into 'the jurisprudence and the jurisdiction' of the existing courts of justice and police establishments operating in the whole of British India, to make reports and suggest alterations, due regard being given to religious opinion and custom prevailing among the different peoples in the British territories in India. In the preparation of the code, the Act thus took an all-India view that might represent the highest common denominator applicable to both Europeans and Indians.

The Commission was appointed in 1834 with Macaulay as president and three civil servants representing the three presidencies.[46] The object of the Commission was to frame a criminal code for the whole Indian Empire. 'This code', according to Macaulay, 'should not be a mere digest of existing usages and regulations, but should comprise all the reforms which the Commission may think desirable. It should be formed on two great principles—the principle of suppressing crime with the smallest possible amount of suffering, and the principle of ascertaining truth at the smallest possible cost of time and money.'[47] In points of both law and procedure Macaulay was thus deeply imbued with the knowledge and principles of English jurisprudence.

On his arrival Macaulay was faced immediately with the task of providing against the mischief or danger that might arise from the removal of restrictions on Europeans in the interior of India under the Act of 1833. For, in their explanatory letter of 10 December 1834, the Court of Directors had asked the Governor-General in Council to make 'laws for the protection of the natives from insult and outrage—an obligation which in our view you cannot possibly fulfil unless you render both native and European responsible to the same judicial control. There can be no equality of protection where justice is not equally and on equal terms accessible to all.

Consistent with the spirit of the Act, the privileged position of British nationals was ended in 1836. By Act XI of 1836 the special right of appeal to the Supreme Court was taken away, and it was laid that no person by reason of place, of birth or descent should be excepted from the jurisdiction of the Court of *Sadr Diwani Adalat*, of the zila and city judges, or the Principal *Sadr* or *Sadr Amins* in the territories subject to the Presidency of Fort William in Bengal.

The Act passed in 1836 met with the approval of all the provincial governments, as of most of the Company's servants. But it excited the European residents of Calcutta into virulent agitation. Macaulay described the opposition as 'expected'. It later subsided and he went ahead with the work of Law Commission.

The Code appeared in the course of 1837 itself. According to Fitzjames Stephen, Macaulay's work was 'far too daring and original to be accepted at once. It was a draft when he left India in 1838'. The draft of Macaulay's Indian Penal Code and the Code of Criminal Procedure which appeared in 1837 were later revised by Drinkwater Bethune and Sir Barnes Peacock. But they did not become law until about twenty-two years after his departure from India. The first Law Commission lingered on for several years after Macaulay, and periodically published bulky volumes of reports without the Governor-General in Council being induced to effect any measure of codification. It was simply allowed to expire.

In these circumstances the Charter Act of 1853 provided for the appointment of a second Commission which, prior to its own recommendations, was instructed to make a diligent and thorough inquiry into the working of the law and procedure already established, and to examine and consider the various recommendations made from time to time by the previous Commissioners on the subject of legal and procedural reforms. The second Commission was appointed in November 1853. It consisted of eight members, including Sir John (afterwards Lord) Romilly, Sir John Jervis, Sir Edward Ryan and Robert Lowe.

The Commission sat in London till the middle of 1856 and presented four reports, including plans for the amalgamation of the Supreme and Sadar courts as well as a uniform code of civil and criminal procedure, applicable not only to the High Courts to be so called on amalgamation, but also to the lower courts in the interior of British India. It also referred to the inadequacy of substantive civil law and submitted its views regarding the best mode of overcoming that deficiency. The outbreak of the Mutiny, however, produced in the meantime a situation which threw law reform into the background. The new policy which flowed from the Queen's proclamation was to recognize the distinctions between eastern and western ideas and institutions in a spirit of understanding, but also to proceed without hesitation in the direction of reforms on a national and uniform principle that would preserve India's unity against the forces of disintegration. The effect of the Mutiny on the statute-books was therefore unmistakable.

The recommendations of the second Commission resulted in important legislation both in Parliament and in the Legislative Council of India. The Code of Civil Procedure was enacted in 1859. This was followed by consideration in the Legislative Council of Macaulay's Indian Penal Code, which was revised and passed into law in 1860, to come into operation on 1 January 1862. A Code of Criminal Procedure was likewise enacted by the Council in 1861. On the recommendation of the Police Commission of 1860, separate police establishments were also created in British India under the Indian Police Act of 1861. By an Act of Parliament passed in 1861 itself, the Supreme and Sadar Courts were amalgamated into the three High Courts of Calcutta, Madras and Bombay, and provision was made for the establishment of another High Court for the North-Western Provinces. By 1861 India had thus acquired a penal code and codes of both civil and criminal procedure, the whole concept underlying each being an all-India and rational approach, meant for all persons, British or natives, all for all places, presidency towns or rural districts. The Codes represented the highest common denominator evolved out of the different categories of legal and judicial systems that preceded them.

How far these Codes were related to the resolution of social conflicts amongst India's social and economic categories is a question that formed no part of the Commission's concern. What is of immediate relevance to our study is to note that the judicial and police establishments which functioned under these codes provided a unified institutional mechanism to enforce sustained political loyalty to the paramount authority. For the Indian Penal Code applied without exception to all classes and conditions of people, irrespective of their religious persuasions or regional differences. The mechanism so devised represented a synthesis of the Queen's and the Company's courts, though heavily weighted on the side of the former in respect of both penal and procedural provisions as well as organizational and ecological matters, which were considered necessary for the promotion of the rule of law as the foundation of the entire system. But, as in the case of social reform and retaining the concept of two Indias, considerations of policy demanded that due regard be given to the rights, feelings and peculiar usages of the people. Consistent with the Amending Act of 1781, no departure was made from the fundamental principle that the Indian people were to remain entitled to their own personal law. The modernization of the legal system was thus sought to be effected from within the framework of indigenous institutions themselves. Modernity proceeded with due respect for tradition.

Separate Legislative Authority

Prior to the Charter Act (1833) the Governor-General in Council at Fort William exercised both executive and legislative functions, although under the Amending Act (1786) legislation became the collective responsibility of the Council, the Governor-General not being authorized to act independently or override its decision, as he could do in his executive capacity. The exercise of legislative functions was even otherwise anomalous and ill-defined. The Governor-General in Council could, for example, legislate for the settlement of Fort William on British-born subjects. But such regulations were not valid unless made with the approval of the Supreme Court and duly registered in it, subject to an appeal to the King in Council. No such approval was, however, needed for the enactment of regulations governing the Company's provincial courts and councils under the Amending Act (1781), or for the levy of taxes and duties within the presidency town itself under an Act of 1813. There was, in fact, a complete want of uniformity and precision in the functioning of the legislative organs of government, which differed little from its executive. As already noted, this flowed from the existence of different legal systems operating under separate presidency governments.

The anomaly of legislative authority in the Indian situation was brought into focus by James Mill, who pointed out that it sprang basically from 'the extraordinary circumstances of there being a class of people in the country, a class in reality of foreigners, not very considerable in point of numbers, but remarkable in certain circumstances and from the power attendant on those circumstances, who are not subject to the legislative power of the government under which they live: who claim exemption from its enactments, and for whom the government has no power of making laws'. In the course of his evidence before the Parliamentary Select Committee of 1832, Mill therefore emphasized equality in the eyes of the law between the subjects of British India.

Mill's suggestion, however, raised a constitutional issue. For it became necessary not only to bring British immigrants within the bounds of Indian laws, but also to reform these laws and to reconstitute the law-making body on principles that might not be repugnant to those of the mother country with its free legislature. This was a difficult task, for there were in India different laws for different classes of people and places. And, above all, there was no example in the country of free legislation, but only of executive-made regulation that was unacceptable to British-born subjects.

Macaulay had expressed himself in favour of continuing the Company as an administrative corporation, a neutral body possessed of experience but free from the influence of party politics of England. In other words, its government in India must likewise be conducted on non-party principles. Following from this was his other suggestion, which was to 'give a good government to a people to whom we cannot give a free government'.[48] His concept of good government was based on the writings of Jeremy Bentham, the exponent of a strong executive functioning within the framework of legislative authority. Macaulay wanted India to have a strong and centralized executive government with some kind of a separate legislature, subject to control by Parliament. He believed that any law enacted by a separate Indian legislature would not only be acceptable to European settlers, but might also eventually lead to the expansion under this system of the public mind of India for better government based on European institutions.[49] But the immediate purpose for which a separate legislative authority was sought to be established in India was to meet the legal exigencies arising from increasing British colonization, which dictated the necessity of one legislative authority for the whole of India.

The Act of 1833 accordingly provided that the Governor-General in Council 'shall have power to make laws and regulations ... for all persons, whether British or native, foreigners or others, and for all courts of justice, whether established by His Majesty's charters or otherwise, and the jurisdiction thereof, and for all places and things whatsoever within and throughout the whole or any part of the said territories, and for all servants of the said Company'.[50] This power was, however, not to extend to the making of any laws and regulations (i) which should repeal, vary or suspend any of the provisions of the Act of 1833, or of the Acts for punishing mutiny and desertion by officers and soldiers in the service of the Crown or the Company; or (ii) which should affect any prerogative of the Crown, the authority of Parliament, the constitution or rights of the Company, or any part of the unwritten laws or constitution of Britain, whereon may depend the allegiance of any person to the Crown or the sovereignty or dominion of the Crown over the Indian territories; or (iii) without the previous sanction of the Court of Directors, which should empower only courts other than a chartered court to sentence to death any British subject born in Europe or their children or abolish any of the chartered courts.

There was in addition an express saving of the right of Parliament to legislate for India, and all Indian laws were to be laid before Parliament to enable it to exercise its power. Any Act made with the final sanction of Parliament was to be treated as an Act of Parliament, which did not require registration or publication in any court of justice. It signified the supremacy of the legislative authority vested finally in Parliament.

The powers of the subordinate presidencies to legislate were withdrawn from them and vested in the Legislative Council of India, which was reinforced by the addition of a professionally qualified legislative member. The subordinate presidencies were authorized merely to submit to the Governor-General in Council drafts or projects of any laws or regulations which they might think expedient, and it was the Governor-General in Council who was required to consider them and communicate his resolutions to the government that proposed the drafts or projects of law. The laws made under the Act of 1833 were known as Acts, which took the place of 'regulations' made under the previous Acts of Parliament. Legislative functions were thus centralized in the same way as executive decisions.

But centralization apart, the new legislative authority also differed in respect of its constitution and procedure, a difference which contributed to the growth of a separate all-India legislature as yet another powerful instrument of territorial integration. Constitutionally, the beginning to a separate legislative body was marked by a change introduced by the Act of 1833 in the form of a fourth 'Ordinary Member' who was appointed not from within the Company's service, but from outside its cadre, a choice made by the Crown. He was to be well versed in law and jurisprudence and, further, distinguished from the other members of the Council in that he was to attend only such meetings as were specifically called for legislative purposes. This inclusion of the legislative member among the Ordinary Members amounted, as Macaulay had suggested, to engrafting 'on despotism those blessings which are the natural fruits of liberty'.

Although the addition of a mere legislative member to the Council of the Governor-General did not make a legislative council in its full sense, the fact remained that, ever since the appointment of Macaulay as the fourth Ordinary Member, a procedural as well as constitutional distinction had begun to be recognized both in India and England, a distinction which conduced to the formation of a separate legislative body. It was a clear case of advance or modernization from within the existing political framework. It was both continuity and change mixed together, a principle which governed institutional developments in other fields also.

The speed with which change proceeded in the legislative field, was, however, accelerated by British immigrants, more especially under Lord Dalhousie when he lent his full support to the colonial interest in India and promoted it by making the legislature its central forum and powerful instrument. On his recommendation, the Executive Council of India came to consist of five members, including the Commander-in-Chief and the Law Member, who, under the Charter Act of 1853, became full member with powers to sit and vote at all meetings. For legislative purposes the Executive Council was to have a representative from each of the four governments of Madras, Bombay, Bengal and the North-West Provinces, in addition to the Chief Justice and one of the judges of the Supreme Court at Calcutta. All six legislative councillors (called Additional Members) so appointed under the Act of 1853 were not to sit and vote in the Executive Council except at meetings held for making laws and regulations. No law or regulation was to be passed unless assented to by the Governor-General.

After the Charter Act of 1853 there developed a clear distinction between the Executive Council and the Legislative Council. This, however, proceeded not so much from any specific statutory provision as from the manner in which Dalhousie chose to work it; for, like the previous statute of 1833, the term employed by the Charter Act of 1853 to designate the legislative organ was 'the Council of India for making Laws and Regulations'. It was his Minute, approved by this Legislative Council in its meeting of 20 May 1854, that made the difference. The Minute distinguished the 'Council of India' in its executive capacity from that for making laws and regulations. Dalhousie designated all six Additional Members as 'Legislative Councillors' and the Council of India, when meeting along with them, as the 'Legislative Council'.[51] In fact, he treated the two Councils as separate and distinct under the Act itself, and it was on the basis of this construction of the Act that he commended to his Legislative Council certain basic principles of legislative procedure which, he thought, were in conformity with 'past practice' and 'the experience gained during the last twenty years of the work of legislation in India'.[52] One of the 'basic principles' was the observance of due formality in the conduct of legislative business, which was recognized as controlled by an authority emanating from no other source but the Legislative Council itself.[53] Another principle introduced was oral discussion and legislation on the basis not only of proposals made by provincial governments, but also those received from individuals, or drafts suggested by persons who were not officers of the government. It meant complete freedom for the introduction of private bills, the only exceptions being those relating to public finance, the constitution of the army and foreign affairs, where no bill could be entertained by the Legislative Council unless transmitted to it by the Executive Council itself. In addition, provisions were made in the Standing Orders for the first, second and third readings of a bill, for its discussion in the Select Committee and the press, and for the admission of the public and press to debates of the Legislative Council, which appointed its own 'Clerk of Council' in place of the Home Secretary who previously played this role. All these rules were fashioned on the model of the British Parliament and one of the Standing Orders even provided for questions being asked in public, although the executive could refuse to answer them. But recourse to refusal could not always be had except on the grounds of going against public interest.⁵⁴ This kind of legislative independence was in fact encouraged by the presence of the judges of the Supreme Court, who were well versed in law and legal procedure.

The growing independence of the Legislative Council proceeded, however, from the increasing strength of European settlers in the country and the exclusive control which the European element exercised over the law-making body in Calcutta. The exclusion of Indians was justified by the argument that, since the 'large and liberal' principles of European countries 'had not yet made progress in Indian society, the European mind ... must lay down the general principles and direct the general movements of government'.⁵⁵ With added support from the Governor-General himself for colonial interests, the Legislative Council naturally emerged as the spokesman of that interest which called for legislative control over executive government to ensure the promotion of free enterprise. The representatives of this interest in the Legislative Council acted as an opposition to the executive government after Dalhousie's time, and not only called for information and papers, but also invited petitions to inquire into grievances. P. W. Le Geyt of Bombay, for instance, moved a bill in 1856 to authorize the Legislative Council to make inquiries, call witnesses to its bar and to compel the production of documents.⁵⁶ The proposed measure was abandoned, but cases occurred where members of the Legislative Council, more especially its judge member or the representative of a provincial government, called for full details on financial schemes whenever proposals were made to levy a tax on income or trade. Members even regarded withdrawal of information as a breach of privilege.⁵⁷

The Mutiny revealed the need for a resolution of the ill-defined, though dischotomous, relationship which had been developing between the colonial movement, which brooked no governmental interference with free enterprise and the exercise of racial superiority, and imperial policy which provided for a strong central government and a law commission to work out a common code of substantive and procedural law so as to ensure the protection of all, irrespective of race, place or position. The anomaly of the Indian situation, however, became manifest when the London and Calcutta authorities began to be noticeably partial to *laissez-faire* until, unlike Bentinck, Dalhousie identified himself completely with the colonial interest he was expected to control. As regards the codification of law, as we has seen, the entire work was laid aside until the Mutiny led to a realization of its importance from considerations of both justice and imperial security, which not only produced in a couple of years a unified system of law and justice, but also modified the machinery of legislation in a manner consistent with the strength of the executive government.

The Mutiny brought to the surface the venom of racial bitterness and Canning, who succeeded Dalhousie, soon became a target of virulent criticism by British settlers. An important reason for European opposition to Canning was his policy of moderation[58] in the course of the Mutiny and his apathy to Christian Missions who were always keen on conversion. The European population of Calcutta had gone so far as to petition the recall of the Governor-General for his 'clemency'. But the Queen's Proclamation, which was issued on the lines indicated by Canning, was expression of an imperial policy marked by restraint and concern for the interest of all, irrespective of racial or religious distinctions. The Mutiny was, in fact, a lesson to the policy makers to distinguish between colonial and imperial interests which, though related and overlapping in certain areas, were yet distinct, the latter taking a total and broad political view of things, while the former represented only a narrow and partial economic view.

There were other constitutional changes which followed immediately in the wake of the Mutiny to make imperialism as the most effective instrument of unity and integration. The government of the British Indian territories, which, under the Charter Act of 1853, was to continue under the Company for an indefinite period, was, for example, taken over directly by the Crown under the Government of India Act (1858), which placed all superior appointments of a political nature in the hands of the Queen, whose powers were in practice to be exercised by a Secretary of State for India, a Minister of Cabinet rank, who was to be assisted by as Under Secretary of State and a Council of fifteen members called the Council of India. The Secretary of State in Council was invested with all such powers as jointly or separately belonged earlier to the Court of Directors and the Board of Control.

The provisions of the Government of India Act (1858) need not be recapitulated here in detail. It is relevant to note, however, that, as the spokesman of the imperial government on Indian affairs, the Secretary of State had a say in all decisions except financial matters, where the concurrence of the Council majority was required for grants or appropriation of any part of the Indian revenues. He could even send a dispatch to India without reference to his Council. As a Minister of the British Cabinet he doubtless owed responsibility to

Parliament, but this responsibility was nominal; for the entire India Office was maintained from the Indian revenues and the establishment was not subject to votes by Parliament. The Secretary of State was, in fact, a Minister of the Crown without receiving his salary from the Crown's treasury. As regards his control over India, it became far more effective when an overland cable was established in 1868, the Suez Canal opened in 1869, and a British submarine cable completed in 1870. The Council of India Act (1869) also enhanced his powers: he was authorized to fill vacancies in the Council, not during good behaviour as before, but for a period of ten years, subject to re-appointment at his own discretion. A majority vote in the Council, required earlier in the appointment of Executive Councillors of the Governor-General and the Governors of Madras and Bombay, were now dispensed with. The Secretary of State as an imperial agency thus came to symbolize the unity of the entire Indian subcontinent.

In the post-Mutiny period the authority of the Governor-General of India was equally enhanced. He had remained out of the Presidency of Bengal during the Mutiny and for some time even afterwards, when he singly exercised in a state of emergency all the powers of government without the assistance of his Council. The strength of the Government of India was also increased by a break in the monopoly which the Bengal Civilians had hitherto held in the constitution of the Governor-General's Council. Canning imparted to the Council an all-India complexion for the first time by appointing as its Ordinary Member a Bombay Civilian, Sir Bartle Frere.[59] By far the most important constitutional change introduced in his time to strengthen the authority of the executive government was the Indian Councils Act (1861), which reduced the powers of the Legislative Council constituted by Dalhousie in 1854 under the Charter Act of the previous year.

The Governor-General's dispatch of 9 December 1859 contained the first set of proposals on the subject of legislative councils. It referred to the 'misleading effect' of the existing rules governing legislative procedure, and suggested that the function of the Legislative Council of India be limited to mere legislation. The dispatch defined the scope of legislation and proposed the creation of separate legislative councils for Madras and Bombay. For Charles Trevelyan, then Madras Governor, had earlier in a bitter attack described the Executive Council as the 'Calcutta Council', and in a Minute of 3 July 1859 raised serious doubt as to whether 'the real government of the whole of India can be carried on by one set of men from one place', especially when 'the South of India differs from the North as much as France does from Germany'.[60] Sir Bartle Frere supported the case of legislative decentralization, but did not favour reducing the powers of the Legislative Council to the extent Canning wanted. The Judge Member was of course to be excluded, but in the aftermath of the Mutiny the necessity of bringing into the legislature certain respectable Indian non-official elements was generally realized.

The bill which Sir Charles Wood (Secretary of State) carried through in Parliament became the Indian Councils Act on 1 August 1861. He made it clear that the Council created under the Act of 1853 had never been intended to be a sort of 'debating society or petty Parliament', which in practice it actually became. Thus, in the enactment of the new measure care was taken to avoid any word which might convey the notion of the separate existence of a legislative council. The new Act of 1861 therefore authorized the Governor-General to nominate additional members, not less than six nor more than twelve in number, half of them being non-officials, and including Europeans and Indians. The persons so nominated were 'to be members of the Council for the purpose of making Laws and Regulations only', and were not to be 'entitled to sit or vote at any meeting of Council except at meetings held for such purposes'.[61] While the Secretary of State could disallow any measure of the Indian legislature, the Governor-General was authorized to make and promulgate ordinances which could have the force of law for a period of six months. Another precaution taken to undo trends from Dalhousie's time was the stipulation that rules of legislative business be framed by the Governor-General in Council in his executive capacity, not by the Council when meeting for legislative purposes. Dalhousie's Clerk of the Council, too, was abolished and replaced by the Home Secretary, who took over his function to lend support to what was claimed as the identity of legislative and executive business. Even the term 'session' was sought to be removed from the rules of business simplified in 1862, and care was taken to see that the words 'Legislative Council' were not used in any official parlance. Wood's whole effort was, in fact, directed against the emergence of anything comparable to the House of Commons, or 'a committee of grievances, or a grand inquest of the nation'.[62] For these were features of a representative government or council which, in the existing state of Indian social development, would only mean representation for European settlers who could not be 'trusted to deal with the natives'.[63] The only alternative left was the 'despotism' of the executive government, a sort of bureaucratic despotism, 'merciful', rule-bound and rational.

But Wood's was after all a pious wish to protect Indians against European dominance, although it was not without reason or justification. The functional and procedural distinction which Macaulay had earlier made to meet the exigencies of increasing European colonization could not altogether be undone. For his basic concept of 'liberty', which he wanted to engraft on executive despotism to produce 'good government', consisted in the separation of legislative functions and procedure as a means of placating immigrants who interpreted it to their own advantage through such standing orders or rules of legislative business as tended to strengthen colonial interests. Though publicly opposed to any measure that might weaken the authority of the executive government, Canning himself in a private letter to the Secretary of State admitted that it was too late to completely subordinate legislation to that authority or to withdraw public attention altogether from the proceedings of the Council.[64] In practice, the erstwhile distinctions between the two councils remained. For making laws and regulations the Council of India continued as the Legislative Council, meeting in 'session' from December to April, and generally on Wednesdays. As the legislative proceedings of the post-1861 period show, all legislative measures were criticized and discussed publicly: and although there were no political parties yet to function as a link between the government and the people for integrative purposes, leaders of public opinion, organized associations and interests concerned were consulted and their opinions sought before enactment.

It is true that public discussion of legislative measures could not yet be deemed to give the Council a character comparable to that of the House of Commons; for, as observed by Sir James Stephen, the discussions hardly played an effective role even in 1871 and the real work of legislation was done by the select committees acting practically under the sanction of the Executive Council. But the Legislative Council was not a mere enlargement of the Executive Council. The form and mode of legislation according to a fixed procedure, separate and distinct from the rules of executive business, made the Legislative Council potentially representative. Contrary to Wood's justification of 'despotism', the European business houses and their Indian collaborators who had already emerged under European patronage, later joined together and pressed for a formal recognition of representative principles. Andrew Yule, a leading European managing agent of Calcutta, was, for instance, actually installed as president of the Indian National Congress in 1888, and the demand for representative government was conceded under the Indian Councils Act of 1892. The experience of the Mutiny dictated the expediency of executive initiative in the work of legislation, but it could not put the clock back. Legislation could not be subordinated completely to the executive government or withdrawn from the press or public attention. Legislative authority in the Indian situation was obliged to steer a course in between the two extremes, namely, executive despotism and parliamentary tradition.

The effect of increasing colonization was noticed in local demands for decentralization, more especially in view of the experience of the Mutiny which revealed the failure of centralization and exclusive dependence of Bengal Civilians for a knowledge of local requirements. This was more so in a situation where the expansion of British territories, the increasing settlement of Europeans, the new technological advancement and the freedom of the press together called for a new adjustment of relationships in the executive and legislative machinery of Central and Local Governments. The establishment of common codes of law and procedure as well as the institution of separate police establishments at an all-India level were other developments which could sustain the forces of unity and integration under central control, despite a reduction in the degree of legislative centralization. The Indian Council's Act (1861), therefore, restored to Madras and Bombay the power of legislation they had exercised before 1833. It authorized the Governor-General in Council to extend, when necessary, the operation of these provisions to Bengal, the North-Western Provinces, the Punjab and any other province which might hereafter be placed under a Lieutenant-Governor, an office created under statutory enactments. He was empowered to create such a legislative council for Bengal as soon as possible, and the creation of similar councils for the North-Western Provinces and the Punjab was left to his discretion. But in no case was the number of 'additional members' to exceed those allowed for Madras and Bombay, where the minimum fixed was four and the maximum eight.

The restoration of legislative authority to local councils was, however, not without certain restrictions, which were necessary for the integrity of the central authority, a lesson based on the knowledge and experience of Indian history. No local council was, for instance, to legislate on such subjects as public debt, public revenues, coinage, posts and telegraphs, the penal code, religion and usage, the armed forces, patents and copyrights, the Indian states and relations with foreign governments. Even measures of local interest were to remain subject to the assent of the Governor-General in addition to that of the local governor. The principle of decentralization was thus recognized in response to local needs and sentiments. But these were sought to be met within the framework of an all-India unitary system, which responded to the call of changed circumstances without any prejudice to the basic unity of the political structure. It was precisely here that the pre-British country governments were found sadly wanting in terms of institutional strength and the quality of resilience as an integrating force that flowed from it.

Civil Administration and The Public Services

The civil administration of the East India Company and the public services organized and reformed from time to time to run that administration were no less important in terms of their contribution to unity and integration. They provided institutional arrangements for the execution of policies which, with the attainment of paramountcy, began to assume an all-India, integrated form and character, which was in large measure reflected in the Charter Act of 1833. The growth of departmentalism and increasing pressure for delegation of authority to subordinate presidencies or provinces had doubtless tended to affect integrated direction at both policy and administrative levels. But, as indicated before, care was taken to regulate change in a manner and to an extent not inconsistent with the requirements of integrated action. That unity of policy which became institutionalized in the office of the Secretary of State and his Council filtered down to executive levels in the districts.

The Unifying Role of Administration

In my works on the administration and public services of India I have examined in some depth and in an historical setting the structure and function of the administrative bureaucracy both before and after 1947. Here we shall merely assess their importance in relation to India's political unity, a point which needs to be recognized in the context of the country's current condition.

The importance of administration as an integrating force under British rule lay in the confidence it created among the people by its determination to establish a reign of law and its capacity to ensure peace and tranquillity. The early years of that rule were, of course, no better than the administration of the country powers, Mughal or

Maratha, an administration where the agencies of government themselves reduced spoliation to a system and engaged in raids to raise revenues and resources from the people in payment for the services they claimed to have rendered to their masters. The reports of the Parliamentary Committees of 1772 on the affairs of the Company are replete with cases of corruption and abuse of power that followed in the wake of Plassey and the subsequent changes of government in Bengal. Warren Hastings was the first Governor to fight against the rampant corruption arising from the exercise of power without responsibility, a state of administrative 'immorality' which constituted a serious challenge to raising a clean and weighty fabric of government that could protect the peasantry and win their confidence. His main contribution in this direction consisted of his attempt to build a system based on political principles and to recognize, despite the existing state of corruption, the potential abilities of the Company's European servants to become an agency for the task of administrative reconstruction.

Soon after assuming charge as Governor of the diwani provinces Hastings noticed that the Company's government 'consists of a confused heap of undigested materials, as wild as the chaos itself. The powers of government are undefined; the collection of the revenue, the provision of the investment, the administration of justice (if it exists at all), the care of the police, are all handled together, being exercised by the same hands, though most frequently the two latter offices are totally neglected'.[65]

Hastings advised the Directors to reform the existing system rather than punish stray individuals for corruption. He pointed out that 'whatever may have been the conduct of individuals or even of the collective members of your former administrations, the blame is not so much imputable to them as to the want of a principle of government adequate to its substance, and a coercive power to enforce it'. His point was that, although the extent of the Company's territorial gain in Bengal was equal in resources to most states in Europe, the constitution of its government was still traceable only in 'ancient charters' meant for profit-motivated commercial operations with no means or power to protect the revenues of a large kingdom from private violence and embezzlement. It was therefore not surprising that many of the Company's servants used this situation to secure 'the advancement of their own fortunes, or that those who were possessed of abilities to introduce a system of better order, should have been drawn along by the general current'. But knowing, as Hastings did, 'the general habits and manners of the Company's European servants, he assured the Court of Directors that 'you will hear of as few instances of licentiousness amongst them as among the members of any community in the British empire'.[66]

This implicit faith in the Company's European servants as an instrument of administrative reconstruction thus induced Hastings in 1772 to place rural districts in the hands of Collectors,[67] invested immediately with the collection of revenue but charged also with executive administration as district officers. The regulations enacted in 1772 to enable Collectors to function as district officers with powers to preside over both civil and criminal duties were the first to define functions and establish some kind of administrative procedure. This arrangement was, however, short-lived. After the formation of a new government under the Regulating Act (1773), the Collectors were recalled by the Council majority. But when administrative re-organization was taken up again in 1781, the functions of civil judge and magistrate were once more united in the office of the Collector, who alone was recognized as the district officer empowered to deal with local complaints.

In his anxiety to recognize the different categories of zamindars as actual proprietors, Cornwallis united the magistracy with the office of civil judge, and also reduced the rank of the Collector by making the judge-magistrate the first officer of the district. The jurisdiction of the Collector was not to extend beyond the collection of revenue, and there, too, his decisions were subject to revision by a civil court. The Collector could no longer make justice executively available to those who could not afford judicial proceedings in a regular court of law. The Cornwallis Code of 1793 tended, in fact, to destroy the security of a status-bound society that was based on a popularly recognized concept of natural justice. It introduced, instead an alternative system which, though designed in theory to promote equality before the law, was at once dissipative and unequal because of the social and economic imbalances among contenders for justice under the Judge-Magistrate system.

But Bengal was not all-India. Metcalfe, for example, was a strong critic of Cornwallis' Permanent Settlement, which he regarded as gravely unjust to the actual cultivators. His settlement of land in Delhi marked an absolute departure from Cornwallis' principles. So did Elphinstone's in the settlement of the Peshwa's territories and Munro's in his raiyatwari settlement in Madras. The methods of all these statesmen differed on account of varying local customs, but all wanted to maintain and strengthen the self-governing activities

The Instruments of Territorial Integration 45

of village communities which Cornwallis had wholly disregarded in Bengal. The periodical mahalwari system, which Holt Mackenzie introduced in the North-Western Provinces, likewise tended under the influence of Bentinck and Metcalfe, to assume the character of either raiyatwari or village settlements. In their anxiety to establish regular contact with village communities all three wanted the Collector to function as the head of district administration. The general trend of opinion about the office of the Collector was that he alone, by reason of his immediate function, was able to establish a direct link with the actual cultivator and look after his welfare as part of the Liberal Indian policy that was emerging in the 1820s. During Bentinck's administration there had in fact emerged a general consensus that the magistracy must stay with the head of district administration and that it should be united with the office of the Collector who had the promotion of agriculture and the care of public revenues as part of his immediate duty.[68]

The popularity of British rule in the rural districts flowed in the early days from the personal approach of district officers, their easy accessibility to the people, which built up the administrative fame of Malcolm, Munro, Edwardes, Elphinstone, Henry Lawrence, John Lawrence and others. Independently of whether they were in their bungalows, the district headquarters, or in their settlement camps in villages, they remained open to direct contact with the people and kept away the intermediaries who traded on the people's ignorance. In 1831, the Collector once more became the head of district administration. Land revenue policy as well as administration was thus geared to win popular approbation and to meet the needs of patriarchal society for purposes of political unity and integration, which the British government wanted to achieve through its administrative agency in the department of land revenue. There were changes between 1837 and 1845 when the magistracy was separated from the office of Collector to enable the latter deal with the resumption of revenue-free estates. But the arrangement of 1831 was again put on a stable footing in 1859.

However, the restoration of the Collector's position as the head of district administration did not signify a departure from respect for law, form and procedure, which Cornwallis had inculcated as a test of administrative morality. In the administration of revenue, the judicial decisions of the Collector thus remained subject to review by a regular civil court. A strict adherence to legal forms and procedures not only applied to judicial administration, but tended to become a general feature of the administration as a whole It conduced to institutional stability, which in turn strengthened territorial unity and integration. I have dealt in separate volumes with the growth of the Central Secretariat, from its small beginnings with a single Secretary and some Assistants at Fort William in 1756 to a full-fledged Secretariat in New Delhi, a Secretariat which successfully managed the conduct of two World Wars before it was transferred to Indian hands in 1947.[69] It is relevant for our purpose to note that the growth of the Central Secretariat was closely related to the evolution of an all-India political and constitutional fabric. As a unified decision-making centre it contributed to unity of action in making geographical India correspond to political India. Despite the growth of increasing departmentalism in vertical hierarchies, the Central Secretariat symbolized an instrument of integrated policy-making and action. This was first sought to be attempted by Wellesley (1798–1805), who provided promotional avenues and created a Chief Secretary to head the Secretariat organization with powers of general control and authority for the proper conduct of business. Other central agencies of control came into being over the years to ensure financial co-ordination and political unity. These were, for instance, the departments of audit and accounts, posts and telegraphs, political residencies and the public works department, with ramifications covering the whole of India from a single centre at the Secretariat. Through its record-based nerve-centre and rule-making power, the Secretariat, Central or provincial, exercised a general control in maintaining unity of action and the execution of policies in districts all over the country.

Civil Service

In pre-British days there was nothing like a modern civil service organized on a competitive basis, a contractual arrangement which provided promotional opportunities according to seniority, conduct and quality of performance. The Mughal mansabdari system was essentially a graded military service, paid generally by grants of jagirs or military assignments. Even those who were employed on the civil side held a military rank without troops, though they were paid according to the rank assigned.

In *The Bureaucracy in India* I have discussed in some depth the problems of civil service organization as an integrating force, which, under its cadre rules, combined the field experience of the districts with the Secretariat experience of policy-making at the headquarters.

A periodical cross-fertilization of the experiences of districts and headquarters not only tended to bridge communication gaps and conduced to understanding, but also helped in keeping together the political structure, despite considerable variations in the regional cultures. However, it took time for such a principle of organization to mature, and it did not happen during the Company's rule, which was mostly a period of patronage, though not unregulated or of public interest. It is not proposed here to dwell on this subject at length; however, we should note certain organizational features of the Company's public services which lent themselves to a competitive system, a pre-condition for the creation of an imperial or all-India service.

Even as a trading organization, the Company's service was organized into the hierarchical gradation of writer, factor, junior merchant and senior merchant, paid according to each grade with promotions based on seniority as well as quality of performance and behaviour. Before entering the service all recruits were required to sign a covenant undertaking to serve the Company with loyalty and good conduct. As each Presidency had its own separate cadre, its President and Council submitted to the Court of Directors a complete list of their covenanted servants, arranged according to their respective grades and denominations, with remarks in each case which helped the Court in the determination of their rank in the service. As the salary happened to be nominal, they were permitted to trade on their own account under certain conditions. This service came to be invested with the administrative and political management of the diwani provinces in 1765 and men like Hastings and Cornwallis endeavoured ceaselessly to evolve out of chaos a public service competent to rule a kingdom as civil servants rather than mere traders in pursuit of profit or individual gain.

While Parliamentary enactments attempted to change the constitution of the Company and Presidency governments in India from 1773 onwards to provide clean administration on an all-India imperial principle, the continuance of the Company's rule and its old parochially-oriented Presidency system operated as a serious constraint on the growing free-trade movement which favoured competition both in trade and government, and which the Company viewed as prejudicial to its monopoly as well as patronage. In his correspondence with Dundas, Wellesley, for instance, made no secret of his design that the Company's patronage be made subject to the Governor-General and that its government in India placed directly under the Crown. His imperial idea found expression in his important Minute of 10 July 1800, where he justified the establishment of a college at Fort William for the training of civil servants in the higher branches of European learning and in the languages of India. 'Duty, policy and honour', he pointed out, required that British India should not be administered 'as a temporary and precarious acquisition', but that 'it must be considered as a sacred trust and a permanent possession' which should be governed by a civil service capable of 'an inexhaustible supply of useful knowledge, cultivated talents and disciplined morals'.[70] The College at Fort William was designed to ensure this supply.

Consistent with this imperial object, Wellesley proposed to bring all the covenanted servants from Madras and Bombay to Fort William to be trained before being assigned to the several Presidencies. Wellesley was clearly thinking in terms of an all-India imperial service, but the Court of Directors willed otherwise. While under their letter of 27 January 1802 they ordered the abolition of the College and reduced it to a small language school for Bengal civil servants only, they asked the Government of Madras in a separate letter of 12 March not to send any writer to Bengal for training. In a long letter of 5 August the Governor-General justified his plan and suspended for a time the execution of the Court's order until final decision. But after his recall, Madras stopped sending writers to Fort William and established a collegiate institution of its own in 1808.[71]

This conflict between the Court and the Governor-General was in fact an expression of a conflict of ideology between two opposite economic systems, namely, mercantalism and free trade, the first representing monopoly and the second competition. The Court was fighting a losing battle. While the Charter Act of 1813 abolished its trading monopoly, that of 1833 put an end to its trading rights itself. But so long as British territorial gains kept on flowing from the Company's government and trade was treated as interwoven with it the right to patronage remained unaffected with the continuance of the Presidency system in the distribution of its covenanted servants. In fact, the establishment of Haileybury College had in the meantime taken the wind out of Wellesley's sails; and, despite the abolition of the Company's trade, its right to patronage was merely modified by a mode of limited competition, but not abolished. Only the control of the Supreme Government over its subordinate Presidencies, which in the past had remained ineffectual despite statutory provisions, was tightened by the Act of 1833 through financial and legislative channels.

The Company's patronage thus remained a problem, an obstacle to the development of an all-India service, with provisions for its disposition according to the requirements of government at the centre and in the provinces. The centralized control of the Government of India over subordinate Presidencies was a step forward in this direction. But since the cadre of covenanted service continued to be recruited separately for each Presidency on the basis of patronage assigned to the 'Chairs', members of the Court and others, the all-India view of the service was lost sight of. The Presidency-wise training in the country's languages and separate regulations too stood in the way of an all-India perspective.

The system of nomination by the Directors had, of course, been considered unsound for many years. The Select Committee of 1832 had pointed out that it was not the best mode of securing civil servants of high calibre and that 'a system of competition' was therefore 'deserving of serious consideration'. At the instance of Macaulay, the Act of 1833 inserted clauses which rearranged the system of appointment to the civil service on a basis of competition,[72] a 'fourfold system'[73] which did not, however, come into effect. The Directors of the Company had then been too strong for him. 'Backstairs influence in Leadenhall Street contrived that the clauses embodying Macaulay's plan lay dormant in a pigeon-hole at the Board of Control, until backstairs influence in Parliament at length found an opportunity to procure their repeal'.[74]

As President of the Board of Control Sir Charles Wood, however, called upon Parliament to enact, as part of the India Bill of 1853, that 'a nomination for the Civil Service of India should thenceforward become the reward of industry and ability, instead of being the price of political support, or the appanage of private interest and family connection'.[75] In the course of the second reading of the bill, Macaulay spoke eloquently[76] in favour of an open competitive examination as the basis of recruitment, an examination which should test the quality of mind of candidates in terms of depth rather than 'hold out premiums for knowledge of wide surface and of small depth', which was the knowledge of 'a mere smatterer'. The choice of subjects for academic competition did not matter—they might be a range from literature to science. Macaulay in fact anticipated that, if recruited on the basis of an open competitive examination that tested the quality of a liberal education, the young civilians would in future 'be inferior to no class of public servants in the world', more especially in terms of 'nicety of honour and uprightness of character'.

The India Act which Parliament passed in 1853 gave its seal of approval to the appointment of civil servants by open competition. Sir Charles Wood entrusted the task of making the necessary arrangements to a committee with Macaulay himself as Chairman.[77] The Committee completed its report in July 1854 and incorporated the principles Macaulay had earlier advocated. It included a complete list of subjects of examination with the proportion of marks allotted to each. The Indian Government adopted it in its entirety, accepting all the recommendations,[78] whether they related to the age of the candidates, the abolition of the Company's College at Haileybury, or the training of probationers during the two years which were to intervene between their first selection and final departure for India. After the approval of the Board of Control, regulations were established in January 1855 for the conduct of the examination. The Board appointed the examiners who conducted the first examination held on 16 July. After the selection of candidates so examined their appointment proceeded from the Court of Directors. When the Secretary of State for India in Council took over the functions of the Board of Control and Court of Directors under the Government of India Act (1858), the advisory function which the Macaulay Committee earlier performed were transferred to the Civil Service Commissioners appointed by the Crown on 21 May 1855. While the Secretary of State in Council exercised the executive power to make regulations for the conduct of examinations and admission to the civil service of India, the Civil Service Commissioners were to advise and assist him in framing these regulations, to control and superintend the examinations, and finally, to grant certificates of fitness for appointment.

It is thus the open competitive examination held under the direct control of a single centralized imperial authority that imparted to the Company's civil service for the first time a unified all-India complexion. Each Presidency or province was, of course, to have its own separate cadre as hitherto but Lord Canning was the first Governor-General to impart to his Council a real all-India character by appointing a Bombay civilian to it. It was the beginning of an end of the monopoly which Bengal Civilians had so far held in the Executive Council,[79] a beginning which later established for integrative purposes an effective link between the districts and Secretariat headquarters.

Contribution of the Indian Surveys

The contribution of the Indian surveys consisted not only in the preparation of maps of routes and in providing a knowledge of local geography for military operation and political control, but also in the collection of data for descriptive memoirs containing, besides a narrative of each survey, carefully digested statistical, historical and antiquarian information for social, economic and cultural purposes. The detailed topographical surveys based on triangulation acted as an effective instrument of cohesion. While the maps facilitated the development of communications as a means of physical control, the study of Indian antiquities brought into focus the importance of Indian art and literature, religion and culture—a new development which led to the emergence of renaissance movements, that collaborated with the Company's Government under Bentinck in the promotion of a liberal social policy. The work of the Indian surveys needs to be studied in depth so that their contribution to India's territorial integrity is properly appreciated, but here we shall merely indicate some of the landmarks.

The task of recovering from the past the ancient languages and literature of India was undertaken in the eighteenth century by Europeans who realized the importance of studying Sanskrit, the language of India's ancient history and culture. The immediate incentive to study Sanskirt originated in the serious attention given to this field by the Company's administration under Warren Hastings; from considerations of policy, he wanted to win the confidence of the people by attempting to govern them through their own laws and customs. No doubt at various intervals and during a hundred years or so before Hastings a few isolated students, chiefly missionaries, had published pioneering accounts on Sanskrit literature and grammar. But their assessment of Hinduism was for the most part hostile and narrow. Hastings was the first to initiate the compilation of a digest of Sanskrit law-books by pundits. And since no one at the time was found able to translate the works into English, they were translated first into Persian, and from Persian English versions were made by Halhed in 1776. It was left to Sir William Jones, Chief Justice of the Supreme Court at Calcutta, to place this study on a firm basis by the establishment of the Asiatic Society of Bengal in 1784. Jones laid the foundations of comparative philology in the course of an address delivered as President to the Asiatic Society in 1786, when he observed that Sanskrit was closely connected not only with Greek and Latin, but also with the languages of Persia and those of the Celts, Germans and Slavs. 'The Sanskrit language, whatever be its antiquity', he said, 'is of a wonderful structure; more perfect than the Greek, more copious than the Latin, and more exquisitely refined than either: yet bearing to both of them a strong affinity, both in the roots of verbs, and in the forms of grammar, than could possibly have been produced by accident; so strong indeed, that no philologist could examine them all without believing them to have sprung from some common source, which perhaps no longer exists. There is a similar reason, though not quite so forcible, for supposing that both the *Gothick* and the *Celtick*, though blended with a different idiom, had the same origin with the *Sanskrit*; and the old Persian might be added to the same family.'[80]

Topographical Survey

About this time the work of topographical surveys began under the guidance of Colin Mackenzie. One of the most indefatigable surveyors and collectors of information, Mackenzie commenced his explorations just after the close of the Anglo-Maratha War in 1783 and worked for more than thirty years in the Presidency of Madras, carefully surveying the peninsula of India along with a group of other officers trained at the Madras Military Institution established at the instance of Mackenzie himself. He worked in Coimbatore and Dindigul, in Nellore and Guntur, and in the Ceded Districts and Mysore. He remained in Mysore for several years after its conquest in 1799. His topographical surveys covered 40,000 square miles, including a general and seven provincial maps. But the most important result of his topographical surveys was a valuable descriptive memoir containing statistical, historical and antiquarian information in several manuscript volumes.[81] He devoted himself to the study of Indian antiquities and visited in the course of his surveys every place of any interest from the Krishna to Cape Camorin, accompanied by a staff of Indian assistants who helped copy and collect records. He got together 30,000 *sassanums* or tenures originally inscribed on stone or copper. According to Markham, the Mackenzie Collection included in fact valuable source material which brought to light the excellence of Hindu attainments in the Deccan peninsula where Islam had made little or no dent on indigenous art, language, literature and culture. It 'consists of 1,568 manuscripts in different Indian languages, 8,076 inscriptions, 2,630 drawings, 78 plans, 6,218 coins, and 108 images'.[82] Mackenzie, in addition, sent to India House some sculptured stonework from Amravati and published several papers of historical and topographical importance.

After his Deccan surveys Mackenzie was appointed Surveyor-

The Instruments of Territorial Integration 53

General of Madras in 1809 and commissioned a couple of years later to join an expedition to Java. He returned to Madras in 1815 and soon proceeded to Calcutta to take over as Surveyor-General there in 1816.

Colonel Lambton was another officer of repute who superintended the surveying of the South Indian regions on a system of triangulation different from that of Colin Mackenzie. Most of the work of surveying was, however, done by officers trained by the Madras Military Institution. Lieutenant Garting and his group, for instance, surveyed in 1811–12 the Portuguese territory of Goa, including its coastal areas, anchorages and rivers. It was done on the principle of Lambton's trigonometrical survey of North Canara, where the work was completed in 1815 with a descriptive memoir in two volumes containing an account of the general aspect of the region surveyed, with details concerning its history, cultivation, water-supply, inhabitants, tenures, trade, routes and road communications. Lieutenant Conner, who surveyed the mountainous principality of Coorg in 1817 on the basis of Lambton's primary triangles, produced his memoir in one volume to illustrate the map.

The surveys of Travancore and Cochin were completed between 1816 and 1821, and those of Malabar during 1825, the latter being connected with the work of Buchanan. The eastern districts of the Carnatic as well as of Ellore, Rajamundry and Guntur were also surveyed by officers of the Military Institution on Lambton's system of triangulation. Apart from historical information and statistical data, their descriptive memoirs were significant for the light they threw on hill tribes. Of these, the memoir which accompanied the survey map of Tinnevelly (surveyed during 1807–13) was by far the most valuable from an historical point of view. For Tinnevelly formed part of the ancient Pandyan kingdom, and the memoir 'described inscriptions on granite walls; the temples and other religious monuments; the climate, population, products; and details respecting the boundaries, the resources of each talook, and the roads'.[83]

Thus, the information collected on South India in the course of scientific topographical surveys not only provided knowledge for military and political control, but, in addition, recovered the source material required for an historical reconstruction of the past—the intellectual background to Hindu resurgence and the restoration of its confidence.

While the topographical survey of Madras was in progress and Lambton's work being completed, the surveyors in the north were going ahead with a survey of the Ganges, the mountainous regions of Nepal and districts of Bengal. Colonel Colebrooke was especially interested in the source of the Ganges which called for a knowledge of the Upper Himalayas and Tibet, a knowledge which was at the time derived from Chinese sources through Jesuit missionaries who had contact with the Lamas of Tibet. But as it was not based on scientific surveys, Colebrooke obtained in 1808 the sanction of the Government for an expedition to complete the examination of the sacred river from Hardwar to its source in the Himalayas, the survey from Hardwar to Allahabad being already completed by Lieutenant Wood in 1800. The Surveyor-General, however, fell fatally ill in 1810 and his mission was executed by Captain Webb and two Lieutenants. Colebrooke, the eminent Sanskrit scholar and kinsman of the Surveyor-General, was, on the other hand, interested in knowing the heights of the Himalayan peaks.[84] As for Nepal, Colonel Crawford had even before the Anglo-Nepalese War attempted a survey of its mountainous range and measured some of the peaks. But it was Lord Hastings who appointed Colonel Hodgson and Lieutenant Herbert in 1815 to survey the regions beyond the sources of the Sutlej and the Ganges, which were bounded on the north by Chinese Tibet.

In 1818, Hodgson and Herbert were both engaged in surveying Garhwal on the basis of the same triangulation employed earlier.[85] Captain Webb, who had earlier surveyed the course of the Ganges from Hardwar to Gangotri, continued his work in the province of Kumaon during 1815–20. It seems the Himalayan regions with all the kingdoms included in them began to engage the most serious attention of the British government as soon as the prospects of an Indian empire became evident at the turn of the century. This perhaps also explains the emphasis laid on the scientific survey of the Indian peninsula, more especially of its coastal and strategic areas, which are important from considerations of both defence and as sources of history.

For a statistical survey of Bengal, Lord Hastings nominated Buchanan who had already reported ably on Mysore and Malabar. In his report on Bengal, he covered such subjects as the people, their religion, agriculture, production, etc. He was aided by assistants and draftsmen, and his work extended over seven years from 1807 to 1814.[86] Captain James Franklin undertook the survey of the important terrain of Bundelkhand, which he completed between 1815 and 1821. He produced a valuable map of the whole region, together with a memoir containing particularly useful information, on its geology. A survey based on routes and cross-routes was also made in 1821 of Bhopal and Bairseah in Central India. The Sunderbuns were surveyed between 1812 and 1818 by two young Lieutenants, Hugh and Morrison. They conducted it despite ceaseless annoyance from tigers and alligators.

Between 1818 and 1819, surveys were commenced in a number of other places also. These included, for instance, Cuttack in Orissa, Bakerganj and the Sylhet frontier in Bengal, Jaunpur in the North-Western Provinces, and Midnapur to Nagpur for purposes of ascertaining the practicability of constructing a road. This was the period when in the Presidency of Bombay Colonel Monier-Williams made by compass and perambulator a careful survey of Gujarat, Cutch and Kathiawar and maps were at the same time compiled from the route surveys in the Deccan. Geographical work was on the other hand, conducted in Malwa, and Captain Tate surveyed the islands of Bombay and Salsette.

Lord Hastings was thus credited not only with the virtual establishment of British paramountcy through conquests and annexations, but also with the promotion of a series of surveys which attempted for purposes of territorial integration to bring into focus India's natural frontiers in the north and provide a knowledge of details respecting the internal history, topography and conditions of its people, a knowledge which was sought to be employed in the interests of internal security and cultural revival for a renascent social policy and change. The emergence of an Indian policy in the 1820s is appreciated in the context of a Hindu cultural revival which, though largely still clinging to orthodox practices, produced a small but distinctive group of reformist intellectuals like Raja Ram Mohan Roy. They sought to use the liberalism of the British Indian policy to form Hindu society on liberal principles inherent in Hindu philosophical traditions. The Muslims, though benefiting from economic developments, formed no part of the new renaissance. For the Muslim nobility and landed class were politically displaced, and, in the absence of leadership, the peasantry remained quiescent and backward. Quick to advance in western education through English, the Hindus stole a march over the Muslims who clung to their classics as an article of faith.

Revenue Survey

The object of the revenue survey was to ascertain all revenue-paying districts and to show the limits of every village. It was conducted by a

Local Government according to its civil requirements and involved a *khesera* or plot measurement of fields by native *ameens* who filled in all the details on the maps for geographical purposes also. A topographical survey, on the other hand, generally produced a map for military purposes. Both were carefully connected with the great triangulation, the revenue survey by a chain and theodolite system of measurement fixing all boundaries of estates and villages, and the topographical by breaking up the large triangles into minor triangulations with sides sufficiently short to give bases for plane-table sketching. The topographical surveyor did not use the chain for political reasons; for independent tribes viewed it as the sure harbinger of loss of territory. While surveying the mountainous regions situated between the sources of the Sutlej and the Ganges, Lieutenant Herbert, for instance, measured a base with wooden staves, fixed latitudes by the stars' zenith distances and longitudes by observations of Jupiter's satellites. The scientific part of measuring a base on a plane table was of course highly creditable, but the work of filling in the details at each trigonometrical station by reference to neighbouring objects lacked accuracy. This defect was later sought to be removed by means of a distinct and separate staff appointed specifically to fill in details on each triangle.

Thus, both revenue and topographical surveys were conducted on the basis of triangulation to extend the geographical knowledge of India as a means of keeping the country territorially integrated. The work of surveying in both fields continued with added vigour and efficiency from 1847, the end of the first Sikh War, when Colonel Thullier took over as Deputy to Sir Andrew Waugh, Surveyor-General at Calcutta. Thuillier brought both energy and talent to the improvement of the entire system of surveying and succeeded Sir Andrew in 1861. Since his appointment in 1847 he had brought out correct maps and reports on the surveys of Punjab, Oudh, Sind and the Lower Provinces as well as other districts put under his charge. The Madras Revenue Survey, too, had been steadily progressing since 1856 under the superintendence of Colonel Priestley. It consisted of cadastral field surveys for every village of revenue paying districts. The maps in each case were well drawn and lithographed soon after the survey to make the results accessible to the public. Rent-free tracts were, however, omitted until the gaps were ordered to be filled in later by means of topographical surveys on a moderate scale.

Regrettably, the revenue surveys of the North-Western Provinces were made with great speed without any regard being paid to accuracy and precision. In their anxiety to complete the settlement of rent-free estates, Collectors had pressurized surveyors to sacrifice everything to cheapness and rapidity of execution. The result was that the maps drawn for revenue purposes omitted topographical features and were confined to the actual definition of village boundaries and the *khesera* measurement of fields by native *ameens*, without any attempt being made to connect them with related features of geographical importance. The maps so drawn thus served no purpose as geographical material and it is interesting to speculate as to whether this omission of topographical and geographical material in the original village maps of the North-Western Provinces led to failures in military operations during the Mutiny of 1857.

A minute revenue survey of the Bombay Collectorates was started in about 1836 by Major Wingate, who was aided by Goldsmid of the Civil Service. The elaborate design of the survey consisted of a cadastral measurement of fields by means of the cross staff and chain. But, as in Madras, the field surveys were not connected with the triangulation of the great trigonometrical survey later extended to the Bombay Presidency. The result was that field surveys could not be reduced to the great trigonometrical points for geographical purposes. Though rivers and streams were partially indicated on the maps, major topographical features, such as hills, remained altogether absent. The problem in fact was one of reducing trigonometrical points to a single scale to serve the purposes of both revenue and topographical surveys. As said before, all village maps of earlier revenue surveys were drawn on a scale different from that of topographical surveys. The progress of science and technology however required connecting both for purposes of locating engineering and other projects of the Public Works Department, their object again being to reinforce the integrating agencies for political ends as well as for social economic development.

Geological Survey of India

The development of the country's economy, which has its own relevance to India's political stability and progress, was also connected with the work of the Indian Surveys. For instance, it was linked closely with the Geological Survey of India. The first official appointment of a geologist was made in 1818; he was instructed in the beginning to join Colonel Lambton of the Topographical Survey and collaborate with him in doing a preliminary study of Indian rocks and minerals. It is not proposed here to discuss the survey here, except to note its importance as a basic source of unity.[87]

The Army in India

Prior to 1895, there were Presidency armies in India. A single, united Indian Army had yet to come into being. However, certain important changes had been taking place since 1864, which cleared the way for the eventual union of the armies under a General Order of the Government of India, dated 26 October 1894. This abolished the Presidency armies with effect from 1 April 1895.[88] The Order divided the army into four commands, namely Punjab (including the North-West Frontier and the Punjab Frontier Force), Bengal, Madras (including Burma), and Bombay (including Sind, Quetta and Aden). There existed, in addition, local corps which did not belong to any of these commands but remained under the control of the Governor-General. These included the Hyderabad Contingent, two regiments of Central India Horse, the Malwa Bhil Corps, the Bhopal Battalion, the Deoli Irregular Force, the Erinpura Irregular Force, the Mewar Bhil Corps, and the Merwara Battalion.

Organizational Development up to 1858

The term 'Indian Army' itself began to be employed loosely only from 1875 onwards, and it was not until after 1903 that it became a stereotyped and officially recognized title. The earliest force which could be regarded as the embryo of the Indian army orginated in Bombay, when that island was ceded in 1662 to Charles II by the King of Portugal. A detachment of the King's troops and a few Europeans who collected locally formed its garrison. Later, Charles offered Bombay to the Company, who accepted the offer in 1668 on a yearly rental of £10 in gold. The King's troops were converted from service of the Crown to service of the Company. Besides a couple of gunners and some pieces of cannon, the garrison of Bombay consisted now of 5 officers, 139 non-commissioned officers, and 54 *Topasses*—people of mixed origin claiming descent from the Portuguese. As there were no troops at Surat, the headquarters of the Company's agency, this small body of troops formed the nucleus of the Bombay army. This, in fact, was the origin of the Indian Army, as the troops located in Madras and Bengal were of negligible strength at the time.

With the union of the old London Company and the new Company of Merchants chartered in 1698, a further change came about in 1708, when the United Company, known as the East India Company, pro-

The Instruments of Territorial Integration

ceeded to take one of the first definite measures to constitute the three Presidencies, each separate and distinct from the other, and each absolute within its own limits. The President of each, who was also the Commander-in-Chief of the military forces of the Presidency, was to be responsible only to the Court of Directors in London. As a natural corollary, the armies of the three Presidencies became as distinct and separate as the Presidencies to which they belonged. The Presidency system thus came to be firmly established in terms of both organization and disposition. Each Presidency army had by the first decade of the eighteenth century progressed from a mere unorganized and ill-disciplined body to a force of small but organized military units. It consisted in those days of Europeans recruited from England or collected locally, and of Indian sepoys. The sepoys 'were mainly armed with their own native weapons, wore their own native dress and were commanded by their own native officers.[89]

Another important occasion which called for military reorganization was the war with France, which temporarily ceased with the Peace of Aix-la-Chappelle in 1748. As neither the French nor British could afford to provide reinforcements from home in the course of the first Anglo-French conflict in the Carnatic, the employment of larger bodies of Indian troops remained the only alternative. Before the outbreak of the next Anglo-French war in May 1754, the Court of Directors took steps to free the Company's artillery from its marine origins and reorganize it on the lines of the European system, with a separate group of five officers and 110 British rank and file, including 100 gunners. The Directors also ordered that promotion in the armies should be by seniority, a principle which could only be waived with the express sanction of the governors of the respective Presidencies. The British armies of India comprised the Company's troops only—European and Indian; and except in the Bombay army, which had adopted the battalion system, the European infantry was organized in separate companies and the Indian troops were little better than an armed police.

The year 1754 marked the arrival of the first Royal troops in India since 1662. Among the reinforcements which arrived at Madras under Admiral Watson were what later came to be known as the First Battalion of the Dorsetshire Regiment. This introuced a new element in the army in India, which came to be divided into the King's troops, the Company's European troops and the Company's Indian troops, three distinct categories of combatants, more especially the Indian counterparts, who needed to be reorganized for effective and coordinated action. The task was undertaken by Clive who, shortly before the battle of Plassey in 1757, started reorganizing the Indian troops under his command by forming them into regular battalions with a small nucleus of British officers placed over them for efficient command. He armed and dressed the men in a fashion not very different from that of the Europeans. The first battalion of about a thousand combatants so reorganized was as follows:[90]

British officers	British non-commissioned officers	Indian officers	Indian rank and file
1 Captain	1 Serjt. Major	1 Commandant	50 Havildars
2 Subalterns	Several Serjeants	1 Adjutant	40 Naiks
		10 Subadars	20 Drummers
		30 Jamadars	10 Buglers
			700 Sepoys

The battalion consisted of ten companies which included all the Indian ranks except the Indian commandant and adjutant. On parade the Indian commandant took his post with the British captain and was followed by the Indian adjutant. The British officers and non-commissioned officers formed the staff of the battalion.

The Indian troops, which had earlier been organized in separate companies, came now to be integrated with a regular battalion, a change which increased promotional avenues. The other change was the introduction of the British element in positions of command and as the battalion's staff. This too contributed to understanding and integration as well as mobility and unity of action. While reorganizing the Indian troops under his command, Clive was careful not to depart radically from the previous system, however. There was still an Indian commandant and the proportion of Indian to British officers was high, although the introduction of British non-commissioned officers was an organizational feature which continued and remained the general rule up to the time of the Mutiny and even subsequently.

The success of the Indian troops so reorganized by Clive led immediately to the raising of a second battalion after the battle of Plassey in Bengal. Madras raised battalions on Clive's model in 1759. Its merit was recognized also by Bombay, which had an auxiliary force consisting of a rabble of Arabs, Abyssinians, Indian Muslims and Hindus attached to the European battalions. These were organized into independent Indian companies in 1760 and then into battalions in

1767. Clive's system in fact grew increasingly popular, and battalions which had been supervised by a small staff of British officers but commanded by Indian commandants, were converted into battalions with British commandants and British company commanders, a development which met all the exigencies that arose from the extension of territories in the course of nearly forty years after the battle of Plassey.

The importance of British officers for Indian troops thus acquired a new dimension. They were selected from the officers of the European regiments, and were chosen not only for their soldierly qualities but also their suitability for serving with the Indian troops. With an increase in the strength of the Indian forces in the three Presidencies, the British officers of sepoys had in fact emerged as a factor of recognized importance, enjoying many advantages. But they did not rank with the officers of the King's troops, for they belonged to the Company's regiments. The result was that they continually found themselves in subordinate positions to younger officers of the royal force. The implied inferiortiy was resented. The result was the reorganization of the Company's armies in 1796, which distinctly improved their prospects of advancement and promotion, although the distinction between the Europeans of the King's and the Company's forces remained until both were merged and transferred to the Crown in 1858.

The strength of the European soldiers in India before the reorganization of 1796 was about 13,000, while the number of Indian troops in the three Presidencies was roughly as follows:[91]

Bengal	24,000
Madras	14,000
Bombay	9,000

The 1796 reorganization reduced the strength of the native armies, but the most important changes were the considerable augmentation in the number of British officers in each of the units of the Indian armies, the creation of separate artillery battalions, and the formation for the first time of double battalion regiments.

Prior to the reorganization, the strength of British officers in a battalion had already increased beyond what Clive had provided in 1757. It had, for instance, a commandant and adjutant in addition to ten subalterns commanding companies—a total of twelve. But the reorganization provided, besides a colonel commanding the regiment, 1 lieutenant-colonel, 1 major, 4 captains, 11 lieutenants and 5 ensigns (excluding 1 British non-commissioned officer)—a total of

22 officers. But the most remarkable feature of the reorganization of 1796 was that, while the establishment of the 22 British officers for each battalion came to approximate closely to the King's or Imperial forces, the dignity and authority of the Indian officers, which had already decreased under Clive's system, declined yet further. The high proportion of British officers to Indian ranks was retained even later under the reorganization effected in 1824. But in comparison to 1796, the proportion of British to Indian troops had declined in 1805, as shown below:[92]

	Britsh Troops	Indian Troops	Total
Bengal	7,000	57,000	64,000
Madras	11,000	53,000	64,000
Bombay	6,500	20,000	26,500
Total	24,500	130,000	154,500

The first general reorganization of 1796 did not remain confined to the infantry, European and Indian, but extended to the cavalry also. The rulers of the Indian states, with whom the Company's government had long been in conflict, favoured the employment of large bodies of horsemen called *bargirs*, who were regularly organized in hierarchies as well as paid and equipped by the state, and *silladars* who had their own horse, dress and equipment. Though partially equipped and scarcely disciplined, the mounted soldier had remained a characteristic feature of the indigenous military system. The Company had always been slow to develop the cavalry arm of the army in India and the few Indian cavalry regiments that existed were organized on a regular basis and officered on the full European scale. Even so, the proportion of Indian cavalry to infantry after the reorganization of 1796 was only about 8 single battalion regiments to 59 battalions, or in the ratio of 1 to 7½. By 1824 this ratio increased to 1 to 5⅔. On the eve of the first World War the ratio had reached nearly 1 to 3½, when it began to fall again.[93]

The increase in Indian cavalry regiments after 1796 was effected by raising the number of regular regiments, and by adding to them a certain number of irregular units organized on the basis of the *silladar* system, a virtual imitation of the system employed in the Indian states. The essential difference between the regular and irregular units was not solely in the principle of their organization, but also lay in the fact

The Instruments of Territorial Integration

that the establishment of British officers was reduced to a minimum in irregular corps. One of the chief defects of 1796 was thus sought to be removed by augmenting the strength of irregular units.

In fact, the rapid extension of the Company's territories between 1796 and 1857 raised some other organizational issues besides the need to add to Indian cavalry units. It involved, for example, a considerable expansion of the spheres of action of the three Presidency armies. And the extent of this expansion became so great that it suggested eventually the expediency of an all-India, unified control in place of the existing Presidency system. The same cause also made it necessary to raise irregular corps and local contingents for a particular service in particular localities. Of these, perhaps the best known were the Hyderabad Contingent and the Punjab Irregular Force, which later became the Punjab Frontier Force. Some of the irregular corps rendered considerable assistance to the British cause during the Mutiny.

At the end of the Company's rule the army in India thus consisted of certain units of the British Imperial Army, known as the King's Royal Troops, the Company's three Presidency armies comprising British and Indian units, and various local forces and contingents. After the abolition of the Company in 1858 the units of the 'British army serving in India' came to form part of the Imperial British Army. This decision necessitated the transfer of the late Company's European troops to the service of the Crown. The distinction between 'Royal Troops' and the 'Company's European Troops' which had existed for more than a hundred years thus disappeared. The Company's European infantry became British regiments, and the Bengal, Madras and Bombay European artillery units were amalgamated with the Royal Artillery.

As an Instrument of Conquest

The British military organization which developed in India on the lines briefly indicated above served during the Company's rule not only as an instrument of conquest but also assisted its civil administration in the maintenance of internal security for purposes of political stability, a function which, though not directly related to wars and annexations, was none the less vital to sustained territorial integrity.

A study of the growth of British paramountcy in India shows that the Company's government was heavily preoccupied with the exclusion of the French in the contest for power; once the French bastions and allies in the country were either destroyed or weakened, the establishment of that paramountcy became a *fait accompli*. The country's segmented society and denationalized governments did not constitute a serious challenge to the British, despite the tradition of chivalry in India. Thanks to the weakness of their social and political systems as well as their outmoded weaponry and military organization, the Indian powers succumbed to the organizational superiority of the British army in India.

The Frenchmen who helped the country powers to raise infantry battalions on European lines were ultimately adventurers and guided by the changing patterns of relationships with England.[94] Besides, they could not ensure a regular supply of military stores, or improve on indigenous manufactures which were for the most part of inferior quality. The English, on the other hand, maintained their regular supplies through their naval control. But much more important than this was their unity of command, which was sadly lacking on the Indian side. The Treaty of Bassein, which Peshwa Baji Rao II concluded with the English in 1802, was an expression of complete Maratha bankruptcy in political terms. The English made full use of it politically to reinforce the justification of the military conquests effected with the cooperation of 'the Army in India', reorganized in 1796 and commanded later under the Marquis of Wellesley by such able generals as Arthur Wellesley (later Duke of Wellington) and Gerald Lake (later Lord) who forced Perron, the French general, to retire from Sindia's service. India's defeat proceeded in fact more from the weakness of its society and polity and its incapacity to unite for combined manoeuvres under able leadership, rather than any lack of courage to fight or face dangers. It was, after all, the Indian troops under British officers who conquered the country for Britain. While Mir Jafar, the chief of Sirajuddaula's army, won the battle of Plassey for Clive in 1757, Baji Rao II, through the Treaty of Bassein, consigned the Maratha empire to the grave in 1802. The higher interest of the country or its people did not matter.

An Instrument of Integration

The final defeat of the Marathas came about in 1818 as an immediate consequence of British military operations against the Pindaris who constituted under their cover a serious menace to peace and internal security. The Maratha states, more especially those of Holkar and Sindia, utilized the services of the Pindaris even in their internal dis-

The Instruments of Territorial Integration 65

sensions. The Pindaris were not a tribe but a military system of bandits consisting of all races and religions. In the words of Captain Sydenham:

> Every horseman who is discharged from the service of a regular government or who wants employment and subsistence joins one of the *darras* (principal divisions) of the Pindaris; so that no vagabond who has a horse and sword at his command can be at a loss for employment. Thus the Pindaris are continually receiving an assession of associates from the most desperate and profligate of mankind. Every villain who escapes from his creditors, who is expelled from the community for some flagrant crime, who has been discarded from employment, or who is disgusted with an honest and peaceable life, flies to Hindustan, and enrols himself among the Pindaris.[95]

Though provided in some cases with matchlocks, the Pindaris were all mounted, armed with spears and distinguished by the secrecy and precision with which they accomplished their object—plunder. Each of their parties consisted of one to four thousand, marching forty to fifty miles a day without tents and baggage. They were not satisfied with only loot and plunder, but destroyed what they could not carry away. Villages were seen in flames wherever they marched, while women were subjected to outrages compared to which torture and death were mercy. The hands of children would frequently be cut off as the shortest way of obtaining their bracelets. When the Pindaris repaired to their lairs, the claim of the government under which they served had first to be satisfied, or in case they pursued their vocations independently, the demand of their chiefs had to be met. Amir Khan, who was originally a Pindari, had risen to power in alliance with Sindia and established even a military state at Tonk in Malwa. Maratha politics and the geography of Central India were both helpful to the operations of the Pindaris, whose widespread depredations extended even to Madras in 1814–15.

These conditions of anarchy and lawlessness, which constituted a threat to security and imperial defence, led the Marquis of Hastings to justify the maintenance of a large army[96] to guard against not only external dangers, but also the disintegrating forces of internal unrest and adventurism. The latter necessarily called for the establishment of a number of military stations throughout the country to assist civil authorities in the exercise of their functions and to help the Indian princes in the fulfilment of their own obligations to keep roads safe for trade and the movement of the Company's army.

As the mounted elements of lawlessness used to enter the Nizam's territories in the Deccan by eluding the regular force stationed on the

Narbada, the necessity of increasing the regular cavalry regiments and adding a certain number of irregular units to them had already begun to be realized after the reorganization of 1796 itself. Several such detached corps were therefore sought to be widely distributed and irregular units organized to meet the exigencies of territorial integration.

Unlike the rule of the Mughals, which was essentially military rule, the Company's army in India remained subordinate to the civil power. Although a gifted soldier himself, Clive, for instance, laid down firmly a principle that the whole army must remain subject to the control of the Company's Governor and Council, who alone were the trustees of the Company and the guardians of public property under a civil constitution,[97] a constitution which conduced to change within the framework of law and legislative authority. Clive put down with determination the mutiny of the Bengal army when it agitated for additional allowances. Disputes arose from time to time, but the authority of the civil government remained supreme and created confidence in the administration of law and order, an important condition for peace and stability.

The development of transport and communications, which had immediate relevance to internal security, was at the outset governed chiefly by military considerations. The projected construction of a railway in India, for instance, was first discussed by the Governemnt of Lord Hardinge, a known annexationist, who agreed to help a private company provided the railway so constructed gave priority to the movement of the Company's troops and military stores. This was one of the conditions imposed by the Government of Lord Dalhousie when it entrusted the construction of railways to a number of private companies on the principle of what was known as the 'guarantee system'. The first railway introduced in India was opened in 1853 and the first train steamed out of Bombay to Thana on 14 April. Apart from strengthening internal security, this was the first tangible step towards industrialization,[98] which led to the augmentation of the country's commercial and agricultural resources. The Government of Lord Dalhousie constituted the postal services[99] as a separate branch in 1854, and the first telegraph line was put up in 1853 from Calcutta to Agra. The development of transport and communications, which proceeded initially on a trunk-line system, connected the hinterland of the Bombay, Bengal and Madras Presidencies with their principal ports and with each other.

As already noted, the army's contribution to territorial integration was particularly recognizable in the field of Indian surveys, which were manned by officers belonging to military cadres. Their work covered a wide range of activity which had a direct bearing on the country's defence and development. Apart from the work of army personnel in such fields as revenue, topographical and marine surveys, their achievement in bringing the hill tribes and 'backward' communities to the path of peace and progress was no less praiseworthy. They learnt the local languages, lived amidst the tribes, won their confidence and studied their institutions to provide guidelines for the government in the formulation of its policy towards the hill tribes on India's frontiers and its hinterland. Lieutenant Macpherson of the Madras army, for instance, raised an irregular Police and Khond Battalion from among the Khonds of Orissa and worked in their midst during 1837–45. He brought them to a life of law and discipline, persuaded them to give up their custom of female infanticide, and, instead of human sacrifice, induced them to sacrifice buffaloes, monkeys and other animals with all the ceremony which earlier attended human sacrifice. In Alexander Mackenzie's excellent study of the work of army and other officers among the hill tribes of the north-east frontier of India, particular tribute is paid to 'the untiring [army] officers of the Survey, for whom no peak is inaccessible, no jungle impenetrable, and no tribe too rude to be faced'.[100] Hall of the Bengal army bacame known for similar work among the Mers living in the hills near Ajmer. He persuaded them to give up human sacrifice, slavery and the sale of women, and raised a Mer corps which remained loyal to the British during the Mutiny.[101]

Chapter II

THE DELIMITATION OF INDIA'S FRONTIERS

The political stability of India proceeded mainly from three related premises—freedom from the dangers of foreign invasion, territorial reorganization for purposes of internal security and development, and institutional stabilizing agencies and arrangements. They called for the fulfilment of certain conditions involving policy dimensions and executive action. The delimitation of India's frontiers to check foreign aggression was sought to be achieved by making the frontiers conformable to her geographical unity and cultural tradition.

The delimitation of frontiers generally involved questions of buffer states and diplomatic relations with adjoining countries. It was inextricably linked with questions of trade, which necessitated political control not only of frontier territories but also of adjoining states, more especially to the buffers which had to remain friendly. For reasons of security, the delimitation of boundaries also called for an administrative reorganization of frontier territories on the Indian side, especially in the north-east and the north-west, a task which, though of local importance, formed part of the total provincial reorganization for security and developmental purposes. The diplomatic and administrative tasks were to a large extent sought to be achieved with the help of the Indian surveys which supplied extremely useful and secret information on foreign topography, trade and military routes, local ethnography and economy, information that shaped both policy and action.

The North-West

Safeguard against Russian Attack

Maintaining Afghanistan as a buffer state remained for most of the period the main concern of British policy in the north-west of India. The boundary of the Company's territory was confined to the Sutlej until 1849 and Punjab and Sind were earlier little known to the English. At the instance of Sir John Malcolm, Lieutenant Burnes of the Company's army had attempted a general survey of the Indus valley on his way to Lahore to meet Ranjit Singh in 1831, and to

The Delimitation of India's Frontiers

offer him presents from the King of England. In 1832 Burnes again made a famous journey to Bokhara via a mission to the Amirs of Sind, whom he persuaded to agree to a more precise survey of the Indus and its estuaries. It was through his reports that the Company's government acquired a knowledge of the trans-Indus territories stretching along the borders of Afghanistan, including Peshawar and other places of strategic importance, which the Sikhs had by then come to occupy.

The Treaty of Tilsit (1807), which brought the French into an alliance with Russia and marked the emergence of the latter as a power to reckon with in the politics of Central Asia, led Lord Minto to send out a series of missions, such as those of Seton to Sind, Metcalfe to the Punjab, Elphinstone to Afghanistan, and Malcolm to Persia. But the most effective step taken was to persuade the Portuguese in India to exclude all kinds of anti-British foreign influence from their territories in India.[1] An agreement to this effect was negotiated with the Portuguese Viceroy by the British political officer in Goa, Captain Courtland Schuyler.

The English fears building up after Tilsit became obsessive by the time Auckland took over. In 1836 Burnes was appointed to proceed on an allegedly commercial mission to Kabul, a mission which in reality was political in its objectives.[2] In fact the correspondence which Burnes carried on with W. H. Macnaughten, political secretary to the Governor-General, related for the most part to Dost Muhammad's demand for the recovery of Peshawar and the Russian mission of Vitkevich sent to Kabul.

Right from 1831, when Burnes did the first survey of the Indus and its surrounding areas, there developed in fact an increasing emphasis on the importance of Sind and the outlying part of the north-west frontier regions. These became in course of time an object of direct British concern, which led Burnes to acquire through his travels to Bokhara an intimate geographical knowledge of intervening countries for regional defence against foreign infiltration as well as the protection of India's immediate frontiers. At the time this was a responsibility of the Sikh ruler who, in the long-term scheme of British strategy, did not figure as a permanent factor in the politics of the north-west. The conquest of Sind in 1843 and the annexation of Punjab in 1849 advanced the British administrative boundary across the Indus, made it coterminous with the territories of the Baluch and Pathan tribes, and eventually brought the Government of India into closer contact with the Khan of Kalat and the

Amir of Afghanistan. The deputy commissioners of the six districts of Hazara, Peshawar, Kohat, Bannu, Dera Ismail Khan and Dera Ghazi Khan were entrusted with the conduct of relations with the tribesmen. A new North-West Frontier Province was later carved out of the trans-Indus tracts under Lord Curzon. It consisted of these settled districts and, in addition, comprised a number of trans-border political agencies acquired during Anglo-Afghan border disputes after the second Afghan War of 1878.

The development of Anglo-Afghan relations, which finally resulted in the second Afghan War, forms no part of our analysis, which is concerned mainly with boundary delimitation for defence against aggression. Certain policy trends which impinged on these relations are, however, relevant for a clear understanding of the problem of Indian defence arising from the Russian advance towards Central Asia and bordering Afghanistan. In 1844, for instance, Russia and Britain agreed that the Khanates of Central Asia, comprising Bokhara, Khiva and Samarkand, should be treated as a neutral zone between the two empires in order to prevent their hostile contact. This understanding, however, became inoperative after 1854. The British in India came to know of this through the confidential reports of their agent, Fateh Muhammad, who travelled in 1855 through Turkestan and visited Khiva, Persia and Afghanistan to ascertain the exact position of the Russians in those places. This report on these and such other areas as Bokhara, Balkh, Badakshan, Merv, Mashed, Herat and Kabul showed that, while Persia was following a regular policy of aggression against Afghanistan, the British position in Central Asia was far from favourable.[3] Russian expansion in fact had begun soon after the Crimean War and by 1864 touched the borders of Khokand, Bokhara and Khiva. The year following, Russia occupied Tashkent. A new province of Russian Turkestan was constituted in 1867 with Kaufmann as its first governor. Bokhara was reduced the same year. Khiva too became a subsidiary ally in 1873. All these dependencies were put under military control, and made subject to the Russian foreign office at St Petersburg. The old understanding of the Oxus being the neutral zone and dividing line between Russian and British interests was thus unilaterally ruled out by Russia, which now insisted on Afghanistan itself being a neutral zone. The undefined northern borders of Afghanistan thus became open to Russian incursions, while its western frontiers of Herat came under constant threat of attack and seizure by Persia with Russian support.

British policy, far from being firm and decisive, was characterized by caution bordering on what is called 'masterly inactivity'.⁴ It could not stop the Russian military advance in Central Asia, which was intended to keep England in check by a threat of intervention in India. This was reflected in John Lawrence's refusal to make an alliance with the ruler of Afghanistan, and, after the death of Dost Muhammad in 1863, his recognition of any contender to the Afghan throne who happened to establish his authority. Accordingly, Dost Muhammad's son, Sher Ali, then Afzal Khan, and then Azim Khan, and, finally, Sher Ali once more, succeeded in establishing themselves as successive rulers of Kabul, and the Indian Government under Lawrence was content with recognizing each in turn. This wait-and-watch policy, which caused instability in the internal affairs of India's buffer state, continued even after Sher Ali had come to stay and despite Lawrence's departure in 1869. The Amir reiterated his desire for an alliance which would bind the English to support him against external attack, with a promise, in addition, that they would never acknowledge any friend in Afghanistan save Sher Ali and his descendants.⁵ Lord Mayo's meeting with Sher Ali at Ambala in March 1869 was doubtless friendly, but it signified no change of policy. His successor, Lord Northbrook, it is true, assured the Amir in 1873 of help, if necessary. But the Duke of Argyll at the India Office rejected his proposal. When Disraeli became Prime Minister in 1874 and Lytton replaced Northbrook in India, it was realized that something tangible should be done to assuage Sher Ali's fears. But by then the failure of the Government of India to give Sher Ali the assurances he sought, had led the Amir to strengthen his relations with the Russians, who were willing to help. In a letter to the Amir in June 1878 Kaufmann thus emphasized that the external relations of Afghanistan required deep consideration and that he was sending Stolietoff to appraise the Amir of all that Russia planned to do. Stolietoff, in fact, carried with him a draft treaty offering terms very similar to what Lytton proposed—recognition of the heir apparent and assistance against external attack. However, Neville Chamberlain, the envoy whom Lytton had chosen for a mission to Kabul to counterbalance that of Stolietoff, was prevented by threat of violence from passing Ali Masjid.

Second Anglo-Afghan War

The Amir's refusal to accept the British representative led to the second Anglo-Afghan War, which broke out in November 1878. The results were quick and decisive. Towards the close of December 1878

Sher Ali announced his retirement into Russian territory, where he died early in 1879. Negotiations were started with his son, Yakub, who signed the Treaty of Gandamak on 26 May 1879 before British forces had even entered Kabul. The new Amir assigned to the British the districts of Kurram, Pishin and Sibi, and, in addition, agreed to conduct his foreign relations with the advice of the Governor-General of India. A permanent British representative was to be stationed at Kabul.

Cavagnari, the political agent who had conducted the negotiations, was accordingly charged with the responsibility of handling a still tense and delicate situation. He reached Kabul on 24 July. Though well received initially, he was murdered on 3 September, which led to a renewal of the armed campaign. Roberts advanced by the Kurram Pass and occupied Kabul with the cooperation of Yakub. But since British opinion was unanimous against the restoration of Yakub, he was removed to India, pensioned and allowed to reside at Dehra Dun till his death in 1923. Abdur Rahman, a nephew of Sher Ali who had earlier fled to Samarkand and Russian protection, returned to Kabul on receving conciliatory messages from the English and entered into negotiations with Lepel Griffin, the new English political agent at Kabul. As a result of the understanding reached with Abdur Rahman, the English retained Pishin and Sibi. The new Amir also placed the management of his foreign affairs under the Government of India who, in return, promised to pay him an annual subsidy. The settlement thus brought to a close a dangerous phase in the Central Asian question. The error of 1873 had to be rectified and the border country of Afghanistan reconciled to exist as a friendly buffer state.

Boundary Delimitation Commission

The aftermath of the second Afghan War, which called for a joint boundary commission, was no less significant. The need for the commission arose from Russian movements in the Merve oasis, a region of strategic value, where the Merve chiefs on the northern borders of Afghanistan were being influenced to seek Russian protection. The Russian War Office was found to have prepared in 1884 a map showing the Merve boundaries stretching southwards and touching the Hari-rud near Herat. It was Granville who not only took up the idea of formally defining the northern boundary of Afghanistan, but also proposed the establishment of a joint delimitation commission consisting of an Indian official, Sir Peter Lumsden, at the head of the British mission, a set of qualified officers to be provided by the Amir, and members of the Russian mission. A joint meeting was proposed to be convened at Saraks on 1 October 1884 on the assumption that Russia would co-operate. But the Russian leaders chose to bid for time so as to first secure control over the entire body of nomad Turkoman tribes; and although they named General Zelenoi as head of the Russian boundary mission, they announced his inability to function on the score of illness, so that nothing could be done till February 1885. The work was further delayed by Russian insistence on Punjdeh being declared independent of the Amir; and when the British proposed to have cases of boundary disputes decided by the commission, the Russians would not agree. They even attacked a body of Afghan troops and drove them out of Punjdeh in March 1885. the Punjdeh crisis became the cause for further delay, and it was some time before an agreement was signed in September 1885 under which Afghanistan was assigned Zulfikar on the north-west, instead of Punjdeh. Even so, the commission continued to function, and the boundary survey it undertook proved to be of great importance in terms of Indian defence strategy.

The boundary survey had in fact started in all seriousness in 1884 itself,[6] and within about two years it made considerable progress. According to the Indian records of the surveys 'an enormous addition to our knowledge of northern and western Afghanistan was made by the labours and researches of the Boundary Commission, which was engaged, in conjunction with a Russian Commission, in laying down the northern boundary of the country in 1884-6'.[7] The Afghan and Russian missions, it seems, operated separately, though from time to time the observations of the respective teams were compared and verified by the Russian staff with a 'Repsold instrument of a much higher class than the small English theodolities'. On the basis of such verification, the topography of the area was codified, whether by Russians or the British officers and Indian sub-surveyors, who not only did their share of boundary work but also extended smaller scale geographical work to a vast area in Persia and Afghanistan over some 110,000 square miles.[8] Other important tasks completed by the Afghan Boundary Commission were the surveys of the Hindu Kush and the routes and passes connecting Afghan Turkestan with Kohistan and Kabul. These surveys were carried out, among others, by Colonel Holdich, Talbot and the sub-surveyors Yusuf Sharif, Ata Muhammad and Hira Singh. Their work marked the end of the geographical researches of the Afghan Boundary Commission in so far as its European officers were concerned. Connected with the work of the Afghan Boundary

Commission, some important explorations of the Pamirs and Upper Oxus valley were carried out by Ney Elias of the Foreign Department, who was already reputed for his adventurous journey across Mongolia in 1872–3. Starting from Yarkand in September 1885, Elias left the plains at the frontier village of Inghiz Yar and travelled through Kashgar, Yangi Hissar and other central Pamir peaks which had earlier been measured by Captain Trotter. The track he followed at more than 14,000 feet was frequently well nigh impassable for a caravan, or even a loaded pony. Passing through Bar Punjah and Ishkashim, Elias traversed the Tajik and Turki regions, and reached Khanabad and Kunduz in February 1886. He was obliged to travel down to the Turkoman country to join the Afghan Boundary Commission, then near Maruchak on the Murghab. Elias took leave of the Commission at the end of April to return to Badakhshan. He was recalled to India via Chitral and Gilgit, arriving eventually in Kashmir in September 1886. The Afghan boundary, too, was formally laid down that year, from the Oxus westward to Zulfikar.

Apart from being extremely important and relevant in terms of India's defence strategy, the empirical data collected from the researches and survey operations of the Boundary Commission made a major contribution not only to a knowledge of geography, but also of geology, social anthropology and the biological sciences.

Geographical Surveys and Explorations During War

Considerations of defence doubtless called for explorations and the gradual opening up of the peripheral regions of British Indian territories. But the task of surveying wild and unsettled regions beyond the limits of India's frontiers was not easy for British officers. A plan for training Indians[9] was therefore introduced, a plan which was greatly developed in the post-Mutiny period, more especially after 1875, was crowned with complete success. Clad in local dress and speaking the language of the natives, Indians could pass on without the risk of detection and do their jobs with the help of the instruments they carried on their bodies.

But when involved in hostilities no such restrictions applied. This was the case when the first Afghan War was declared. The surveys, even though hurriedly conducted, contributed to the extension of geographical knowledge about Afghanistan. Their results were at the time never combined and compared, but they were nevertheless helpful. So long as British policy considered it unsafe to extend survey operations beyond the actual line of British control, the survey officers could not do more than fix on their maps all the prominent points on the hill ranges beyond, which were visible from the frontier attained in the course of the war. But they also filled in details from local information or through the secret agency of native explorers.

On the declaration of war with Afghanistan in 1878, however, the errors of 'the hurried surveys' of the first Afghan War were sought to be rectified. Surveyors were duly attached to each of the four columns invading the country, even though for purposes of geographical exploration they were grouped in three divisions. Much of the survey work in southern Afghanistan was done with Quetta as its headquarters and embraced altogether about 5,000 square miles. The Kuram valley was done by Captain Woodthorpe who was attached to General Roberts. Woodthorpe accompanied the first advance in November 1878, and carried on his work in the midst of the fighting. He corrected several errors in the old maps, and sketched a good deal of the topography from the highest point of the Sufed-Koh range. As many as three officers were attached to the northern column: they were obliged to conduct surveys under heavy fire from the fort of Ali Masjid to Jalalabad, a continuous survey which found Peshawar to be nearer Jalalabad than previously suspected. While surveying north of the Khyber Pass, the party was attacked by the Memands in the country south of the Kabul river, but soon after this encounter the leader of the party, Scott, visited the Sekhram (Sikaram) peak, the highest point of the Sufed-Koh range, and obtained observations on numerous distant peaks, which later helped in the work of the Afghan Boundary Commission. The task completed by the surveyors of the northern column covered altogether an area of about 2,200 square miles, extending from Forts Michni and Jamrud on the British frontier near Peshawar to the Surkhab river west of Gundamak, and including a little of the northern and most of the southern portion of the basin of the Kabul river. A good deal of information was in addition gathered from across the Kabul river in the Lughman plain and neighbouring tracts.

The reports of these surveys conducted during the Afghan Wars, the political negotiations with Russia over the question of the northern frontiers of Afghanistan and the researches carried on by the Afghan Boundary Commission, equipped the Government of India with a stock of knowledge and information about the routes and topography of its near and distant frontier regions, which could be used for defence purposes to ensure political stability in the country.

The formal recognition of the Afghan boundary from the Oxus westward to Zulfikar in 1886 was followed by six years of comparative quiet until disputes arose in respect of the Pamirs in 1892, when British officers were arrested in territory which they asserted was not Russian. Russian agents not only visited Chitral, but even entered territory under the actual occupation of the Afghans. In the middle of 1892 Russia sought in fact to establish its dominion over the whole of the Pamirs. Despite a proposed delimitation commission and discussion on the controversy in progress, no solution was in sight because of the persistent claims of the Russian War Office to the whole of the Pamirs. It was only on 11 March 1895 that an agreement was signed by which Afghanistan was to surrender territory north of the Punjab, while Bokhara surrendered that part of Darwaz which lay south of the Oxus. This settlement left no further room for dispute on the Afghan boundaries and marked a gradual relaxation of Anglo-Russian tension, although their mutual distrust did not, of course, disappear altogether.

The Indian Marine Surveys for Naval Defence

The survey of the coasts of India was of great importance not only for navigational purposes, but also for naval defence. The operations of the Indian navy were not confined only to the Indian coasts, but extended to the Red Sea, the Persian Gulf, the Arabian and African coasts, the Burmese and Siamese waters as well as the China Sea and, in fact, other regions connected directly or indirectly with India's external defence or commerce. The Indian navy was, however, abolished in 1862, and no alternative arrangement was made to continue the survey work for which the service had acquired its good reputation. After a period of inaction an efficient, though small, department was organized in 1875 under the superintendence of A. D. Taylor who corrected the errors of earlier surveys, took steps to remove hindrances to navigation on the Indian coasts, and extended the scope of the marine survey to areas considered even remotely relevant to India's strategy.

Even in an earlier period, the Indian navy was not only concerned with the defence of India's coastal regions for which it carried on the surveys of distant seas and water courses, but also involved in securing the political supremacy of the British Government in India over such remote regions as were in any way linked with the security of the country. The question of controlling by means of the Indian navy the politics of Egypt from Bombay, for instance, formed the subject of correspondence between Dr Beke, an enterprising traveller in Abyssinia, and the British Foreign Office under Palmerston. The correspondence occurred in 1846 and 1848, and on the basis of the information collected earlier by the Indian Marine Surveys Dr Beke emphasized the importance of Abyssinia as a springboard for political ascendancy against the French in Egypt; he suggested the expediency of using the principal Abyssinian port of Massawa as the nearest point for the transport of troops and military supplies from Bombay. For while the distance from Bombay to Suez by sea was 3,050 miles, that from Bombay to Massawa was 2,090 miles. In other words, Massawa was situated at no more than two-thirds of the entire distance that the British forces and supplies would need to travel by sea, and control over Massawa with the consent of Abyssinia would mean a saving of time and cost.[10] In the absence of the Suez canal, which was opened in 1869, political control was thus sought to be established in Abyssinia, so that Massawa, which was only about 390 miles from the British territory of Aden, could be used not only for India's security against foreign attack by sea, but also to counteract hostile action by a foreign power in Egypt or any of the East African countries along the Arabian Sea or the Indian Ocean. Steps were also taken at about the same time to ensure naval security in the Persian Gulf areas.[11]

The Survey of India

Organizationally, the office of the Surveyor-General of India at Calcutta was not reconstituted earlier than 1875, when General H. Thuillier retired. The Government took this opportunity to call upon his successor, Colonel J. T. Walker, to reorganize the department, involving an amalgamation of the three branches of the Survey—the Great Trigonometrical, the Topographical and the Revenue. Up to that time, they had been virtually separate departments, each with its own cadre of officers and establishments consisting of both European and Indian surveyors, with a separate Superintendent attached to each. The duties which each had to perform happened to be essentially distinct. The Trigonometrical Survey was, for instance, required to furnish the basis on which all surveys of interior details were to rest. It supplied the framework within which interior details were filled and connected together. The Topographical and the Revenue Surveys were both to furnish the interior details. But while the former did the surveying by plane-tabling and covered the whole country, including the Indian states, the latter was confined to the survey of the richer British revenue-paying districts on a larger scale with chains instead of plane-tables. The Topographical Survey, though originally intended for the primary general survey of India, had for strategic and other reasons to undertake in many cases detailed surveys on a larger scale, and the Revenue Survey had in addition to its own special functions been largely employed on the topography of hill districts on a trigonometrical basis. The amalgamation of the separate departments thus proceeded from the functional intermingling which had already started. It was, however, a sound measure which provided unity of control and direction. The efficiency of the survey teams which accompanied the several military columns in Afghanistan in 1878–9 and the work of the Afghan Boundary Commission during 1884–6 were all a result of the coordinated efforts made by the reorganized department.

The Northern Himalayan Frontier

The Himalayas, which constitute the northern boundary of India, served as a frontier between Central and South Asia, containing Nepal, Sikkim and Bhutan on their southern fringe, and Tibet under Chinese suzerainty in the north beyond their crests. The southern Himalayan states were over the years drawn into the sphere of British India and alienated politically, through treaties and agreements, from their erstwhile Tibetan and Chinese influence. But Tibet, which was contiguous for nearly a thousand miles with British territory from Kashmir to Burma, would not permit easy contacts, even for purposes of trade and commerce.

The whole range of the Trans-Himalayan frontier was dotted with three main spheres of contact with Tibet. These were Nepal and Sikkim in the middle, the Assam Himalayas to the east and Ladakh with adjoining Kashmir in the west. And since Tibetan relations with China remained over the years in a state of flux, any attempt of the Indian Government to delimit its Trans-Himalayan frontier on the Tibetan borders gave rise to a clash with both until their interests became mutually exclusive, which meant an opportunity for the British to combine with Tibet to settle its scores with the Chinese and neutralize their influence and power in Tibet.

Central Tibet

Earlier Approaches

The Tibetan tendency to remain secluded and their refusal to receive any Englishman proceeded from their fear of the Company's government, which they considered to be a great power prone to war and conquest. This feeling was conveyed by the Tashi Lama at Shigatse to George Bogle, who had been sent on a mission to Tibet by Warren Hastings in 1774. In the course of his talk with Bogle the Tashi Lama made it very clear that, if the object of the English was to promote trade between the two countries, 'I wish the Governor [Warren Hastings] will not at present send an Englishman', but that 'the Governor would rather send a Hindu'.[12]

Markham's account of Bogle's mission to Tibet makes it evident that Warren Hastings wished to secure his northern borders on a long-term basis and promote through trade and commerce a stable, neighbourly intercourse with Tibet. The Governor-General's concern for a secure northern frontier beyond the Himalayas was clearly reflected in Bogle's conversation with the Tashi Lama, but nothing came of his mission and he returned to Bengal without achieving his objectives. Hastings nevertheless appointed Bogle to go on a second mission to Tibet. However, this did not materialize, for Bogle died in 1781; as a substitute, Captain Samuel Turner was appointed in 1782 to pursue the matter. None of these missions could induce the Tibetans to open trade relations or to grant consular facilities to the Company, but Hastings' sense of realism and foresight remained undisputed.

Without going into the legal and political question of Tibet's international status, the British in the nineteenth century acted on the belief that the Dalai Lama's government was subordinate to the Chinese emperor and that any British overtures to the Tibetans would require prior Chinese approval. This belief flowed from the reports of both the earlier missions and Manning's individual visit to Lhasa in 1811. The approach of the British government in India was pragmatic. It looked upon Tibet as a potential market for British and Indian goods, as a potential land route from India to the Chinese interior, and as a source of gold and wool. And as the physical situation of Tibet and its influence were of considerable importance in the politics of the Himalayan states of Nepal, Sikkim and Bhutan, the Government of

India was anxious to establish within the framework of Chinese suzerainty diplomatic relations with Tibet so as to be able to secure, in addition to commercial advantages, a stable northern frontier beyond the Himalayas. It wished to create a British sphere of influence in Tibet without questioning the validity of the *status quo* or entering into any controversy over the legal definition of actual relationships.

New Approaches towards Tibetan Frontiers

Besides being founded on the experience of the earlier missions, this pragmatic approach to Anglo-Tibetan relations was related to the steps which the government of Lord Hastings (1813–23) had taken to strengthen British positions in the Himalayan regions as well as the gradual submission of the Himalayan states to British power in India. The Anglo-Nepalese War, which resulted in the annexation of a long strip of territory along the foothills of the Himalayas, marked the beginning of a series of surveys conducted for topographical and geo-political purposes in the Himalayas, covering not only Kumaun and Garhwal, but extending to several higher reaches beyond the sources of the Ganges and the Sutlej.

An important offshoot of the Anglo-Nepalese War was the extension of British control over Sikkim, adjoining the Tibetan border itself. Sikkim was afflicted with internal dissensions between the Bhutiyas and Lepchas in the 1760s when, with the support of the Lepchas, Phuntshog Namgyal ascended the throne after the death of Gyurmi Namgyal. With the mediation of Tibet the Bhutiyas and the Lepchas came to terms. But the state was soon subject to frequent invasions by Prithvinarayan Shah of Nepal on the west and by Desi Shidar of Bhutan on the east. Bhutan attacked Sikkim in 1770 and occupied the entire territory east of the Tista river, while Nepal took most of its territory west of the Singli ridge. With the intervention of Tibet, Nepal and Sikkim settled their boundary disputes in 1775. But the Gorkhas broke the settlement and occupied the entire lower Tista basin in the south during 1788–90. Chhogyal Tenzin Namgyal, who had ascended the Sikkim throne in 1780, took shelter in the north in search of Tibetan aid. But since Tibet itself was involved in conflict with Nepal, nothing could be done before 1792, when the suzerain Manchu emperor came to the Dalai Lamas's aid and imposed severe terms on Nepal. The Tibetan government, however, appropriated Tenzin Namgyal's ancestral estates in central Tibet which had earlier been granted by the Sixth Lama. Tibet also took the Chumbi Valley, and thus made the Cho La and Jelep La ranges the northern and eastern boundaries respectively of Sikkim. On the western side the Gorkhas continued to hold Sikkimese territory up to the Singli ridge. But since the Nepalese were settled in strength in the lower basin of the Tista river, Bhutan realized the potential threat which Gorkha power posed and started co-operating with Sikkim to contain Nepal.

The British came into contact with Sikkim under Tsugphu Namgyal, the son of a Lepcha mother, who succeeded Tenzin Namgyal in 1793. He shifted his capital to Tumlong in the north of Sikkim in 1814 and sought British help against the perpetual Gorkha menace. The strategic importance of Sikkim induced the Company's government to accept Tsugphu Namgyal as an ally in its war against Nepal (1814-15). Under the Anglo-Nepalese Treaty signed at the conclusion of the war, Nepal surrendered to the British the Morang strip at the foot of the Sikkim hills, which had originally been part of Sikkim, though later annxed to Nepal during 1788-90. Later, on 10 February 1817, the British agent at Purnea concluded a separate treaty with the representatives of Chhogyal Tsugphu Namgyal at Tetuliya. The British restored to the Sikkimese ruler the whole of lower Sikkim, bounded on the east by the Tista and on the west by the Mechi river as well as the Singli range. It became a barrier against any further attempt on the part of the Gorkhas to extend power. But Sikkim was reduced to the status of a protectorate. The Company's government virtually became its paramount lord, with the right to arbitration in disputes with neighbouring or other states. Sikkim undertook to allow free passage to merchants and to provide troops on requisition.

Security against foreign invasion had largely been achieved, but the internal feuds between Bhutiyas and Lepchas did not stop. This gave a handle to the Lepchas in Nepal to carry on raids into western Sikkim, which led to British intervention in 1828 and the withdrawal of the Lepcha infiltrators. But it was about this time that the English officers deputed to explore the region learnt of the Darjeeling hill tract, a suitable site for a sanatorium which, in addition, could serve as a base for the pursuit of British political and commercial interests in the eastern Himalayas. Chhogyal Tsugphu Namgyal was thus persuaded to present Darjeeling to the Company's government, which acquired it in 1835 and sanctioned an annual compensation of Rs 3,000 to Sikkim in 1841. The presence of the British so close to Sikkim and the development of roads and other communication facilities aroused fears in Tibet, which then restricted the Chhogyal's visit to

Lhasa and curtailed the grazing rights of his people on its borders. The opening up of Sikkim was in these circumstances stopped in 1847 by the Chhogyal's Bhutiya minister, who even arrested Archibald Campbell, the first Superintendent of Darjeeling (1839–64). The march of a small military force into Sikkim, however, led to the release of Campbell who stopped the annual subsidy and annexed Morang as well as the entire hill tract around Darjeeling.

In 1857 Anglo-Sikkimese relations became seriously strained on account of Sikkim's acquiescence to the use of its territory by the rebels of the uprising in India. In 1860 Campbell laid siege to that part of Sikkim between the Raman and Ranjit rivers. Chhogyal Tsugphu Namgyal fled to Tibet, where he died in 1863. His son Sidkyong Tulku, who took over in 1861, accepted British terms, which included trade and extradition clauses, the right of the Government of India to construct roads through Sikkim to the Tibetan frontier, banishment of the anti-British Bhutiya minister who had fled to Tibet, and removal of the seat of government from Tumlong to Gangtok in the eastern part of Sikkim. The treaty of 1861 lifted all restrictions on free intercourse and permitted the Government of India to complete its projected survey of the whole country up to the borders of Tibet, which thus came to be a buffer state on the frontiers of Sikkim. Like the other princely states of India, the ruler of Sikkim came to be designated a 'Maharaja' and his government as the Sikkim Darbar. The new title, however, did not continue for long.

The virtual establishment of British paramountcy under Lord Hastings and the growth of British control over Sikkim as an aftermath of the Anglo-Nepalese War were doubtless new conditions which encouraged a renewal of effort in the direction of establishing the projected relations with Tibet as a natural frontier beyond the Himalayan crests. But it was the British entry into Bhutan after military action that immediately prompted the Government of Bengal to induce the Indian Government to resume the old threads of negotiations which had begun with Bogle's mission of 1774.

Of the three Himalayan states, Bhutan had by far the closest links with Tibet in both cultural and political terms. It was contiguous with the territories of the Assam rajas who were said to have kept under their possession that narrow tract of fertile tarai land which stretched along the base of the Bhutan hills from west to east, and varied in breadth from ten to twenty miles. This land, which sloped downwards to the plains, was a source of contention between the Assamese and the Bhutiyas; the inhabitants of the barren hills, who, taking advantage of the indifference shown lately by the Assam rajas on account of the malarious and deadly character of the tract, held practical command of the border, and in course of time established what they considered their right over the whole of the disputed tract. As Bhutan bordered the most westerly districts of Assam, the British occupation of that province in 1826 brought them into peculiar revenue relations with the Bhutiyas of the tarai tract and the neighbouring highlands.

There were along the frontier of Bhutan as many as eighteen passes (*duars*), eleven on the frontier of Bengal and Cooch Behar, and seven on that of Assam. The passes and the fertile tract below them were made over to the Bhutiyas by the Assam government in consideration of an annual payment of tribute. A regular source of trouble, however, was that while the tribute was paid in kind, the value was fixed in specie. During the time of the later Ahom rulers, who were known for their weakness, it was the Bhutiyas who got differences settled to their own advantage. But since the British would not relent in financial matters, the attachment of the passes and the land below followed non-payment of any arrears or dues, a development which resulted in confrontation and the kidnapping of some British officers.

Captain Pemberton, an authority on eastern frontier matters, was deputed in 1837 to meet the Deb and Dhurm Rajas of Bhutan and to settle not only terms of commercial intercourse, but also the tribute payable for the duars and tarai land. His impression from what he saw of the country was that the central government of Bhutan was powerless to control the Penlows or local authorities of the outlying districts. The peace of the frontier, he added, was often disturbed, especially by 'the Tongso and Paro Penlows, the Governors respectively of East and West Bhutan'[13] who incited their subordinates to stir up violence. The troubles on the frontier of Assam continued until 1841, when the whole of the duars on that front were attached in return for Rs 10,000 to be annually paid to Bhutan as compensation. This measure added 1,600 square miles of territory to Assam.[14]

The Bengal section of the duars, however, remained subject to frequent Bhutiya outrages. Inquiries showed that the Raja and the Tongso Penlow were both involved. When warnings failed to produce any permanent effect on the Bhutiyas who continued their raids year after year, the Government of India, before taking any drastic action, decided to send a mission to the Deb Raja of Bhutan to explain to him the real situation arising from the annexation of one of the estates. Sir Ashley Eden, Secretary to the Bengal Government, was accordingly appointed in August 1863 to negotiate some stable arrangement for the better conduct in future of relations between Bhutan and the Government in India. Consistent with his instructions, Eden exercised restraint and forbearance, as Pemberton had done before. He encountered many difficulties on his way to Poonakha, the capital city, only to meet with insult and inconvenience. He had in fact to buy his safe return by signing under protest a treaty which called for the surrender of all the Assam and Bengal duars.

The extraordinary patience shown by the British in dealing with Bhutan flowed from their desire to use Bhutan as the medium of communication with the Dalai Lama in pursuit of their old policy towards Tibet. Access to the Dalai Lama had become more elusive with the increase in his importance during the First Opium War (1839–42): not only had Manchu power been shaken in the war, but Tibet had also withstood the Dogra invasion of its territory (1841–2), despite the lack of Manchu help and troops.

War followed the disavowal of the treaty which Eden had signed. The Bhutiyas were deprived of all the lands they had held below the hills, and the Bengal duars were formally annexed. A sum of Rs 25,000 was, however, allowed to be paid annually as compensation for all eighteen duars as a token of goodwill towards the Bhutanese aristocracy. In 1872–3, the boundary between Bhutan and British territory was duly fixed. It stretched from the Manas river in the west to the Deosham river in the east, where the territory of independent Bhutan ended. Any collection made forcibly by Bhutiyas from traders operating in the duars so annexed was to be deducted from the next payment stipulated by the treaty concluded in 1865.

Despite provocations and the war with Bhutan care was nevertheless taken to see that restraint was used to reduce tension. This was not without effect. In 1874, Sir Richard Temple, the Lieutenant-Governor of Bengal, received a visit from the Deb Raja and exchanged friendly greetings. When civil war broke out in 1876, the Poonakha Jongpen and Paro Penlow, who had earlier quarrels with the British, took refuge in Kalimpong near Darjeeling. The Government of India gave them not only asylum but also pensions. The importance of Bhutan as an immediate approach to Tibet and China was never overlooked.

The Trans-Himalayan route surveys, which were sought to be completed secretly, were another step towards the enduring Anglo-Tibetan relations originally visualized by Warren Hastings. The first of these surveys, arranged by Captain T. G. Montgomerie of the Great Trigonometrical Survey of India, took eighteen months

The Delimitation of India's Frontiers

to complete, from January 1865 to June 1866. It was a route-survey from Nepal to Lhasa, and from Lhasa, north of Assam, to Gartok, north-east of Simla, beyond Lake Manasarovar, by the great road called the Jong-lam, which covered along the top of the Himalayan range a distance of about 800 miles and kept more or less close to the Brahmaputra river. The second was again a route-survey from Leh in Ladakh to Lhasa and back to India via the north-east frontier of Assam, a survey completed during 1873-5 and drawn up by Captain (afterwards Sir Henry) Trotter. Both the operations were the work of Pandit Nain Singh, an Indian from Almora district, whose countrymen had always been permitted by the Chinese to travel and trade in Ngari Khoraum, the upper basin of the Sutlej. Nain Singh had been trained at the headquarters of the Great Trigonometrical Survey in India by Colonel J. T. Walker who, on his departure for England, put him and his brother (Kishen Singh) under Captain Montgomerie.

The original plan was for Nain Singh to advance direct from Kumaun via Manasarovar to Tibet. But this was given up and Nain Singh was advised to go to Kathmandu and proceed from there to the great road between Manasarovar and Lhasa. The instruments taken for the route-survey consisted of a large sextant, two box sextants, prismatic and pocket compasses, thermometers for observing the temperatures of air and boiling water, a pocket watch and a common watch. Assuming the dress of a Ladakhi and adding a pig-tail to his head to complete his disguise, Nain Singh passed through Shabru and Kerun Shahr in the company of a merchant and reached Tala Labrang, where he first caught sight of the Brahmaputra river, which he crossed to arrive at the Trādom monastery along with a party of traders going westwards to Manasarovar. He stayed there till he met a Lhasa-bound Ladakhi merchant employed by the Kashmir Maharaja. On 2 October 1865, the merchant's headman consented to take Nain Singh on to Lhasa. Starting with the Ladakhi camp, the Pandit had to move eastwards along the great road, and it took more than a couple of weeks before he reached a town called Janglache on the Brahmaputra. Shigatse, where Captain Turner had met Tashi Lama in 1783, was about five days march (85 miles) from this point down the river. Most of the Pandit's companions went by boat; but since he had to do the route-survey and count paces to determine latitudes from astronomical observations, he went by land and arrived at Shigatse on 29 October.

The road, which had hitherto been more or less close to the Brahmaputra, went far south of the river from Shigatse. On 25 December, the party reached the large town of Gyantse on the Pen-nang-chu river. They crossed to Lake Palti which, according to the Pandit's observations, was about 13,500 feet above the sea. From the basin of Lake Palti the party crossed the Kan-pa pass, reaching the Brahmaputra once more. They sailed to Chu-shul, where they again left the river and followed its tributary, the Lhasa river, in a north-easterly direction, reaching Lhasa on 10 January 1866. Nain Singh remained there till 21 April 1866.

The Pandit left Lhasa with the Ladakhi party on 21 April and marched back by the great road as before, till he reached Tra-dam monastery on 1 June. From Tra-dam he followed that road to Manasarovar, over elevated tracts 14,000 to 16,000 feet above the sea. Crossing the Ma-yum pass, the watershed between the Brahmaputra and the Sutlej, he reached Tar-chen on 17 June. At Tar-chen the Pandit and his Ladakhi companions parted with mutual regret, the Ladakhis going north towards Gar-tok, while Pandit Nain Singh proceeded towards the nearest pass to British territory. He crossed the Himalayan range on 26 June 1866 and met his brother who, at the Pandit's request, was sent to Gar-tok to carry on a route-survey to that place.

Pandit Nain Singh's account of his route-survey contained a number of meridian altitudes of the sun and stars taken for latitude at 31 different points, including a number of observations at Lhasa, Shigatse and other places. It included an elaborate route-survey extending over 1,200 miles, and defined the road from Kathmandu to Tra-dam as well as the whole road from Lhasa to Gar-tok, fixing generally the entire course of the Brahmaputra from its source near Manasarovar to the point where it was joined by the stream on which Lhasa stood. It also contained observations of the temperature of the air and boiling water by which the height of as many as 33 points came to be determined, in addition to other observations taken at Shigatse, Lhasa, etc., indicating their climatic conditions. There were, in addition, notes on what was seen and information on a variety of topics culled by Pandit Nain Singh in the course of his expedition.

The explorations which Pandit Nain Singh conducted during 1865–66 earned him the present of a gold-watch from the Royal Geographical Society of London. He did equally excellent work as part of a group of Indians who explored and surveyed the headwaters of the Sutlej and Indus in 1867, and served with a mission deputed to Yarkand in 1873. Finally, Nain Singh volunteered to make a fresh expedition from Leh in Ladakh to Lhasa by a much more northerly route than the one previously followed.

The Delimitation of India's Frontiers   87

With suits of lamas' clothing secretly made and packed, the Pandit and his companions left for Leh on 15 July 1873 and reached six days later. They stayed there for a couple of days, then proceeded to Chugra, three marches further on, where they found a summer encampment of shepherds, the last inhabited spot on the road to Yarkand. 'At night under cover of darkness the Pandit and his three men cast off their old garments and donned their lamas' clothes. Before morning they were all well on the road'.[15] They followed for the first day the Chang Chenmo route to Yarkand by the Marsmik pass (18,420 feet high), but left the Yarkand road and turned east through the Kin pass, reaching Niagzu on 29 July 1870.

Niagzu was at the time recognized as the boundary between Tibet and Ladakh,[16] the right bank of the Niagzu or Rawang stream belonging to the latter, and its left bank to the former. The country from Niagzu to Noh was for the most part uninhabited, except for a few solitary tents belonging to Noh shepherds and a hut at Gunnu chauki which was occupied by a frontier guard. From Noh the Pandit toiled for several marches along the Tibetan plateau. The first seven miles to the east contained the eastern termination of a series of lakes known to geographers as the Pangong. The Pandit's survey discovered all these lakes, extending from Noh to Lhasa; until then the only lake that was known was Tengri Noh Lake to the extreme east.

During his brief stay at Lhasa the Pandit made himself familiar with two of the roads that led to Peking from Lhasa. The one generally used and believed to be open all the year round went due east from Lhasa to Chiamdo, the capital of the Kham country. It then took a southerly direction and passed through Ba-tang and the Chinese province of Szechuen, crossing *en route* numerous snow-covered passes across the ranges which divided the streams from Tibet flowing into the Yangtse. The distance from Lhasa to Peking was about 500 miles by this route, covering 136 caravan marches. The other was a northern route via Nag Chu Kha and across the headwaters of the Yangtse, leading to two alternative roads to Koke Nor and thence to Peking. This route had earlier been followed by the Abbe Huc who took fifteen days to reach Lhasa from Nag Chu Kha. The Pandit, as previously planned, took the highway to Peking and, on the first night halted at Kombo Thang, only two miles out of Lhasa. The following day he reached Dre-chen, a town on the bank of the Lhasa river, but wheeled round under cover of dark and began his return journey to India along the Brahmaputra and the Assam Himalaya, passing through Tsetang and Tawang to Odalgiri, a British frontier town. Pandit Nain Singh reached Calcutta on 11 March 1875.

In addition to the general information supplied to the Government, Nain Singh made a careful and well-executed route-survey of the whole line of country traversed in the course of this exploration. This route-survey covered '1,013 miles from Luking (the west end of Lake Pangong) to Lhasa, and 306 miles from Lhasa to Odalgiri. Of this total distance of 1,319 miles, throughout which his pacings and bearings were carefully recorded, about 1,200 miles lie through country which has never previously been explored. Numerous lakes, some of enormous size, and some rivers, have been discovered; the existence of a vast snowy range lying parallel to and north of the Brahmaputra river has been clearly demonstrated, and the positions of several of the peaks have been laid down, and their heights approximately determined'.[17] The unique service which Pandit Nain Singh provided this time was to survey the route between Lhasa and Assam via Tawang, about which 'next to nothing' had earlier been known. His brother, Kishen Singh, covered, on the other hand, the whole of the 'lake regions on the Tibetan plateau to the north-west of Lhasa and extended the scope of his explorations to Mongolia'.[18]

The work of Pandit Nain Singh and his brother provided in fact a stock of knowledge and information about Tibet which put the Government of India in a position to go ahead with the delimitation of its North-Himalayan boundary along the whole length of the Tibetan frontier, from Kashmir and Ladakh to Assam and Burma. Its friendly relations with Kathmandu, its political hold over Sikkim and its territorial acquisition of the eastern duars between Assam and Bhutan were all new inputs to help in the attainment of a legitimate natural frontier. Warren Hastings had envisaged such a frontier, but it was Lord Hastings who first began taking effective measures to give it concrete shape in both geographical and political terms.

Towards Boundary and Trade Disputes with Tibet

A century after Warren Hastings the Government of India again made a real effort to settle questions of boundaries and trade with Tibet. In the intervening period the government had acquired a full topographical knowledge of Tibet and its neighbouring regions and, in addition, created such political conditions on its frontier as might make that effort a success. It is not true, as Younghusband avers, that for 'a century they were content to let things take their course'.[19] There were, in fact, continuous efforts to realize the goal that Hastings had envisaged. What had stood in the way was the British preoccupa-

The Delimitation of India's Frontiers

tion with the task of paramountcy; and once it was for all practical purposes accomplished, the Trans-Himalayan frontier soon acquired the importance it deserved.

In 1873 the Government of India began to take stock of its position in the context of its new strength. The occasion for a stock-taking was provided immediately by a letter from the Bengal Government which pressed 'for an order of admittance to Tibet' in order to renew 'the friendly intercourse' which had 'existed in the days of Bogle and Turner'. But who was to be pressed for the order to establish commercial and diplomatic relations with Lhasa—the Chinese Government at Peking or the Tibetan Government of the Dalai Lama at Lhasa? History as well as the reports of Bogle and Turner suggested the need to negotiate with the Chinese who, as the account of Pandit Nain Singh also confirmed, not only had officers posted on frontiers to check entry into Tibet, but also maintained a Resident (*Amban*) at Lhasa to determine its foreign relations in consultation with his central government at Peking. In addition, this view was said to have been reinforced by the inquiries of the Lieutenant-Governor of Bengal who observed that 'the prohibition to intercourse with Tibet is part of the Chinese policy of exclusion imposed on the Tibetans by the Chinese officials and enforced by Chinese troops stationed in Tibet'.[20] The letter therefore made it clear that the Chinese should be pressed for permission to enter Tibet.

Anxious to remove an embargo on the export of tea from Bengal and to find in addition a market in Tibet for Manchester and Birmingham goods and Indian indigo, the Bengal Government would not wait for any attempt to ascertain the Tibetan view of the Chinese, or an examination of the Tibetan-Chinese relationship in terms of its international implications. It straight off advocated an immediate course to settle the boundary and trade matters direct with the Chinese. This was done on the assumption that the Dalai Lama was subordinate to the Chinese emperor whose prior approval would be required for any overtures to the Tibetans.

Though possibly a matter of local importance, the Government of India could not, as with Anglo-Afghan relations of the same period, settle the Tibetan issue by local action. It had to enter into voluminous correspondence with the home authorities which involved diplomatic references between London and Peking, a time-consuming process that could not produce the quick results of the kind possible through local initiatives before the Red Sea cable was laid in 1870. In this particular case the Chefoo Convention was concluded in 1876. The

Convention was an outcome of the respect shown by the British Foreign Office to 'the symbols' of Manchu sovereignty in Tibet. The British minister at Peking, Sir Thomas Wade, persuaded the Chinese to agree in principle that the British should be allowed to send a commercial mission to Tibet under Chinese protection. The Chinese in fact could not refuse; but their acceptance was so worded as to make the dispatch of a mission subject to the political situation in Tibet as interpreted by the Chinese Resident at Lhasa—a condition which essentially recognized Lhasa's attitude as decisive. The Tibetans in fact dreaded English power no less than the Chinese, and the assessment of the Bengal Government that the Chinese alone were opposed to British entry into Tibet was a gross oversimplification. Even so, the Chefoo Convention was the first document to grant the English permission in principle to enter Tibet. However, the Convention failed to ensure any protection. The Government of Bengal was in the meantime left to go ahead with the improvement of road communications inside the British frontier and to further as much trade as possible within its limits.

A mission was planned to be sent to Tibet in 1886 under Colman Macaulay, a Secretary to the Government of Bengal. The Chinese emperor granted the necessary passports and the mission appeared to be set to move from Darjeeling through the Chumbi valley to Lhasa, when international considerations[21] stood in the way and the Government of India countermanded the whole affair. For when the British mission was finally assembled, the Chinese had little difficulty under the terms of the 1876 Convention in demonstrating that the Tibetans would not welcome it and might oppose its passage through their territory. In return for postponing their Tibetan scheme the British were compensated with Chinese recognition of their annexation of Upper Burma, a region which the Manchus had long treated as part of the sphere of their subordinate states. This transaction was formally approved in the Anglo-Chinese Convention of 24 July 1886, under which the British tacitly agreed that in future they would only establish diplomatic relations with the Tibetans through the mediation of the Chinese.

The Chinese approval of the projected mission to Lhasa and its subsequent abandonment created a situation that demanded immediate demarcation of the Anglo-Tibetan boundary. This situation arose in the first instance from the rigid reluctance of the Thirteenth Dalai Lama to accept the Chinese right to dictate his foreign policy. Thus, when the Tibetans learnt that the Chinese had approved the dispatch of a British mission to Lhasa, they resolved to oppose its advance. This determination was reinforced by the mission being abandoned, a move which the Tibetans interpreted as a lack of will to pursue the approved plan. In June, they sent a detachment into Sikkim, which they now claimed as their own. They crossed the Jelap-la, the pass from Chumbi into Sikkim. At the village of Lingtu in Sikkim which lay on the main road from Darjeeling to the Tibetan border, they set up a military post to bar the passage of the Macaulay mission, and refused to retreat even after it had been given up. And when in October 1886 the Chinese rebuked the Tibetans for their opposition to a mission which the emperor himself had authorized, the Tibetans assumed a more defiant attitude towards the Chinese, closed the passes from Chumbi to Sikkim, and reinforced Lingtu.

The situation emerging from the Tibetan action became further confused as the Maharaja of Sikkim was inclined to be more loyal towards the Dalai Lama and the Manchu emperor than towards the British. The Chinese themselves showed every inclination to deny that Sikkim was in fact controlled by the British. There were, in addition, cases where many Sikkim villages near the border paid dues to Tibet also. The Sikkim Maharaja himself exchanged insignia or marks of honour with both sides without the knowledge of the Government of India. Under the existing pro-Chinese British policy there seemed no way of ascertaining what the Tibetan claims were, let alone of rebutting them, without reference to China. And in case a reference was made to China for a withdrawal of the Tibetans on the score of their being within the territorial limits of Sikkim, it would involve discussion of a boundary, the very mention of which might give rise to a specific assertion of China's suzerainty over Sikkim, which had at all cost to be avoided. If, on the other hand, the Tibetans were allowed to challenge without question the existing status of Sikkim, it would have an adverse effect on the security of British treaty relations with Nepal and Bhutan.

Direct military action against the Tibetans was in these circumstances the only way out. For the Darjeeling merchants too started complaining of the alarm caused by the Tibetans amongst the teaplanters of Bhutan and Sikkim. In December 1887 an ultimatum was therefore sent to the Lingtu garrison to withdraw by 15 March 1888. In February, Lord Dufferin, the Viceroy, repeated the warning to the Dalai Lama. The Chinese were anxious that the Tibetans should withdraw from Lingtu. For they realized that they could not retain their influence at Lhasa in the presence of the British. The

Chinese Amban too advised the Tibetans to quit Lingtu. But when the Tibetans refused to listen there was near unanimity among British policy makers that they must be forcibly expelled from Lingtu.

In March 1888 a force of 2,000 men under Brigadier-General Graham drove out the Tibetan garrison without difficulty, though a couple of months later the Tibetans unsuccessfully attempted a surprise attack on the British at Gnatong. Another attack in September led to the Tibetans being pursued into the Chumbi valley itself, which brought both the Amban at Lhasa and the Sikkim Maharaja to their knees. While the former was reported to be on his way from Lhasa to the Tibetan frontier of Chumbi, the latter expressed his readiness to come down to Darjeeling for a talk with the Lieutenant-Governor of Bengal. The authorities in Peking, on the other hand, came to realize that only by negotiating with the British could they hope to preserve any influence in Tibet. The British did not refuse to talk. A. W. Paul, the British Political Officer who had accompanied the expeditionary force, was authorized to hold conversations. But in no case was he to raise questions of the Sikkim frontier involving an examination of its status, which was to be settled as part of a treaty with the Maharaja, not in negotiation with the Chinese Amban at Lhasa. Paul was soon joined by H. M. Durand, the Indian Foreign Secretary, who was assisted by Ney Elias.

In the talks with the Amban it transpired that, while the Chinese were prepared to accept *de facto* British control over Sikkim, they insisted on the preservation of its *de jure* dependence on China and Tibet, which meant that the Maharaja must continue to pay his traditional homage to the Amban and be permitted to retain the rank and insignia conferred on him by the Chinese emperor. But to Durand the homage was no trifle. It symbolized suzerainty, and if he gave way on Sikkim, he would in future be obliged to recognize the same position not only over Bhutan and Nepal but also Kashmir and its feudatories, such as Hunza and Nagar—a principle which could be extended to other Himalayan states too. The Chinese could then proceed to claim suzerain rights over Darjeeling and the Bhutan duars, which the British had acquired from Bhutan's feudatories. Anxious to break the influence of the Tibetans both in Sikkim and Bhutan, Durand did not budge from his position and threatened to end the negotiations with a warning that, if nothing came out of them, the British would not hesitate to raise their reputation in the Himalayan states by occupying the Chumbi valley up to Phari, the gateway to Shigatse and Lhasa, a move which was supported by Lord Lansdowne, who succeeded Dufferin as Governor-General.

However, the negotiations were saved from complete collapse by both Chinese and British initiatives. The Chinese, for instance, realized that unless they secured an agreement right then, they could not prevent direct Anglo-Tibetan contact without Chinese participation in future disputes. Thus, when Durand broke off negotiations, the Chinese announced that James Hart, the Irish Secretary, had been instructed to proceed to Gnatong to assist the Amban in his discussions with Durand. The British Foreign Office was more concerned with the future of Anglo-Chinese relations than the Indian border. So it viewed the Viceroy's initial refusal to talk with Hart as 'unreasonable' and 'hasty', and suggested that it would be 'more prudent to keep the negotiations alive' by making some concessions to the Chinese. The talks were accordingly resumed when revised proposals were submitted by James Hart on behalf of the Chinese government.[22] The proposals met to a large extent the objections raised earlier by Durand on questions pertaining to the Sikkim-Tibet border and the status of Sikkim. With certain modifications, the proposals were accepted by both China and India, on the assurance given that 'China will be quite able to enforce in Tibet the terms of the Treaty', the result of a 'Convention between Great Britain and China relating to Sikkim and Tibet', signed on 17 March 1890.

The Anglo-Chinese Convention of 1890[23] provided that 'the boundary of Sikkim and Tibet shall be the crest of the mountain range separating the waters flowing into the Sikkim Teesta and its effluents from the waters flowing into the Tibetan Mochu and northwards into other rivers of Tibet. The line commences at Mount Gipmochi on the Bhutan frontier and follows the above-mentioned water parting to the point where it meets Nepal territory'. The frontier thus finally demarcated was as hitherto insisted upon by the Government of India.

As regards the status of Sikkim, the Convention recognized that the British Government 'has direct and exclusive control over the internal administration and foreign relations of that state'. It was specifically laid down that 'except through and with the permission of the British Government, neither the ruler of the state nor any of its officers shall have official relations of any kind, formal or informal, with any other country'. This put an end to the earlier claims of the Chinese and Tibetans to receive homage from the Sikkim Maharaja without reference to the British. The 1890 Convention reaffirmed by implication the principle of exclusive British supremacy over all the border states under British protection. But this was sought to be achieved without alienating China and Britain and China engaged reciprocally to respect the boundary as defined and to prevent acts of aggression from their respective sides of the frontier.

As Tibet lay on the other side of the great Himalayan wall, the Convention did not permit Britain to climb over. On the disputed question of pasturage, too, no further concessions were settled. The possibility of increased facilities for trade across the Sikkim-Tibet frontier was raised at the Convention of 1890, but it remained subject to further discussion.

The Joint Commission appointed by China and Britain comprised A. W. Paul, the British Political Officer for Sikkim, and James Hart, the Irish Secretary to the Amban at Lhasa, Sheng Tai, who had signed the Convention of 1890 and took a leading part in the proceedings of the Joint Commission. Consistent with their inherent political objective as part of their trade negotiations, the British demanded 'nothing short of free trade and free travel for all British subjects throughout Tibet, without payment of duty on all articles of trade except arms, ammunition, military stores, liquors and narcotic drugs'. For purposes of trade they wanted a free mart to be established well inside Tibet at Phari and its environs where British subjects, trading or travelling, should be entitled to acquire land and build houses, shops and other edifices. The Government of India was in addition to be free to appoint 'an Agent', with assistants if necessary, 'to reside at Phari or elsewhere in Tibet', and 'the Chinese authorities shall afford the said Agent and his assistants, if any, facilities for their movements and communications'.[24]

Sheng Tai, the Chinese Amban, indicated his difficulties arising from the intransigence of the Tibetans and said that, while his Government was in favour of opening a trade mart at Yatung, a place just inside the Tibetan border, it was opposed to the selection of Phari on the direct route to Lhasa. He also expressed his inability to arrange for the free travel of British subjects beyond Yatung. As for the right of British subjects to acquire land and build houses, the Amban pointed out that, as Yatung had been chosen as the trade mart, the Tibetan authorities had already erected buildings which could be hired there by British subjects, as also the rest-houses built along the road. But they could not be permitted to purchase or rent land, or erect buildings.

The negotiations dragged on for three years. While the British kept insisting that Phari be selected as the trade mart, Peking had to accept the opinion of the Tibetan Council, who were determined to see that 'the mart should not move an inch beyond Yatung'.[25] And when London was persuaded to accept Yatung as a temporary mart subject to a change of trade regulations after five years, Peking and Lhasa were bent upon excluding Indian tea from Tibet, a move which the Calcutta government resisted. But since both the British and Chinese governments considered friendly relations between Tibet and Britain of basic importance, Regulations regarding trade, pasturage and communications were finally signed at Darjeeling on 5 December 1893.[26] The Regulations so drawn up were to be appended to the Sikkim-Tibet Convention of 1890.

A trade mart was accordingly established at Yatung and was to be open to all British subjects from 1 May 1894. The Government of India was free to send its officers to reside at Yatung to watch over the trade, but British subjects could not acquire land or erect buildings on Tibetan soil, even though when trading at Yatung they could travel freely between the frontier and Yatung. They could reside at the mart, rent houses and warehouses and the Chinese undertook to provide suitable buildings for the officers appointed by the Government of India to reside at Yatung. Between the frontier and the mart there were rest-houses built by the Tibetan authorities at Lang-jo and Ta-chun where British subjects could break journey and stay on payment of rent, and the Chinese undertook to give them 'efficient protection' in the transaction of their business. The principle of free trade was recognized for a period of five years in articles other than those already prohibited or permitted only on conditions approved by the governments on either side. As regards the import of Indian tea, the Regulations provided that 'trade in Indian tea shall not be engaged in during the five years for which other commodities are exempt'. But even when permitted for import thereafter, the tea could be imported into Tibet at a rate of duty not exceeding that of Chinese tea imported into England.

Though still not conceding all the British objectives, the Regulations of 1893 doubtless mark a clear advance in the penetration of Tibet. The British advance in Tibet under the Regulations was within the overall framework of Anglo-Chinese *entente*. They used considerable restraint to preserve it and denied independent international status to Tibet. In fact, the Chinese Amban at Lhasa was recognized as the proper agency for communication with the Government of India.

However, the Tibetans recognized neither the Convention of 1890 nor the Regulations of 1893. In matters of trade the Tibetans were not only obstructive but even issued orders against the free movement of goods, their officials at Phari at the head of the Chumbi valley charging 10 per cent on all goods passing beyond that point. Likewise, the demarcation of the frontier between Sikkim and Tibet was not allowed to proceed smoothly. The Tibetans demolished boundary pillars fixed at the Jelap and other passes along the frontier. While the Political Officer in Sikkim and Nolan, the Commissioner of Darjeeling, both wanted the Chinese Government to be told that they should carry out the treaty engagements or leave the government in India free to hold the Chumbi valley as a security to enforce the terms of the treaty, Sir Charles Elliot, the Lieutenant-Governor of Bengal in fact asked the Government of India to proceed on these lines; but bound as the Indian Government was to act on the policy formulated in London for imperial purposes, it continued to use forbearance and deal with the Chinese, who lacked the power to carry out their engagements, and not with the Tibetans.

Under Lord Curzon a change occurred in the whole approach of the Government of India towards Tibet, not so much because of the question of frontiers or trade as the Russian spectre appearing on the Tibetan plateau in the guise of the Russian Buriat, Dorjiev by name, who had acquired a position of great importance in the monastic hierarchy of the Thirteenth Dalai Lama.[27] In 1900 and again in 1901, Dorjiev visited Russia on what were reported in the press as embassies from the Dalai Lama to the Tsar; and although the Russian foreign minister denied the diplomatic character of Dorjiev's special missions, the fact remained that he was received by the Tsar as an envoy extraordinary. Curzon concluded that Russian influence would soon be established in Lhasa to an extent which might suggest to the Himalayan states, more specially to Nepal, the advantages of a policy of playing off Russia against Britain to their own benefit.

The obvious remedy which occurred to the Government of India was the establishment of a British agency at Lhasa as a counterweight to Dorjiev and an unobstructed channel of communication between Calcutta and the Tibetan capital, preferably through a British representative permanently stationed at Lhasa; for there was no other means of communicating with the Dalai Lama who refused even to receive letters from the Government of India.[28] The solution which Curzon proposed to the home authorities in his letter of 8 January 1903 was therefore to dispatch a mission with military escort to Lhasa, which might make the Dalai Lama acknowledge the role of the Government of India. But the home government did not accept Curzon's proposal and all that the Viceroy secured in 1903 was permission to send the proposed British mission to the town of Khambajong, roughly twelve miles into Tibet from the Sikkim border, the object being merely to discuss with both Chinese and Tibetan representatives the pending questions of border and trade disputes.[29] Curzon acted on this, for he knew that, once the Khambajong talks broke down, it would be difficult for the Cabinet to prevent the advance of a mission deeper into Tibetan territory. This was precisely what occurred. Colonel Francis Younghusband, a confidant of Curzon, was appointed to lead the mission and, as already anticipated by Curzon, the proceedings at Khambajong proved fruitless and the advance into Tibet became indispensable.[30] Brodrick, who had succeeded Hamilton as Secretary of State for India in September 1903, permitted Younghusband to advance deeper into Tibet, to Gyantse. This took place in the first half of 1904. There was some armed Tibetan resistance, culminating in May with an attack on the headquarters of the mission outside Gyantse, which provided the justification for Younghusband to march to Lhasa itself. He entered Lhasa in August 1904, the Dalai Lama fled towards Mongolia and Younghusband obtained on 7 September 1904 the treaty called the Lhasa Convention. In the absence of the Dalai Lama it was negotiated with the Tibetan Council and members of the Lhasa monastery and National Assembly, and was a treaty to which the Chinese Amban declined to affix his signature.

Under the treaty, two more trade marts, in addition to Yatung, were to be opened at Gartok in Western Tibet and Gyantse on the direct road to Lhasa from Darjeeling through the Chumbi valley. A trade agent was to be appointed to each of the new marts, the one at Gyantse subject to being invested with diplomatic and political responsibilities too. A separate article to the Convention in fact stipulated that the Gyantse agent could from time to time visit Lhasa and carry on such business as might be assigned to him by the Government of India. He was for all practical purposes to function as the British representative in the Tibetan capital, though not actually stationed there. Another important provision of the treaty stipulated that the Tibetans would pay an indemnity of Rs 75,00,000 in seventy-five annual instalments, and that the Chumbi valley would remain under British occupation until this sum had been duly paid. In addition, the Tibetan authorities were required to enter into direct relations with the British and refuse permission to other powers to establish agencies in the country.

The Lhasa Convention thus did not merely warn the Russians to keep away from Tibet. It amounted cumulatively to making Tibet a British protectorate and clashed with the Anglo-Chinese Convention of 1890 which had recognized China's status as Tibet's overlord. Besides, it ran counter to the assurances extended earlier to the Russians by Lansdowne from the British Foreign Office, which characterized the mission as no more than a police action designed to secure the execution of treaty obligations. The whole operation proceeded, in fact, from an absence of unity in policy-making. Concerned immediately with the defence requirements of India, the Indian Government had suggested to the London authorities a policy that best secured its political stability and border security. The latter, on the other hand, viewed the Indian situation in global imperial terms, not necessarily consistent with India's immediate interest. Confusion was thus a natural consequence and the home authorities found it necessary to instruct Lord Ampthill, acting Viceroy during Curzon's absence on leave, to cancel the Separate Article allowing the Gyantse Agent to visit Lhasa, and to reduce the indemnity from Rs 75 lakhs to Rs 25 lakhs, payable in three annual instalments so that the British occupation of Chumbi might be reduced from seventy-five to three years. Explaining his action to the Government of India in a letter of 2 December 1904, Brodrick pointed out that British interference in Tibet was desirable only 'to exclude' the influence of any 'other power', and that once this had been achieved by Younghusband, Tibet should, as in the past, remain in a 'state of isolation'. This policy betrayed an utter lack of appreciation of efforts since the days of Warren Hastings to provide a natural trans-Himalayan frontier for India and friendly trading relations with Tibet.

The Lhasa Convention nevertheless gone India a number of advantages in Tibet which were retained. It defined the boundaries, excluded Russian influence and placed trade relations upon a stronger footing through trade-marts, the trade agents' escorts, the Gyantse telegraph, British-built rest-houses on the Chumbi-Gyantse road and the special economic privileges arising from the new Regulations of 1908, which refined and codified the mechanism of commercial contact with Tibet.[31] The enforcement of the treaty obligations was in itself an achievement of no mean order. But, above all, as Younghusband said later, 'no feeling of race hatred was left behind the Mission'; for 'after the Treaty was signed the Tibetans were better disposed towards us than they had ever been before. And this I consider to be incomparably the most important result of the policy which the Government of India had [under Curzon] so unswervingly pursued'.[32]

Politically, however, it was the Chinese who gained most from the exclusion of Russian influence in Tibet. The repudiation of the Special Article at the instance of the home authorities in Britain left the way open for the Chinese to step into the absent Dalai Lama's shoes and assert themselves as invested with full sovereignty. Reporting from Peking in November 1904, the British Minister in China observed that, according to rumours then current in Peking, the Chinese were planning to declare Tibet a province and an integral part of their empire.[33] Britain's pro-Chinese attitude received further support in late 1905 from the Liberal Cabinet of Campbell-Bannerman, with John Morley as Secretary of State for India and Lord Minto as Viceroy. A believer in non-interference, Morley regarded a Chinese Tibet as the logical consequence of keeping both Russian and British influence out of Lhasa, as Tibetan politics could not stand by itself. Morley recognized that, left to its own choice, any Indian government, including even Minto's, tended to fill a Tibetan power vacuum. Its relations with the Panchen Lama, its trans-Himalayan explorations, and its security measures along the northern frontier, though singly of seemingly minor significance, contributed cumulatively to considerable influence in Tibet, an influence potentially capable of transforming Tibet into almost a protectorate. The treaties concluded by Britain with China on 27 April 1906, and with Russia on 23 September 1907 were therefore designed to block British India's policy towards Tibet with diplomatic and political checks. The British Government engaged under the former not to annex Tibetan territory or to interfere in the region's internal administration. Under the Anglo-Russian Treaty (1907), Britain and Russia also agreed to respect Tibet's territorial integrity and abstain from interference in its internal administration. But as regards the 'suzerain rights of China' both Britain and Russia agreed to maintain the *status quo* in respect of the external relations of Tibet. In the context of the exclusion of both British and Russian influence in Lhasa, this restoration of formal Chinese authority over Tibet not only repaired the damage caused by the direct Anglo-Tibetan relations established by the Lhasa Convention of 1904, but encouraged the Chinese to integrate the whole of Tibet with their empire. China doubtless reaffirmed under the Treaty of 1906 its support of the Anglo-Chinese Convention of 1890 and the Trade Regulations of 1893. She also later codified and put on a well-defined footing special economic privileges which were extended to the British in 1908. But the British in India, as the whole range of their relations with Tibet had in the past shown, were concerned with far more important matters than mere questions of trade marts or the fulfilment of treaty stipulations. In political terms, Morley's non-interference and his recognition of the exclusive right of China to interfere in Tibetan affairs put the clock back to the great prejudice of the Indo-British case. The result was the entry into Lhasa of Chung Ying's military column in February 1910.

Tibet and the Assam Himalaya

The entry of Chinese military force into Tibet created a new situation which boded political trouble on India's Himalayan boundary where the Chinese, instead of the Russians, would now be well placed to intensify intrigues in Nepal and Bhutan and, in addition, infiltrate into the Assam Himalaya. In the past the Assam Himalaya had remained an administrative backwater, with British influence barely touching the foothills by 1910. If allowed to fill the power vacuum created by the revision of the Lhasa Convention, the Chinese could well extend their empire through the Himalayan barrier to the very edge of the Brahmaputra plain. This possibility the Indian Government could not bring itself to accept.[34]

The outbreak of the Chinese Revolution in 1911 and the collapse of Manchu authority in Central Tibet presented the Indian Government the prospect of a new approach to the Assam boundary question, an approach based on direct Anglo-Tibetan agreement without Chinese participation. This was reinforced by a change in the Tibetan attitude towards the British since 1905–6. The Indian Government, too, had been exploring the possibilities of training projects in India for young Tibetans, and during his stay in Darjeeling in 1910–12 the Dalai Lama himself supported, with the co-operation of the Indian Government, a plan to send some Tibetans to England for training.[35] This friendly exchange of views, which flowed immediately from the Chinese attack on Tibet in 1910 and was reinforced by the Republic's plans to reduce Tibet[36], induced the British in 1912 to think in terms of excluding the Chinese permanently from all direct contact with Tibet, a radical departure from Morley's concept of non-interference. The Chinese exclusion from Central Tibet, however, meant the creation of something like an autonomous, if not fully independent, Tibetan state with defined limits and guaranteed defences against Chinese reconquest—a condition for the survival of a viable non-Chinese Tibet, which could not be ensured except with active British support and protection. In other words, it involved a direct Anglo-Tibetan arrangement, which meant substantially a return to the Curzonian approach to Tibetan policy.

But the dispatch of another Younghusband mission was ruled out in 1912, for in view of the Anglo-Russian Convention of 1907 there could be no interference with the internal administration of Tibet. A compromise solution had to be sought which, by closer direct Anglo-Tibetan contact, could guarantee the protection of Indian interests along the Assam-Himalayan border without a total denial of Chinese suzerainty in Tibet. The reasons were plain and simple: while the limitation of Chinese rights without the support of armed might was bound to be unstable, the application of force to create a non-Chinese Tibet not only called for consultation with Russia, demanding concessions in both Persia and Afghanistan, but also involved the uncertainties of an armed conflict.

The Simla Conference, which formally opened on 6 October 1913 for solution of the Sino-Tibetan dispute, was used as the occasion for direct Anglo-Tibetan discussions on the Assam-Himalayan border. The British delegation to the Conference consisted of Sir Henry McMahon as plenipotentiary, assisted by Charles Bell and Archibald Rose as Tibet and China advisers respectively. Lonchen Shatra, the Chief Minister of the Dalai Lama, was the Tibetan plenipotentiary, while Chen I-fan (or Ivon Chen) represented China, which had agreed to negotiate with a Tibetan representative in return perhaps for British recognition for the Chinese Republic established in 1911. As McMahon's Commission put it, the Conference had been assembled because 'a state of war now exists between the Government of China and the Government of His Highness the Dalai Lama', rendering ineffectual the Anglo-Chinese Convention of 1906, and because 'His Holiness the Dalai Lama has invoked our good offices to remove all causes of differences between his Government and China'. As the Chairman of the Conference, McMahon confessed that the British role was in principle no more than that of the 'honest broker' mediating between Tibet and China.[37]

The question of the Assam-Himalayan border with China which was of immediate interest to the Indian Government was not placed for discussion on the Simla agenda. For it was feared, as in the case of Sikkim, that once the Himalayan border was made subject to negotiation, it might embarrass the British, as Nepal and Bhutan would also then come under discussion, and the Chinese could use the opportunity to reassert their ancient claims in these Himalayan states. Besides, the accepted British border in the Assam Himalaya ran merely at the foot of the range. The Government of India therefore did not want to discuss it with China and decided to use the Conference as an opportunity for separate bilateral discussion directly with the Tibetan representative. There was full justification for direct Anglo-Tibetan relations, for India's geographical position and her extended frontier forced upon the Government of India a closer relation with Tibet than could be claimed by any foreign power.

The output of the direct Anglo-Tibetan discussions that took place between McMahon and the Tibetan representative at Simla, consisted in the main of the new Tibet Trade Regulations signed on 3 July 1914 and the McMahon line resulting from an exchange of notes[38] on 24–25 March 1914, by which McMahon obtained Tibetan agreeement to a boundary alignment along the Assam Himalaya which has since become widely known as the McMahon Line. The unsigned Simla Convention, which was initialled on 27 April 1914, was another important result of the conference. But the parties involved in it were not merely Britain and Tibet but also China, its main task being to resolve through British mediation the Sino-Tibetan differences.

The new Tibet Trade Regulations of 1914 did not differ materially from those of 1908. Besides providing well-defined facilities for the conduct of trade at the trade marts of Yatung, Gartok and Gyantse, they permitted the lease of lands to British subjects and made institutional arrangements for the removal of complaints. Their basic feature, however, lay in the exclusion of the Chinese as a party responsible for the execution of the new Regulations. In view of the full internal autonomy granted under the Simla Convention to Tibet, the protection of British subjects and their trade in Tibet became a direct Tibetan responsibility.

As regards the McMahon Line, its genesis as an Indian frontier alignment in the Assam Himalaya is traceable to the Chinese military action in Tibet and China's movement in the Mishmi country along the Lohit river flowing towards the extreme north-east of India on the borders of Burma. In October 1910, Lord Minto proposed to the home authorities a plan 'to gain a buffer' between British and Chinese territory by advancing northwards along the crests of the Assam Himalayan range from the eastern edge of the Tawang Tract to the Irrawaddy-Salween divide in Burma. Minto did not include in his proposed boundary scheme the Tawang Tract and the area down to the foothills, which the Government of India then regarded as

The Delimitation of India's Frontiers

firmly Tibetan. In the absence of a full geographical knowledge of the Assam Himalaya except of the Lohit Valley, the home authrorities took no decision.

Lord Hardinge repeated Minto's suggested alignment and recommended in September 1911 a policy of loose political control over the Indo-Tibetan border in Assam. But to acquire a knowledge of the area he sent the Abor Expedition which was accompanied by its related ventures, the Miri Mission and the Mishmi Mission, the object in each case being to determine the most suitable alignment for the new boundary. However, the Tawang Tract, which cut right through the barrier of the Himalaya was ignored once again, and for the same reason as by Lord Minto. It was the Indian General Staff which, from considerations of defence, decided for the first time in June 1912 that, notwithstanding the terms of the Anglo-Russian Convention of 1907 which prevented British occupation of any part of Tibetan territory, something must be done about Tawang. Their argument was that the existing boundary demarcation was south of Tawang and running westwards along the foothills from near Udalgiri, a British territory, to the southern Bhutan border; this provided a dangerous wedge between the Miri country and Bhutan, through which the Chinese could exert pressure on Bhutan. The General Staff therefore asked for rectification. The boundary suggested was one from the knot of mountains near longitude 93° and latitude 28°.20', to the Bhutan border north of Tsona Dzong in a direct east and west line with the northern frontier of Bhutan. The suggestion so made by the General Staff involved the British occupation of not only Tawang and the Monpa-inhabited districts to the south, but also the Tibetan administrative centre of Tsona Dzong to the north of Bhutan.[39]

McMahon remained for a time reconciled to the decision of the Indian Government to abide by a foothill border in the Tawang area. Later, in November 1913, the new boundary alignment was sought to be moved up along the ridge crossed by the Sela Pass, a few miles south of Tawang monastery, not the whole of the Tawang Tract. It was only in February 1914 that the new boundary was moved still further north, following the alignment of the final McMahon Line and including in the British territory the entire Tawang region, as the General Staff had earlier recommended for reasons of defence. McMahon, who had earlier rejected the General Staff's view, was later brought to agree to this final boundary alignment on the advice of Bailey who, along with Marshead, had by the time come down through Tawang in an adventurous journey along the Tsangpo-Brahmaputra Valley. He furnished first-hand knowledge and accurate information about the Tawang Tract and its strategic importance.

It is unnecessary to go into the details here of the geo-historical, cultural, linguistic and ethnic considerations which went into the making of the McMahon Line.[40] Though vague and susceptible to varying interpretations, its political case in terms of India's defence along the Assam Himalayan frontier was clear enough. Considerations of political stability and territorial integrity along the north-eastern frontier of Assam dictated the necessity of bringing within the limits of British India such areas as the Tawang Tract, the upper Siang and Siyom valleys and the Lohit between the Yepak river and Kahao, containing Mishmi and the Abor and Miri tracts, which traversed several watersheds within the same given limits. It is true that these areas did not become British-administered territory, for McMahon's boundary involved the nominal transfer of territory from Tibet to India, not its annexation. But they were more than British-protected regions. The McMahon Line represented a boundary below which the Indian Government would not tolerate the influence of any power other than Tibet, and even Tibetan influence could be accepted only when it was not obstructive. The areas included in the boundary in fact became a British sphere of interest with loose political control exercised under advanced military outposts on the border.

A constitutional difficulty, however, appeared in the course of discussions in Simla. McMahon realized that some kind of Chinese consent should be obtained for the bilateral Anglo-Tibetan agreement on the McMahon Line already concluded in March 1914, an agreement relating to the Assam-Himalayan border which formed no part of the Convention agenda and of which the Chinese had not formally been informed. It was not easy to secure Chinese approval, though McMahon did make an attempt to obtain it. And the Simla Convention, though initialled on 27 April 1914, remained unsigned.[41]

The Chinese refusal to sign was no doubt related to the Convention's decision to divide Tibet into an 'Inner' Zone with Chinese control and an 'Outer' Zone, with full Tibetan autonomy, a decision which followed the Russo-Chinese declaration of October/November 1913 (during the Simla Convention itself), under which China had agreed to the division of Mongolia into 'Inner Mongolia', where it exercised full sovereign rights, and 'Outer Mongolia', towards the Russian frontier, which, though under *de jure* Chinese suzerainty, was recognized as fully autonomous. For the purpose of the Convention the borders of Tibet were accordingly shown on its map in red, while the boundary between Outer and Inner Tibet was marked blue. Under the Convention both Britain and China were to recognize Chinese suzerainty as well as full Tibetan autonomy in Outer Tibet where China was to abstain from all interference in administration. The Chinese Government was not to send troops into Outer Tibet nor station any civil or military officers there. The British Government too had to observe the same principle, except as provided in the Lhasa Convention which permitted a British agent at Gyantse to visit the Tibetan capital with a suitable escort as and when necessary. The Chinese could maintain, as they did in the past, a high official at Lhasa with a similar escort. But in no case was China to convert Tibet into a Chinese province, nor was Britain to annex Tibet or any part of it. As regards Tibetan rights in Inner Tibet, the Convention provided that nothing in it was to prejudice the existing rights of the Tibetan government, which included the power to select and appoint the high priests of monasteries and retain full control in all matters affecting religious institutions.

The division of Tibet into Inner and Outer zones, with a boundary alignment between the two to suit Anglo-Tibetan interests, was perhaps the weakest link of the Simla Convention. Though hardpressed by internal feuds and external machinations, the newly-born Chinese Republic would not agree to abandon, for example, a territory like Chamdo on the Burmese frontier, which the Chinese still held and from which the Tibetans were unlikely to dislodge them merely because of McMahon's boundary demarcation between Inner and Outer Tibet. The Chinese in fact feared that once they accepted the Convention it would limit their freedom of action in Outer Tibet. In 1915, they even resiled on the Convention's decision about the Inner-Outer Tibet boundary and offered to evacuate Chamdo in case the text of the Convention could incorporate a few changes to emphasize Chinese suzerainty and permitted them to station trade agents at Chamdo, Gyantse, Shigatse, Yatung, Gartok and other places which might in future be opened to British trade. But the British refused to open the Tibetan question and the Convention remained unsigned because of British intransigence. After the surrender of Chamdo to Tibet in 1918 the Chinese again offered to sign the Simla Convention with the modifications proposed in 1915. The British once more refused to take advantage of the offer and banked on the autonomy of Outer Tibet which they considered to have as much freedom as they themselves wished to allow it.[42] This was the legacy which the Republic of India inherited and gave up in 1954 through an agreement with China.

Western Tibet

Western Tibet stretched along the gutter carved out of the high mountains by the Indus and the Shyok, all passing through in their upper reaches through the Karakoram range at a height of 23,000 feet and above. It bordered on tracts where for hundreds of years Chinese, Tibetans, Kashmiris and Mongols fought each other for control over parts of the Indus valley in this controversial region. Between the seventh and tenth centuries Chinese and Tibetan forces repeatedly fought each other for control of the passes to the north of the Indus and set up temporary frontiers across the mountains. In the high valleys of the Indus and the Shyok to the south of the Himalayan crests were situated Baltistan, Kashmir and Ladakh, all bordering on Western Tibet. Kashmir in the tenth century was predominantly Hindu, and Baltistan and Ladakh predominantly Budhist; but by the end of the fourteenth century Kashmir itself came under Muslim rule and Baltistan, too, was conquered and became Muslim at the turn of the fifteenth century.

The Muslim rulers of Kashmir invaded Ladakh several times through the Zoji La, but failed to make any impression. Instead, the Ladakhis conquered Baltistan. Their victory, however, was short-lived and in the sixteenth century Ladakh was invaded by a Mongol army from Kashgar in Central Asia which crossed the Karakoram and raided Western Tibet and the adjoining territory of Ladakh. Finally, however, the Tibetans and Ladakhis inflicted a crushing defeat on the invaders who retreated with heavy loss in men and materials.

By the time the Mongols returned again to invade Ladakh from Western Tibet, Shah Jahan, the Mughal emperor of India, came to the rescue of the Ladakhis on the latter's appeal for help. A Mughal army came over the Zoji La and, reinforced with troops from Skardu, the capital of Baltistan, defeated the invaders in a battle at Basgo, some twenty miles up the Indus from Leh, the capital of Ladakh. In return for services rendered, the Mughal emperor insisted that the king of Ladakh embrace Islam, build a mosque and allow his empire a monopoly of the Tibetan shawl-wool trade. Nothing, however, came of the arrangement so proposed by the Mughal sovereign and

Ladakh continued to live its own life in terms of religion and trade with Tibet. Though Muslim, Baltistan too lived in peace with its neighbouring state along the borders of Western Tibet.

A new situation emerged with the conquest of Kashmir by Ranjit Singh in 1819, a point of time when William Moorcroft, the British explorer in the service of the East India Company, was carrying on trade negotiations with the Ladakhis. In the midst of these negotiations Ranjit Singh sent envoys to Ladakh in 1822 demanding payment of tribute as the new overlord of Kashmir, and rumours reached Leh of Russian movements on the northern borders of Ladakh. This prompted Moorcroft to accept the possibility of a Russian force actually crossing the Karakoram to invade Ladakh as an ally perhaps of Ranjit Singh. When the Ladakhis on their own asked for British protection, Moorcroft warned Ranjit Singh not to interfere in Ladakh and forwarded their petition to Calcutta. The Company's government, however, saw the situation quite differently from Moorcroft, and treated the Sikh state as a convenient buffer against Afghanistan and the unruly tribes to the west of the Indus. The Company went to the extent of stopping Moorcroft's pay and allowances, which forced him to wind up his affairs in India and proceed to Afghanistan, where he died not long after.

Years after Moorcroft had left Leh, Zorawar Singh, one of Ranjit Singh's lieutenants, crossed the mountains and marched towards Tibet. By 1840 both Baltistan and Ladakh were under Sikh rule. One of his columns marched up the Indus valley and, disguised as pilgrims to Kailash and Mansarovar, advanced into Western Tibet in the summer of 1841. The Tibetans, however, mobilized their troops and in the winter of that year the Sikh army was defeated and Zorawar Singh killed near Mansarovar. By 1842 the Sikhs had retreated to Ladakh.[43]

Interested in the supply of fine Tibetan wool to India, the British accepted Gulab Singh Dogra, a Sardar of the Lahore Darbar, as the ruler of Baltistan and Ladakh. Though formally owing allegiance to the Lahore government, Gulab Singh remained neutral in the first Anglo-Sikh War and even acted as an intermediary between the two parties. The Sikhs were forced to give up their claims to a large area south of the upper Indus, including Kashmir, and agree to paying a huge cash indemnity. But since the sum demanded by the Company could not be fully delivered, a separate agreement was signed at Amritsar with Gulab Singh, to whom Kashmir and its adjoining areas were transferred for a much smaller sum. The British wanted to weaken the Sikh state and protect the northern frontiers of India by installing in Kashmir a more or less reliable ally—but not by annexation which involved the risks of direct rule. This in fact formed part of a general pattern of British policy to secure the protection of frontier regions along the Himalaya. Gulab Singh was accordingly confirmed as the independent overlord of Kashmir as well as the strategically important Trans-Himalayan provinces on the Tibetan borders that had already been conquered.

The arrangement which the British made with Gulab Singh in 1846 proved to be of great advantage to them. The son continued his father's policy of friendship, and his Kashmiri troops, like the Gurkhas of Nepal, fought side by side with the British in the Mutiny of 1857. But despite their treaties with Gulab Singh, the British concern for regular and safe carriage of Tibetan wool to India remained undiminished. This concern was reinforced in the post-Mutiny period, when Moorcroft's fear of Russian invasion not only seemed less fantastic, but his old dream of developing trade with Central Asia beyond the Karakoram began to acquire new dimensions of a geopolitical nature, calling for the security of all trade routes from India to Western Tibet and China, more especially through the Indus and other river valleys in the mountains.

The first task of the Indian Government was to obtain full knowledge of the geography of the region, which in turn could not be acquired except by some kind of control over the Kashmiri provinces bordering on Western Tibet. As the security of Gulab Singh's frontiers on the north-east of Ladakh was of immediate importance, the British sent surveying parties during 1846-7 from Leh to the north to try and determine with Sino-Tibetan agreement where exactly these frontiers lay. But the parties concerned refused to agree to a precise demarcation of their frontiers on the ground that the ancient boundries between Ladakh and China as well as Ladakh and Tibet were well-known and observed for many years.

Although some surveying had been done by the Company in Ladakh and Western Tibet, the geography of the high river valleys and mountain passes of the Karakoram had mostly remained a closed secret. The situation of Kashmir and its provinces of Baltistan and Ladakh on the frontiers of Western Tibet was thus of immense importance in acquiring the requisite geographical knowledge of the space between India and Central Asia. The status of the so-called 'independent' state of Kashmir thus needed to be duly modified in the light of British relations with Indian states as a whole. In 1852,

The Delimitation of India's Frontiers 109

Gulab Singh had already been obliged to agree to the appointment of 'an officer on special duty', a loose arrangement intended to control the Maharaja rather than the promotion of trade. On complaints from Punjabi traders, a British agent was sent to Leh in 1867 for the summer to reduce the customs levied by Kashmir. His appointment was sanctioned by the Viceroy as 'an extreme measure' and a mere 'temporary arrangement'. But he stayed on. An 'officer on special duty' was in the meantime also appointed at Gilgit, two hundred miles down the Indus from Leh. He was to watch the southern end of the Karakoram passes.[44] Even so, Kashmir continued to be at least one friendly state between British India and the Sino-Russian territories across the Karakoram in Central Asia. It is from Ladakh that Nain Singh started his major explorations and the knowledge acquired by him and his brother, Kishen Singh, provided a firm foundation for the further study of these crucial regions. The work of the Afghan Boundary Commission, which commenced in 1884, added further to geographical knowledge of the Central Asian region to the west of Mongolia and Chinese Turkestan. And the exploration of Ney Elias from the Pamirs to the Oxus basin, completed at about the same time, provided information on the adjoining Russian possessions.

The main object of British policy at this period was to check the advance of Russia towards the Indian subcontinent which had more or less been going on clandestinely since 1807. Moorcroft's early warnings began to engage the attention of the Indian Government, more especially because of China's incapacity to defend its frontiers with Russia in Central Asia. The difficulty, however, was that even from the material collected by Nain Singh and other surveyors it was not clear whether India could in fact be invaded from the north,[45] and, if so, the precise route by which the invasion could take place. In 1889, Younghusband was dispatched to find an answer. Accompanied by a small party of six Gurkhas, he went to Kashmir and from there proceeded northwards, crossing the Himalaya, the Indus and then the Shyok. It was there that he heard of Russians being interested only in scientific exploration. At the headwaters of the Shyok the local tribes were frequently pestered by raids from Hunza further to the west. On Younghusband's arrival their chieftain asked him to accept their allegiance in return for protection, a request which met with a favourable response. Younghusband later reached Hunza and expressed on behalf of the British sovereign his displeasure at the raids, negotiating with the local chief an arrangement to stop them. Further north, towards the Karakoram mountains, Younghusband actually met some Russians who, though hospitable, made no secret of their intention to invade India. Two years later, on a second exploratory mission to the Pamirs, he again met a group of Russians who forced him to agree to leave in 1892. They claimed the whole of the Pamirs for their own dominion, although Afghanistan owned part of the Pamirs under British influence. A commission of delimitation had already been proposed, but its progress remained slow and tardy and it was not till 1895 that an agreement was finally signed. Under this agreement, Afghanistan was obliged to surrender territory north of the Punjab, while Bokhara had to transfer to Afghanistan that part of Darwas which lay south of the Oxus. The eastern portion of the Afghan territory of Badakhshan thus came directly under Russian control.

Russia and China were thus both situated on the northern frontiers of India beyond Kashmir. India's defence was vitally connected with the kind of relations Britain was obliged to have with them. The Younghusband mission to Lhasa and his Lhasa Convention of 1904 represented a climax in Anglo-Russian relations which, after the Japanese victory over Russia in 1905 and the changing pattern of European politics that followed, normalized into the Anglo-Russian Convention of 1907 which recognized the principle of non-interference in Tibet. As already noted, the British Government took care to observe that principle and, despite the claims of India's geographical situation and cultural tradition, did not agree formally to annex what was considered to be Tibetan territory. The boundary alignment in the Assam-Himalaya was settled by the McMahon Line negotiated between the British Government in India and the Tibetan plenipotentiary.

As regards the Chinese territory of Sinkiang adjoining Western Tibet, the position of the English always remained shaky because of the danger of potential Russian designs on Sinkiang. Out of concern for India's defence, the Indian Government was interested in the extreme north-eastern frontier of Ladakh, a desolate tract known as Aksai Chin. In a note presented in March 1899 to the Peking government the British extended, against Russian claims, its recognition of Chinese rights to a portion of Aksai Chin. The note remained unacknowledged, but the British considered themselves bound by it until Chinese troops entered Lhasa in 1910, leading to an open Sino-Tibetan rupture which the British took advantage of. With the increasing danger of Russian control over Sinkiang, the

The Delimitation of India's Frontiers

Aksai Chin plateau acquired in the eyes of British strategists the importance of a buffer between potential Russian territory and the passes leading from the Karakoram to the Indian plains. In 1912, Lord Hardinge urged that Aksai Chin must be kept out of Russian hands in any negotiation for readjustment of the Kashgar-Kashmir border, an arrangement which might be made a precondition for the recognition of a Russian protectorate or annexation of Sinkiang.

The Simla Convention, which proceeded on the division of Tibet into Outer and Inner Zones, suggested the way out. The red line of the Convention map, which signified the outer frontiers of Tibet, was carried in the north-west to a point on the north bank of the Karakash river, a point which outflanked Aksai Chin; and since Tibet lay south of the red line, it could not but be inferred that some part at least of Aksai Chin, if not all, was in Tibet with its boundary stretching further to the north-west. According to the final Simla Convention the territory within the extreme western end of the red line was to be treated as wholly Tibetan, and therefore friendly to the British under the separate Anglo-Tibetan notes of 24–25 March 1914 and the final Convention map of 27 April 1914, on which the McMahon line was shown crossing the Karakash below Kizil Jilga in Sinkiang. This, as the Indian Government after the transfer of power claimed, was really the traditional and established boundary between India and Tibet, a boundary which formed part of the territory of Ladakh adjoining to it.

It is not within the scope of this work to examine in depth the legal validity or otherwise of the McMahon boundary either on the Assam-Himalayan border to the east or the Ladakh-Tibetan border to the west. What emerges clearly from an analysis of historical events is that the British in India remained for nearly a century seized with the problem of the country's defence against Russian expansionism. Concerned with the security of India's natural frontiers, the Government of India wished to avoid delay and suggested measures to meet local exigencies by a resort to immediate local action. But since imperial considerations called for India's direct political interest being treated in a global context, the Indian Government was obliged to fall in line with the views of the London authorities. Local action became subject to moderation and restraint. It was this restraint which prompted Russia to gain considerably, for example, in the heart-land of Central Asia, more especially on the northern and eastern frontiers of Afghanistan. It was Russian pressure on Sinkiang again that became a cause for concern to the Government of India on its Ladakh border, a border which was susceptible to attack through the Karakoram passes.

Functioning under constraints imposed by London, the Indian government could not at will impart political reality to its trans-Himalayan boundaries. The whole of the Himalayan crests and their northern fringe on the Tibetan side from Kashmir to Burma remained a subject for negotiation, for any resort to war or conquest was more an exception than a general rule in British frontier policy. The issues which determined India's trans-Himalayan frontiers were political and not legal. They involved considerations of defence and international politics and should, as such, be judged politically rather than legally as some contemporary scholars have mistakenly attempted to analyse them.

The North-East Burmese Frontier

India's north-east frontier with Burma came into being after the annexation of Upper Burma in 1885, an act which provided a wall to the east for the protection of the whole of Assam and the hill states which in the past were often subject to incursions from that side.

The Burmese king, Mindon, who succeeded his deposed brother Pagan in 1853 and ruled over Upper Burma till 1878, tried without success to recover Pegu. He recalled his troops and respected the new frontier set up under the treaty of 1852. Mindon was succeeded in 1878 by Thibaw, one of his junior sons.

The idea of a flourishing market for Manchester goods in Upper Burma and Yunnan, the frontier province of China, had been suggested to the Company by Baron Otto des Granges. In a short survey of the countries between Bengal and China, he indicated the great commercial as well as political importance of the Burmese town of Bhamo in the upper Irawaddy and the practicability of direct overland trade between Calcutta and China. A copy of this survey was forwarded by the Company to the Governor-General in Council on 20 October 1847. The survey pointed out that the direct distance between Calcutta and Yunnan, the Chinese frontier, was 540 miles, which could be covered in three stages, namely, Calcutta to Sylhet, Sylhet to Manipur via the state of Cachar, and Monfos, the first Burmese frontier town, to Bhamo via Ava and Katha. Though mostly untraversed by Europeans, the survey indicated that the routes to Bhamo in Burmese territory were of known configuration and presented no great difficulty to travellers.

The Delimitation of India's Frontiers 113

The survey described Bhamo as 'the most important town of Upper Burma, it is the emporium of its trade with China and annually twice at the beginning, and at the end, of the dry season a Chinese caravan arrives here, selling all the goods here, whilst only few merchants proceed to Ava. This market has been frequented since the earliest centuries and formerly even to such greater extent than now, since the comparatively recent invasions of the Burmans have interrupted the trade. Marco Polo, the famous Venetian traveller, who, as an envoy of the Mongol Khublai Khan, visited these countries at the end of the thirteenth century, is the first who gives us some information of this market and of the road leading from here into Yunnan'. In spite of the interruptions caused at times by the unsettled state of the countries since Burmese ascendancy, the trade transacted with 'Yunnan and neighbouring provinces of China was still considerable'. It extended through the Mishmi country and Singphos to Assam.' The survey now suggested extending the trade to Calcutta, the centre of British trade in India. The idea was to establish through the Bhamo-Yunnan channel direct trade between India and China, a trade which might fetch bullion and raw silk from Yunnan in return for the export of English woollen cloth and opium to China. The object of recommending the shortest direct route from Calcutta to Bhamo was to induce the Chinese to come down to Sylhet and develop it into an *entrepot* for purposes of Sino-Indian trade. While land transport[46] from Bhamo to Sylhet was to devolve on the Chinese, the Calcutta merchants were to go to the suggested mart by water, a distance of 250 miles.

The plan so suggested, however, remained inoperative. The Sikh and the Burmese Wars of Lord Dalhousie both intervened. Mindon on his part did, of course, everything possible to foster trade between Bhamo and Yunnan and even removed a governor of Bhamo for obstructing English officers. But the wild tribes north of Bhamo were subject to neither Burmese nor Chinese rule. Mindon was, in addition, a supporter of royal monopolies and the largest dealer in all kinds of produce in his dominions. Trade treaties were thus first made in 1862 and again in 1867 between the King of Burma and the Government of India; these opened Upper Burma to British trade. The overland direct trade between Calcutta and Yunnan, which had been suggested in 1847, did not, however, come into operation. Instead, British steamers plied regularly from Rangoon to Mandalay after 1868, reaching Bhamo in 1869. British officers, on the other hand, visited Yunnan a number of times after the Treaty of 1867. In 1875, however, they turned back when Margary, of the Chinese consular service, who had travelled overland from Shanghai to meet British officers in Bhamo, was murdered by a Chinese rabble.⁴⁷

The unsettled state of Burmese affairs, which emerged not a little from Thibaw's acts of massacre, was viewed as a serious hindrance to British commercial interests. Non-official opinion in British Burma therefore came out unanimously in support of annexing Upper Burma. But the Government of India saw no reason to justify intervention on the score of internal misrule, more especially when statistics did not show any link between trade and misgovernment. What forced the English to act was the peaceful penetration of the French in Upper Burma, a move highly prejudicial to British interest in both commercial and political terms. They had made a trade treaty with the Burmese Government in 1873. But since the Burmese wanted a full alliance with permission to import arms, France had declined ratification of the treaty, particularly since their possessions at the time were limited in the region to Cochin China. However, by 1884, France had Tonkin and was steadily advancing towards Upper Burma. In early 1885 the Burmese were also assured of permission to import arms, and this development caused concern amongst the British and dictated the expediency of firm action by them.

The situation caused more concern to the British when the Burmese envoy in Paris negotiated the establishment of a bank at Mandalay as a source of financial relief to Burma. But unable to wait till the French bank materialized, Thibaw turned to the Bombay-Burma Trading Corporation, a British firm which extracted timber over half his kingdom. The Corporation lent him £100,000, and a further sum of £220,000 early in 1885. But encouraged by the French promises, the Corporation was accused by the Burmese of having failed to pay its employees and defrauding the Burmese crown of royalties in the Yamethin forests. The Corporation was condemned to pay £230,000. The English did not act in haste. The French in the meantime suffered reverses and failed to move westward to lend support to their Burmese ally. They in fact withdrew from Upper Burma. When the Burmese refused to submit the Corporation's case to the arbitration of the Viceroy (Lord Dufferin), British troops reached Rangoon on 13 October 1885. The draft ultimatum to the Burmese, which was recommended locally and endorsed by Calcutta, was approved in London and received back on 19 October. It directed Thibaw to receive a permanent British Resident, to submit the Corporation's case to the arbitration of the Viceroy, to entrust his foreign relations to British control and to assist them through trade with

Yunnan. Thibaw's rejection of these terms reached Rangoon on 9 November and on 28 November he was made a prisoner in his palace under British protection. The annexation of Upper Burma followed the rejection of the British offer of making Burma a protectorate.

The kingdom of Upper Burma was pacified during 1885–90. But it was barely half of Upper Burma. Its greater half consisted of tribal areas, such as the Chin hills, where Burmese rule had never been perceptible, or the Shan states where it was ineffective. The remotest Shan state submitted in 1890, while fighting continued against the Chins till 1896. With the complete restoration of peace, however, Burma under British rule served as a wedge between China and the Assam Himalaya, guaranteeing the security of India's eastern as well as her extreme north-eastern frontiers.

Chapter III

TERRITORIAL REORGANIZATION FOR SECURITY AND DEVELOPMENT

The delimitation of India's frontiers in accordance with the legitimate claims of her geography was a primary condition for protection against foreign aggression. The administrative reorganization of the territory so delimited and occupied, more especially in the sensitive areas of the north-west and the north-east, was designed to provide for internal security and material development, not only in terms of efficient military arrangements and communication, but also of social and economic policy. Security and development were in fact two important additional elements in the reinforcement of political stability, which territorial reorganization was intended to accomplish.

In a well-argued memorandum[1] of 27 November 1849, the year of the annexation of Punjab, C. I. Napier, the 'General Commander-in-Chief', made a broad assessment of the problem of internal security which, though emerging immediately from that annexation, involved larger issues of India's total defence arising from a long stretch of tribal frontier on the borders of Punjab and Sind in the north-west and the hill tribes of Assam on its Tibetan and Burmese frontiers in the north-east. Napier's emphasis was, however, exclusively military in approach, confined to the maintenance and distribution of the army with special reference to Punjab and the hill tribes of the frontier regions. He acted under a belief that the army alone was the panacea, not only for the entire belt of India's external enemies on all her frontiers, but also for internal foes, especially among unruly hill tribes. On the frontiers, he felt that no civil administration in India could 'produce among the people peace' or secure their 'attachment to our rule'.[2] Though willing to accept the professional recommendations of the Commander-in-Chief on the distribution of the armed forces for defence and security purposes, the Governor-General attached no value to Napier's comment on civil government and his policy of political management to secure peace in the sensitive areas.[3] Political stability continued to be treated as a function both of security, a military dimension, and development, a policy dimension of civil administration. An attempt will be made in this chapter to review territorial reorganization, with emphasis on the reorganization of the north-west frontier of Punjab and Sind as well as the north-east frontier of Bengal and Assam.

The North-West Frontier Province

It took nearly fifty years after the annexation of the Punjab for the north-western frontier districts to be severed from that province and formed into a separate North-West Frontier Province in 1901. This reorganized territorial unit was designed from geographical, ethnic and political considerations, to function under the immediate direction and control of the Government of India. The aim was to ensure speed, care and efficiency in the handling of its problems without the intervention of the Lieutenant-Governor of Punjab who could not get rid of local prejudices in the management of tribal affairs involving major problems of imperial defence. This period of nearly fifty years can be divided into two phases, the first from the annexation of Punjab in 1849 to the outbreak of the Second Afghan War in 1878, and the second from 1878 to the formation of the North-West Frontier Province in 1901.

The initial change made by the British on the Punjab frontier was the taking over of frontier districts from the Sikhs, with their ill-defined administrative boundary, including both settled areas and border tribes. Besides the district of Hazara in the north, these trans-Indus frontier districts were, from north to south, those of Peshawar, Kohat, Bannu, Dera Ismail Khan and Dera Ghazi Khan, the last being on the borders of the Baluch country. The Sikh frontier administration had for the most part been loose and ill-organized. It could not realize any revenue or rule without resort to repressive measures or even frequent incursions. In fact, the Sikhs had in the past waged a continuous war against the border tribes and even against the so-called settled districts. The British thus inherited from their Sikh predecessors a virtual state of anarchy and disorder. The arrival of the English army in Peshawar in 1849 under Sir John Gilbert was therefore hailed as deliverance from the 'hated Sikhashahi'.[4] Gilbert halted his squadron at a place called Shadi Bagiar near Jamrud below the Khyber defile, which marked the limit of Sikh rule.

First Phase—Close-Border Policy up to 1878

The Company's army took over Peshawar and other frontier districts as part of the annexed province of Punjab. The north-western boundary of the new province was drawn along the foothills, the line which the Sikhs had claimed by virtue of conquest and revenue. No attempt was made at the time to advance deep into the highlands, or even to secure the main passages through the mountains such as the

Khyber or the inlets into the mountains provided by the Khurram, Tochi and Gumal rivers. Only in the central district of Kohat did the English penetrate deep into the low Khatak hills. By and large, British rule stopped where the hills began: these had marked the Sikh administrative border, dividing the settled districts from the tribal areas beyond the foothills. This Sikh administrative line was recognized as the first north-west frontier, the Indus being rejected, for a variety of reasons,[5] as of little or no value as a true frontier of India for purposes of imperial defence.

The five northern trans-Indus districts from Peshawar to Dera Ghazi Khan, together with one cis-Indus district in the far north, Hazara, became the frontier districts of the new Punjab province. These were organized in 1876 into two Commissionerships, the Commissionership of Peshawar consisting of the districts of Hazara, Peshawar and Kohat, and the Commissionership of Derajat comprising the districts of Bannu, Dera Ismail Khan and Dera Ghazi Khan. After the Mutiny, the provincial cadre of Punjab did not in fact remain confined to the officers of the Indian Civil Service. It became a mixed cadre known as the Punjab Commission which included a number of military officers seconded to civil work. The deputy commissioner of each district belonged to this mixed civil and military cadre, a principle which applied equally to the Political Department, later known as the Political Service, which became invested with the administration of the North-West Frontier Province, with its jurisdiction not confined to the administrative border of the old settled districts but extended far deeper into tribal belts from Chitral in the north to British Baluchistan in the south. In all other matters, however, the Punjab of 1849 remained much the same as any other part of British India. There was no change in the law governing civil and criminal justice, revenue and police, public works and other administrative details. The general tenor of British civil administration continued unaffected to make law-based justice available to a people who had been afflicted by Sikh misrule. Moreover, as the Punjab frontier happened to be too long and rugged to be defended by the military alone, a resort to the political management of the tribes was considered to be essential for peace and security. Deputy commissioners alone continued to deal with tribal tracts and conduct relations with the chiefs, until all six districts were formed into two commissionerships in 1876, a change which was followed by the system of political agencies adopted in 1878.

In the first phase up to 1878, the credit for organizing a new frontier to a new province in a region which, though geographically contiguous with the rest of India, had its own ethnic peculiarities, went to a group of pioneers who established out of chaos an equilibrium which provided stability to the Indian subcontinent. They included men like John Nicholson, Herbert Edwardes, Mackeson, Robert Warburton, James Abbott and Sir Robert Sandeman, who were all held by the Pathans as deliverers. They were a chosen band of officers who had earlier tried their hand in dealing with the Pathans between the two Sikh wars, officers who enjoyed the confidence of Dalhousie and remained dedicated to the task they were comissioned to accomplish. Their difficulties were doubtless enormous: at least since Mughal times there had been no tradition of firm order even in the plains; and there had been no exact limit to which the authority of the new British government could run. The Pathans were, from every point of view—ethnic, linguistic, geographical, historical, different from the Punjab Muslims. There was only one thing in favour of the newcomers: they were better liked than the Sikhs, but this preference could soon evaporate.

In view of the existence of a wide expanse of tribal territory beyond the British-administered border which acknowledged neither Kabul nor Calcutta as suzerain, each deputy commissioner of a settled district was invested with powers to deal with the tribes on his own border and occupy such trans-border territory as formed a no-man's land, usually of unknown extent. Thus, while the deputy commissioner of Peshawar dealt with the important and powerful Mohmand tribe, the Kohat deputy commissioner handled the Orakzais. Bannu dealt with some of the tribes of Waziristan, and Dera Ismail Khan with the Sheranis and Bhitannis. In their relations with the trans-border tribesmen the deputy commissioners made use of intermediaries who invariably happened to be Khans or notables of border villages. This mode of penetration into the trans-border tribal territory formed part of what came to be known as the 'close-border policy' for the political management of the tribes.

The close-border policy was in the main based on a couple of sanctions. These were the organization of the Border Security Force and provision for written agreements and allowances. The Border Security Force did not form part of a regular army establishment. It was an irregular force, a kind of border police, more mobile than regular soldiers and acting under the civil authority. It was a considerable body of militia, eventually named as the Punjab Irregular

Force, later known as the Frontier Constabulary, who functioned as a mobile security force and not as policemen meant to investigate or control crimes. It provided protection against armed raiders and plunderers of the trans-border tribes. In case of serious trouble the border police relied on the assistance of the regular army who, at times, did the counter-raiding and led military expeditions into tribal territory as punitive measures. This, however, was the negative aspect of the close-border policy. Its positive side consisted of written agreements negotiated with every tribe that secured, on paper, everything that the government needed. A typical agreement provided for the services required of a tribe, such as the security of a border, control of raiders, protection of communications, a clause binding the tribe to deny sanctuary to outlaws and a guarantee for the payment of an annual allowance subject to good behaviour and performance of the services stipulated. It was customary for the agreement to be reached in open *jirga*, the tribal council where the maliks and elders of the tribe would affix their seal of approval or thumb impressions. These were all formalities intended to invest an agreement with the weight of tribal consensus and a promise to ensure perpetual peace, a promise which was more an exception than the general rule.

The punitive expedition was, however, not the only remedy. Apart from the stoppage of allowances, there was the blockade, which meant an exercise of economic pressure by exclusion of a tribe from markets, land or grazing in the neighouring district. There was yet another remedy called *baramta*, which meant the seizure of persons, animals or property belonging to a tribe or individual at fault. In the early days of the frontier administration, however, allowances were an important element in the system of political management. Later, their importance doubtless faded; for much larger benefits were made available through service in the army, in civil irregular corps, or even as tribal police. Even so, money payment was an inducement to good behaviour, more especially on the part of the tribes subject to no legal code, no recognized and settled administration. In the first phase of the north-west frontier policy it was the threat of withdrawal of allowances and trade facilities or employment that restrained the tribes from depredations on their more peaceful and settled neighbours.

With its lawyers, appellate courts and Western concept of crime and punishment, the British judicial system was, however, viewed by the Pathans as virtual anathema. It imposed penalties not justified by local custom. Under the law of evidence it acquitted a person whom

Territorial Reorganization for Security and Development 121

everybody knew to be guilty. All this annoyed the tribes and led to frequent violations of law in a society oriented emotionally to custom. In 1872, the introduction of the Frontier Crimes Regulations relaxed the rigidity of the existing judicial system by providing settlement of blood-feuds and come other disputes involving women in accordance with customary methods. The magistrates were authorized to withdraw such cases from ordinary courts and submit them for arbitration by a *jirga*, a tribal jury, which was not bound by the law of evidence. But since the Frontier Crimes Regulations worked more as a supplement than a substitute, the remedy satisfied neither law nor custom. It was a failure. The Pathans of the Punjab frontier districts did not get fully reconciled to their civil administration, much less to the close-border policy intended to effect deeper penetration into tribal territories.

Second Phase—Forward Policy from 1878

The building of a north-west frontier for India involved from 1878 onwards the creation of a strategic line of defence against Russian pressure in Central Asia.[6] The first step in the direction of the new policy was, however, taken in 1876, not exactly on the north-west frontier but from Upper Sind into Baluchistan. After the conquest of Sind, Sir Charles Napier set up a military administration there, selecting his officers not from the civil service but from the soldiers who had crushed the powers of the amirs and used uncompromising repression to put down the marauding incursions of the Bugti tribe from the Kachhi hills and the Dombkis from the Kachhi plains. The Upper Sind border under Napier's military rule was duly protected against tribal raids. To the north on the Sind border lay Dera Ghazi Khan, one of the settled frontier districts of Punjab to the west of the Indus running southward in continuation of the Daman of Dera Ismail Khan. The difference between the two settled districts was that while Dera Ismail Khan was partly a Pathan district overhung to the west by the mountains of Waziristan, a stronghold of Pathan tribes, Dera Ghazi was held in the main by Baluch tribes inhabiting also the extension of the Takht-i-Sulaiman mountains to the west. Unlike the well-armed and warlike Wazir Pathans, the Baluch tribes were less armed and more amenable to their chiefs. Under the close-border policy the deputy commissioner of Dera Ghazi Khan maintained relations with the Baluch beyond his border in much the same way as his brother officers did with the Pathans further north. All the

Baluch and even the Marris and Bugtis who lived adjacent to that border owed a vague allegiance to the Khan of Kalat situated in the Shal valley on a plateau about seventy miles south of Quetta, a town on the same plateau which represented a line of ethnic division at the head of the Bolan Pass, the country to the north being the Pathan belt while that to the south, Baluch and Brahui. A close-border policy was followed along this line to gain territorial extensions by both the Bombay Government responsible for Sind and the Punjab Government responsible for Dera Ghazi Khan. The British had in fact in the course of the First Afghan War acquired a full knowledge of this tract when their officers and troops marched with Shah Shuja's contingent up the Bolan and past Quetta to Kandahar and Kabul.

In 1876 the Government of India was considering the rival views of the Bombay and Punjab governments concerning control of this part of the border with ethnic overlappings. Robert Sandeman, the deputy commissioner of Dera Ghazi Khan, who was an officer of the Punjab Commission with much experience of that border, expressed himself strongly in favour of a proposal for agreement with the Khan of Kalat, which, while guaranteeing the authority of his state and the security of his territory, would provide for the stationing of a British garrison at Quetta. Guided by an immediate concern for the defence of India, the Indian Government laid aside the conflicting claims of the provincial governments and dispatched Sandeman on a mission of reconciliation to the Khan of Kalat, which resulted in the Mastung Agreement and the treaty of 1876. In return for an increased annual subsidy the Khan granted permission for the stationing of troops, and the construction of railway and telegraph lines through Kalat territory. This marked the end of the close-border policy under which deputy commissioners acted with the assistance of middlemen in an effort to gain territory in the trans-border tribal region. According to C. C. Davies, 'Sir Robert Sandeman's tribal policy was one of friendly and conciliatory intervention. Casting all fear on one side, he boldly advanced into their mountain retreats and made friends with the tribal chiefs or *tumandars*. Recognizing that the British side of the question was not the only side, he never condemned the action of a tribe, until he had fully investigated its grievances.' This had been impossible under the earlier system 'which prohibited officers from entering into independent tribal hills'.[7] Sandeman became a legend with the Baluch: he never withheld allowances, and always emphasized that the guilty alone should be punished and not whole tribes.

Territorial Reorganization for Security and Development 123

The net result of the negotiations which Sandeman concluded with Kalat during 1876–7 was that Quetta with its environs, together with the Bolan Pass leading to it, were leased to the British on a perpetual quit-rent, to remain under the nominal sovereignty of Kalat, but administered by the Government of India in accordance with local custom.[8] The treaty made with the Khan laid the foundation of the Baluchistan Agency, for Major Sandeman was on 21 February 1877 appointed Agent to the Governor-General with his headquarters at Quetta.

The occupation of Quetta and Bolan preceded the outbreak of the Second Afghan War in 1878 when the success of the two-pronged advance through the Khyber and the Kurram led to the abdication of Sher Ali and the conclusion with his son Yakub Khan of the Treaty of Gandamak in 1879, which ceded not only the Khyber and the Kurram tracts, but also Pishin, Sibi and Loralar (Bori), Pathan territories lying to the north and east of Quetta. The result of this cession was to carry the Indian frontier across the Khojak range to Chaman near Kandahar. Within ten years a broad-gauge line was carried to the frontier at Chaman by tunnel through the hills. Baluchistan thus became the first point of advance in the pursuit of the new forward policy. Before Sandeman left in the 1880s he even pushed forward from Pishin into the long Zhole valley, with headquarters built at Apozai, later known as Fort Sandeman. The object was to have a lateral communication opened with the trans-Indus frontier districts by an attempt to link the Gumal Pass with the Zhob river, a tributary of the Gumal river in Waziristan, a stronghold of the Mahsud tribe who had yet to be quelled. Allowances were sanctioned by the Viceroy (Lord Lansdowne) for the Gumal tribes, and in 1890 a great joint *jirga* of all the tribes was held by Sandeman at Apozai. The tribes, who were different from those of Quetta, were eager to grab the money, but the Gumal remained closed to the British for a time. Sandeman's success with the Baluch and the Brahuis could not be repeated in Waziristan in the north.

In Baluchistan, however, the ten years which preceded Sandeman's death in 1892, were marked by tremendous development activity. 'Communications', according to Davies, 'were opened out in every direction, irrigation schemes were taken in hand, forests were developed, and arrangements made for the collection of land revenue. In the administration of justice the indigenous system of *jirgas*, or councils of tribal elders, has been developed under British administration. Local cases are referred to local *jirgas*, while more important disputes are placed before inter-district *jirgas*, or before the Shahi *jirga*, which meets twice a year, once at Sibi and once at Quetta'.⁹

In the course of the Second Afghan War, the forward move in Baluchistan was acompanied by a permanent advance into the Khyber Pass, a Pathan region which revolved around the main Peshawar valley. Relations with all the tribes were till then conducted by deputy commissioners on the old close-border through middlemen. In view of the geographical and political importance of the Khyber range, however, a special arrangement was made by appointing Robert Warburton, the son of an Afghan mother, who had no faith in the old middleman system. He was able to move freely under tribal protection between the Pass and the Kabul river over the plateau leading to Landi Kotal. The highlands of Afridi Tirah, a forbidden land for Europeans, however, could not be reached except with the army. Warburton had under him one Aslam Khan, whose father had been in the service of Shah Shuja, a great name among the Afridis. It was Aslam who secured for Warburton the Afridi loyalties and helped him raise an Afridi corps which later came to be known as the Khyber Rifles.

The occupation of the Kurram valley represented still deeper British penetration into tribal territory. The north-eastern Pathan tribes were separated from those of south-western Waziristan by the line of the Kurram river which was reached by the Miranzai valley stretching westward from Kohat throuth Hangu to Thal. The metropolis of the north-east was Peshawar, while that of the south-west was Bannu. The Kurram line, which extended backward to Sikatam as part of the Sufed Koh range on the eastern borders of Afghanistan had been occupied as an advance base during the Second Afghan War and ceded later by Yaqub Khan under the Treaty of Gandamak in 1879. It was, however, left unoccupied at the end of the war, with the result that a state of chaos emerged from constant sectarian strife between the Turis, who were Shias, and their Sunni neighbours. To obviate the danger of any Afghan intervention on its immediate western border, a loose form of administration was set up in 1891-2 on the Sandeman model, with Roos-Keppal as the first Adjutant of the Kurram Militia and then as Political Agent, a development which led to an historic act of state, the fixing in 1893 of the famous frontier known as the Durand Line.

The British advance on the Kurram line was of course viewed by Amir Abdurrahman as a serious forward move which tended to affect the legitimacy of his suzerainty over the hill tribes situated to the north-east as well as south-west of the great Kurram divide.

While the tribes which inhabited the north-eastern region included the Kafirs, Mohmands, Bajaoris, Swatis and Afridis, those who dominated the south-west in Waziristan were the Darwesh Khel Wazirs, Karlanri Pathans and Mahsuds, who all spread over the Tochi, Wana and the Gumal below Thal, with their southern country passing the highway from Khorasan into Sind by Kandahar and Quetta down the Bolan Pass which formed the Baluch-Pathan frontier. Though known for their spirit of independence and constituting 'some of the finest natural fighting material in the world'[10] these tribes could not combine against a common enemy. However, 'excepting the Kafirs, all of them would appeal to the Amir as arbitrator in their disputes, adviser in their military ventures, supporter and provider of refuge in distress, and spiritual guide in their religious counsels'.[11] The British penetration of the tribal valleys and the development of communications in all directions naturally became a cause for concern to the Amir on the north-west frontier of India, more especially because of the extension of the Sind-Peshin railway to New Chaman on the far side of the Khojak range, a move which the Amir regarded as an attack on his preserves and a violation of the Treaty of Gandamak 'which placed the boundary between Afghanistan and British Baluchistan at the foot of the Khojak mountains on the Kandahar side of them'. This boundary at the Old Chaman terminus was carried further towards Kandahar for another seven miles or so and fixed at New Chaman with its pretty little cantonment which, however, created irritations and led to the belief that 'a boundary was necessary between Afghanistan and the independent tribes' on its eastern frontier.[12] The task thus involved was not only political, but geographical and ethnic also.

A proposed mission to Kabul under Lord Roberts did not occur. Next, Sir Mortimer Durand, the Foreign Secretary of Lansdowne, was nominated as envoy. He left Peshawar for Kabul in September 1893 to enter into a boundary agreement with Amir Abdur Rahman. A distinguished company of officials left for Kabul to assist him, but no survey officer was permitted to accompany the mission; for its immediate declared object was the determination of an ethnic boundary, and it was feared that the presence of a survey officer might make the Amir suspect that geographical and territorial issues were in fact at stake. But it was an error to think that non-specialists could give an authoritative opinion on ethnic or other boundaries from maps which were to illustrate the line of demarcation through 1300 or 1400 miles of boundary.[13]

The reception accorded to the mission was excellent and, since

Durand was a good Persian scholar and the Afghans proved friendly, everything went well. Stated briefly, the boundary line was agreed upon from Chitral and the Baroghil Pass up to Peshawar. From Peshawar up to Koh-i-Malik Siyah (the trijunction of Persia, Afghanistan and Baluchistan) the boundary was adjusted in a manner which transferred to Kabul Wakhan Kafiristan, Asmar, Mohmand in Lalpura, and a small portion of Waziristan called Birmal, while the Amir renounced his claims to the railway station of New Chaman, Chagai, the rest of Waziristan (including the Tochi valley, Wana and Gumal), Biland Khel (opposite Thal on the right bank of the Kurram river), Kurram, Afridi, Bajaur, Swat, Buner, Dir, Chilas and Chitral.[14] The tribes thus came to be divided both geographically and politically. Even so, the agreement was finally signed after weeks of waiting. The Amir signed the agreement, but omitted to sign the maps which were supposed to be illustrative of the agreement. The Amir's subsidy was increased by Rs 600,000 annually and the mission left Kabul in November 1893.

It could not be claimed that the Durand Line represented India's precise international frontier; for it started with recognition of the necessity of demarcating an ethnic boundary between Afghanistan and the independent tribes situated on its eastern frontier. The agreement demarcated for the first time a tribal belt under British control between Afghanistan and the administered border of India on its north-west frontier. The old situation of a no-man's land of uncertain extent ceased to exist. The authorities on both sides could now act with a degree of firmness and precision. The agreement in fact did not describe the line as the boundary of India. But since no boundary, tribal or otherwise, could be demarcated without reference to maps indicating geographical situations, the agreement was taken as representing the Amir's dominions and the line beyond which neither side was expected to exercise interference. The reason for the absence of a really well-defined territorial frontier was that the British Government did not intend to absorb the tribes into their administrative system. All that they wanted for reasons of security was to extend their own authority and exclude the Amir's from the territory east and south of the line, an object which was duly accomplished.

Difficulties, however, arose where a sector agreed on the map was never demarcated on the ground. This perhaps was because geographical watersheds and tribal boundaries did not always coincide, for example, on the watershed between Kunar and Bajaur through Mohmand country across the west, a situation which applied to the Khyber sector stretching up to the Sikaram peak of the Sufed Koh range to the west. The Durand Line was conceived as following the Kunar-Bajaur watershed as far as it was defined towards the Kabul river, leaving a considerable belt of Mohmand territory on the Afghan side, while including on the side of British India a number of upper Mohmand clans of such other sections as were not in political relations with Peshawar and drew no allowances. The allegiance of the different clans of the Mohmand tribe was thus divided politically. The failure of demarcation on the Khyber side and the Sufed Koh was of little or no consequence because of the existence of the most obvious natural features. But in the Mohmand country it became a cause for unrest and uprising.

The demarcation on the ground was carried along the frontier of Waziristan, which could only be done after heavy fighting with the turbulent Waziri tribes. In 1894, a few months after Lord Elgin's arrival in India, it was decided to form a military post at Wana, a barren plain north of the Gumal river at the south-western corner of Waziristan. It afforded a key to the backdoor of the restless and virulent Waziri tribes in the same way as Zhob did for the Pathans of the Baluch border. Strategically, it would add a British presence between Jalalabad and Quetta and command the major trade-route between Ghazni and India, a link in the line of Zhob communications with the settled frontier districts. A British brigade was sent to occupy Wana, which was also considered essential for a delimitation of the western boundary of Waziristan.[15] Fighting was a natural consequence of the British attempt to control the Wazir country from within, but the Tochi valley and Wana were both occupied and formed into the separate Agencies of North and South Waziristan, respectively.[16]

The most extensive advance was made in 1895 with the formation of the Malakand Agency including Dir, Swat and Chitral, an Agency which carried India's frontier almost as far as the Pamirs and the upper bends of the Oxus. Under the Durand Agreement this advance was effected partly to subdue the warlike tribes and partly as a counter-move to the Russian advance in the Pamirs. Chitral too was of equal importance in terms of imperial defence against Russian pressure. Situated in the far north of the Malakand Agency, it formed part of the upper valley of the Kunar which ran down to join the Kabul river near Jalalabad, the Kunar itself being separated from Russia only by a few miles of the tongue of Wakhan situated on the borders of the Russian territory of eastern Badakhshan.

Communication with Chitral had earlier been carried on only

from the Gilgit side, not from Malakand, a shorter route, though for long unpenetrated and unknown to the British before 1895. The first officer who was placed in charge of the Malakand Agency was Harold Deane, later the first Chief Commissioner of the North-West Frontier Province formed in 1901. Because of its importance in the direct approach to Chitral through Swat and Dir, the Malakand Agency was from the very beginning placed under the control of the Central Government which, on the model of Baluchistan and Kurram, provided for a sort of loose administrative control for the conduct of local tribal affairs. Dir became a treaty state, but Swat remained for the time in a condition of tribal unrest and little was done to strengthen British control.

The final pattern of the Agencies, which was more or less complete in 1895, consisted, from north to south, of Malakand, Khyber, Kurram, North Waziristan, and South Waziristan. It did not cover the whole of the tribal belt. Such important tribes as those on the Hazara border, the Orakzais and the Bhitannis were still being dealt with by the deputy commissioners of districts. The Adam Khel Afridis of the Kohat Pass were likewise managed jointly by the deputy commissioners of Peshawar and Kohat.

The tribes, more especially in Waziristan, had been watching with concern the steady penetration of their preserves, a development which the Amir of Afghanistan, too, had reason to view with distrust and dislike. For the independent tribes interfered very little with him, and might from religious and other allied considerations be brought under his sovereignty in due course of time. The demarcation of the Durand Line, the setting up of the five Political Agencies, the passage of troops in all directions through tribal territory, and the garrisoning of such tracts as were not only strategically important but also agriculturally fertile and productive were regarded by the tribes as threats to the independence they had so long enjoyed.

A mere spark was needed to set the whole frontier ablaze. This was supplied by a village in the upper Tochi valley of North Waziristan, when the Political Agent went there in June 1897 with a military escort to choose a site for a levy post. Though initially received with geniality, the visitors were suddenly attacked. All the officers were either killed or wounded, but the troops managed to escape. This was followed a fortnight later by a serious uprising in the Malakand region. By August the blaze had spread to the Mohmand country and a fortnight later to the Afridi and Orakzai belt, leading to the fall of the Khyber posts. The Samana forts had in the meantime been attacked, the garrison in one case being wiped out completely, and the Kurram was threatened. The southward spread of the conflagration was, however, checked; for the controlling operations were well under way in the Tochi valley where the Mahsuds, the most formidable and turbulent of all the tribes, did not rise in 1897, having been chastened in 1894.

In the face of a common threat, the tribes nonetheless found some sort of union in their wars against the British during 1897-8, when Upper Swat, Bajaur, Buner, the Mohmand country and Tirah were for the first time traversed from end to end before peace was restored in the spring of 1898. The Khyber was reoccupied and order re-established on both sides of the Kurram divide. The building of new roads and the construction of more forts began with the restoration of peace. But the tribal wars called for a deeper review of the entire north-west frontier policy, given the magnitude of the problem of security and development.

Closely connected with these policy dimensions was the question of the so-called 'scientific frontier' which since the seventies and eighties of the nineteenth century had been engaging the attention of the British Government. The concept was not without problems. It was, for instance, recognized that a 'scientific frontier' would be well-nigh impossible to demarcate if, besides having a geographical unity, it also had to satisfy ethnic, political and military requirements. According to C. C. Davies, it was even 'utopian' to seek a zone 'which does not violate ethnic considerations by cutting through the territories of closely related tribes, and which at the same time serves as a political boundary'.[17] This was the basic drawback of the Durand Line, a drawback that flowed initially from a policy of demarcating a boundary 'between Afghanistan and the independent tribes', which meant a disrtibution of tribal zones between the Amir and British India. A 'scientific frontier', on the other hand, signified a line of defence against invasion from Central Asia, a purely political objective with strategic and military implications. It was generally agreed on the basis of the experience of the Second Afghan War that the best frontier for imperial defence would be the Kandahar-Ghazni-Kabul line.

This line, it was pointed out, was shorter and more easily manageable than any other frontier that could be suggested. While its northern flank was almost fully protected by a series of impenetrable mountains, its southern flank had an equally impassable desert. And if this line was connected with the main Indian railway system, troops could without difficulty be concentrated on either flank. A rapid concentration and dispatch of troops to any part of the line was what was most needed and infinitely more important than the construction of expensive fortifications which a mobile enemy could avoid. The Liberal idea of 'retirement' from advanced positions in the 1880s necessitated the abandonment of the 'scientific frontier'. But the question of whether or not to advance to this line still remained open. The advance made in the 1890s proceeded in the same context. It was in this context that Lord Curzon came out to India as Viceroy in 1899 and approached the whole frontier problem from an angle which resulted in the establishment of the North-West Frontier Province in 1901.

Even long before Curzon took over as Viceroy certain anomalies had developed in the organization of the North-West Frontier as part of Punjab, anomalies which flowed from tribal territory being brought under a vague executive control distinct from the administration of the erstwhile settled districts. In the earlier days, when relations with the tribes were under the close-border policy, the anomalies were not so pronounced. Even so, the pressure of Russian advance in Central Asia had induced Lytton in 1877 to suggest a scheme which gave the Central Government more direct control over frontier policy and administration. The Second Afghan War led to the idea being shelved for a time. But it was not given up. What was needed with the extension of control over tribal country was not only soundness of policy in the context of the Russian advance, but also speed in the execution of that policy. The intervention of the Punjab Government as a medium of communication between the district officers on the North-West Frontier and the Government of India was a factor which tended to affect both. Considerations of security and defence therefore called for a removal of that obstruction.

Despite Lytton's proposal being for a time shelved, its implementation had already begun before Curzon's arrival. Quetta, for instance, remained under central control right from the beginning of the occupation of Baluchistan. The logic of this became all the more convincing with the occupation of the Khyber, parts of Waziristan and Malakand. Already in 1895 the states and tribes brought under the British sphere of influence as a result of the great move forward over the Malakand, had come to be managed by an officer placed directly under the Calcutta Government. The soundness of the logic of central control over the hill tribes of frontier territory was further reinforced by the rising of 1897-8, which made Curzon review the policy pursued in the 1890s.

The position, as Curzon found it, was that more than 10,000 troops were stationed across the administrative border in the Khyber, in Waziristan, in the Malakand area and elsewhere. These advanced positions were not only far removed from a base, but also unconnected with lateral communications. The result, as the rising of 1897 had shown, was the danger of loss of life before assistance could be made available. Moreover, while the Indian Government was still going ahead with the construction of expensive fortifications and the dispersion of forces in tribal territory, the British Government after the troubles of 1897 decided not to split Punjab; for the commissioners of Peshawar and the Derajat were to act under the orders of the Punjab Government in matters of ordinary administration, although in their dealings with the tribes beyond the administered border they were to remain directly under the Government of India. Curzon was opposed to both the administrative and political arrangements. He emphasized that the policy pursued in the 1890s should be replaced by one of non-interference, comparable in many respects to the old 'close-border' system. It meant withdrawal of British forces from advanced positions and their concentration in British territory behind the tribal country, which was to be left for defence by tribal forces, with the support, if necessary, of the British troops stationed nearby as a safeguard. As regards the existing political arrangement, Curzon insisted on direct central control over the entire trans-Indus territory. 'The Viceroy', as he wrote on 5 April 1899, 'is responsible for frontier policy, yet he has to conduct it not through the agency of officials directly under him, but the elaborate machinery of a provincial government.... The result is that in ordinary times the Punjab Government does the Frontier work and dictates the policy without any interference from the supreme government at all, but that in extraordinary times the whole control is taken over by the Government of India acting through agents who are not its own, while the Punjab Government, dispossessed and sulky, stands on one side, criticizing everything that is done.'

Not prepared for any reduction in the extent of his jurisdiction, Mackworth Young, the Punjab Lieutenant-Governor, proceeded to resist the central move with 'disputatious arguments' which drove the Viceroy to a fit of despair. 'I cannot work under this system', he wrote, 'cannot spend hours in wordy argument with my Lieutenant-Governors as to the exact meaning, purport, scope, object, character, possible limitations, conceivable results of each petty aspect of my Frontier policy. If they deliberately refuse to understand it, and haggle and boggle about carrying it out, I must get some fairly intelligent officer who will understand what I mean and do what I say.' The controversy nonetheless persisted for a time, but, as Ronaldshay (later Marquess of Zetland) points out in his *Life of Lord Curzon*,[18] the Viceroy, with a view to 'utterly abolishing an opponent' recorded on 13 September 1900 a strong minute which made the London authorities yield with grace. They agreed to the creation of the North-West Frontier Province, which came into being in November 1901. The new province was created without any prior reference to Mackworth Young, the Lieutenant-Governor.

The new province was divided into two parts: the settled districts of Hazara, Peshawar, Kohat, Bannu and Dera Ismail Khan, which were separated from Punjab to form the province proper; and the trans-border tracts which lay between the administrative and Durand boundaries. In addition to the five Political Agencies of Malakand, Khyber, Kurram, Tochi and Wana, the trans-border tracts included tribal areas controlled politically by the deputy commissioners of the adjoining settled districts. Between Hazara and Dera Ismail Khan there was, however, only one trans-Indus tract which was not separated from the Punjab administration. This was the trans-riverain tahsil of Isa Khel, a non-Pashtu speaking Pathan tract which remained with Punjab. The new province was placed under a Chief Commissioner who combined in his office administrative responsibilities for the districts with political control of the tribal belt as an Agent to the Governor-General, to be appointed by and responsible to the Government of India. In addition, there were to be a revenue and a judicial commissioner.

The staff of the Chief Commissioner included officers of the political department of the Government of India in addition to members of the provincial and subordinate civil services and officers of the police and various other departments.[19] The civil and judicial administration of the settled districts was more or less comparable to that elsewhere in British India, each of the five districts being placed under a deputy commissioner who was assisted by the usual staff of tahsildars, naib tahsildars, kanungos and patwaris. The judicial commissioner constituted the highest judicial tribunal, with the two divisional and sessions judges of Peshawar and the Derajat acting as his subordinates. The revenue administration of the entire settled area vested likewise in the revenue commissioner, Michael O' Dwyer.

As regards the arrangement for the trans-border tracts, the withdrawal of British troops from advanced positions and their replacement by tribal militia under British officers were duly completed.

And as it was feared that a tribal militia might break down if called upon to function as regulars, arrangements were made for their protection and support by flying columns and light railways instead of more fortifications. By 1904 the new arrangements had come into operation along the whole frontier from Chitral to Baluchistan. Regular forces were to be found in Chitral alone because of its proximity to Russian territory. British troops were withdrawn from all other advanced positions and stationed in the military cantonments of the settled districts, their take-off base in an emergency.

However, it was not possible to entirely separate the administration of the five settled districts from the political control of the adjoining Political Agencies. For while British subjects constantly visited the independent territory of hill tribes, the latter lived for a time on both the sides of the trans-border tracts, where residents were connected by ties of race, marriage and association, both territorial and social. This close connexion between the plains and the hills affected the administration of justice in the settled districts. The practice of blood-feuds between tribe and tribe, clan and clan, family and family, which resulted in frequent murders and other serious crimes in tribal country, filtered down to the plains also. To curb lawlessness among the hill tribes from Chitral to the Kabul river, the British dealt with the important chiefs and rulers, such as the Mehtar of Chitral and the hereditary chiefs of the numerous Khanates into which Dir and Bajaur were divided. Further south, between the Kabul and the Gumal, the controlling power was, however, not a chief but the *jirga*, which the frontier officials used in negotiations with tribesmen. The Frontier Crimes Regulation of 1901, which superseded the Punjab Frontier Crimes Regulation of 1887, empowered the deputy commissioner to make both civil and criminal references to *jirgas* of three or more persons convened according to tribal custom.

The creation of a separate province on the north-west frontier of India thus proceeded from two main premises, international and local. While the first involved questions of India's defence against foreign aggression, especially against increasing Russian pressure in Central Asia, and called for a delimitation of the frontier for strategic considerations, the second was a local problem of security against lawlessness arising from variations in ethnic and geographical conditions. Both were linked at most points and the policy pursued could not but take into account the related implications of both in terms of reactions in Afghanistan, the operating catalyst situated between the tribal country on the Indian frontier and Russian territory in Central Asia. After a series of changes in the shades of policy a separate North-West Frontier Province was established under central control to meet the contingencies of both defence and security. The administrative and political arrangement which Curzon made by a withdrawal of British troops from advanced positions was clear recognition of the fact that although central control alone could ensure perspective, speed and resilience in the handling of tribal affairs on a sensitive frontier, its army was not the only answer to the problem of ethnic and geographical difficulties. Attempts had to be made to approach the local problem politically and modify institutional functioning even in the settled districts to make it more or less universally acceptable to tribal sentiments.

This political approach to local problems affected the conduct of international relations. The scheme of defence against Russia was sought to be settled not so much by military strategists as by diplomatists. Both the Forward and Stationary Schools of policy-makers came to agree on the need to build a strong and friendly Afghanistan as a buffer between India and Russia. Amir Habibullah, the successor of Abdurrahman, for example, remained faithful to the British alliance when the war broke out in August 1914. The entry of Turkey in November caused great excitement among the frontier tribes and intrigues on the Indian side of the Durand Line. Punitive expeditions were sent against the Mohmands and the Mahsuds. But since the Amir remained firm in his friendship with Britain, there was no tribal rising.

Assam and the North-East Frontier

As in the case of the north-west frontier, the problem of Assam and its north-east frontier, too, was one of defence against foreign attack and security against local tribes. While prior to the annexation of Upper Burma the north-east frontier remained exposed to Burmese invasion, a new threat to peace emerged later from the Sino-Tibetan side in the twentieth century when the Chinese determined to bring Tibet under their control. The local problem of insecurity arising from the raids of the various hill tribes caused a concern which not only called for constant military alertness but also involved a periodical review of tribal policy to effect conciliation and promote development.

Assam was acquired from the Burmese who, under the Treaty of Yandabo (1826), surrendered the province to the East India Company. David Scott, the Commissioner of Rangpur and Agent to the Governor-

General for Assam, organized the area into five districts. These were Kamrup (Gauhati), Darrang (Tejpur), Nowgong, Sibsagar (Jorhat) and Lakhimpur (Dibrugarh), Goalpara, a perpetually settled tract belonging more properly to Bengal as part of Rangpur, was formed into a separate district under Regulation X of 1822, and came to be administered as part of the Garo hill tract in Assam on account of constant Garo raids on the plains below.

Physically, Assam proper consisted of the two river valleys of the Brahmaputra and the Surma. It was bounded on the north by the lower spurs of the eastern Himalayas, and to the south and east by a number of hill tracts, inhabited, as in the north, by several independent and warlike tribes. Many of them claimed a definite share of revenue from lands near the hills in the plains, while others claimed tribute and services supposed to have been assigned to them by the erstwhile Assam authorities. From the correspondence that transpired on the nature of British relations with the hill tribes in Assam in the 1840s, it appears that the tribes were allowed by the former Assam rulers to collect a definite share of revenue called *posa* or blackmail. The custom of the hill tribes of 'drawing their supplies from the plains and receiving a share of the revenue' was recognized even by the Company's government as 'having long been sanctioned'. It was therefore continued for 'purchasing their goodwill and forbearance' towards British subjects to ensure 'happiness, security and prosperity'.[20]

The researches of Alexander Mackenzie, a Bengal Civilian, who is known for his authoritative *History of the Relations of the Government with the Hill Tribes of the North-East Frontier of Bengal*, confirm the earlier findings of Assam's revenue and judicial officers:

It mattered of course little to us, whether these claims had their basis in primaeval rights from which the Shan invaders [Ahoms] had partially ousted the hillmen, or whether they were merely the definite expression of a barbarian cupidity. Certain it was that such claims existed, and that they had been, to some extent and in some places, formally recognized by our predecessors. The engagements under which the Native Governments lay were transferred to us.... But we are met to this day by difficulties arising from the indefinite nature of the connexion subsisting between the Ahom sovereigns and their savage neighbours.[21]

Hill Tribes on the Northern Border

Beginning from the west on the northern border of Assam were the people of Bhutan, with the Bhutias of the Tibetan dependency of Tawang coming next, a dependency which stretched as far down as the British possession of Udalgiri forming part of the foothills or duars of Bhutan proper on the western frontier of Assam. Situated eastward from the Balipara Frontier Tract, with Tawang on its northern edge, to Walong, its north-east border close to south-eastern Tibet along the Burmese frontier, were a number of hill tribes, such as the Akas, Daflas, Miris, Abors, Mishmis, Singphos and Khamtis.

The source of trouble with the Bhutias was the British occupation of the most westerly districts of Assam, which bordered on the foothills or duars of Bhutan proper and its neighbouring highlands on the Tawang side. As the duars were a tract of fertile land, their possession was important both for the hillmen and the inhabitants of the plains under the Assamese government. The decaying authority of the Assam rajas, however, could not provide for efficient protection to the indigenous cultivators or establish undisputed dominion over the soil and its products. The Bhutia highlanders of both Bhutan and the neighbouring Tawang tract thus held practical control over the duars on the Assamese side. They emerged as a serious security problem after the introduction of British rule which proceeded to tighten its control in the administration and collection of land revenue. The difficulties became more pronounced because of the Bhutias being divided into a couple of separate clans, one of which even disowned allegiance to the Raja of Tawang or the supremacy of Lhasa. They were all sought to be pacified on payment of monthly allowances, subject to their 'promise upon oath that they will neither collect, directly or indirectly, or permit their followers to do so, the former most objectionable contributions, which were denominated *posa,* and by us blackmail'. The payment of a regular allowance led the Bhutias to 'acknowledge themselves dependent upon the British Government' and induced them to promise 'to assist in preventing the encroachments of others upon the plains'.[22]

The Akas were a warlike tribe situated east of the Bhutias. They were divided into two groups, the Kapachor and the Hazari-Khawas. Though small in number, they were comparatively formidable in power and influence. They had long been exacting tribute from the Char Duar Bhutias of Tawang and resisted its Deb Raja with impunity. The Chief of the Kapachors, the 'Taggee Raja', was so powerful that in 1842 he became 'a pensioner of Government upon an allowance of 20 rupees per mensem', with all his former crimes being 'pardoned upon condition of his promising solemnly to refrain from acts of aggression upon the subjects of the British Government'. Despite this agreement, however, the chief never visited Darrang to receive his pension, until summoned directly by the Assistant Agent to the Governor-General.[23] The Hazari-Khawas had earlier been the more powerful of the two Aka clans, but were later reduced by the Kapachors. Even so, they were entitled to the *posa*, a share in revenue which the Kapachors were not permitted to collect as they were treated as dacoits.

The Daflas, who inhabited the mountains to the east, were not a particularly warlike tribe. Arrangements had been made with their *gaums* or chiefs, who were annually paid a fixed sum as their allowance in lieu of *posa*, although care was taken to maintain outposts of armed police along the frontier to guard against any projected raids. Many of this tribe had taken up land in the plains along the foothills and become tenants under the government, a tendency which was encouraged not only among the Daflas, but also with other tribes.[24]

Beyond the Daflas to the east were the Miris and Abors who were closely connected and distributed locally to the north of Dibrugarh in Upper Assam, across the Subansiri river, forming part of the Sadiya Frontier Tract. They claimed the whole country between the low hills below the eastern Himalaya and the Brahmaputra. Their independent ways were a standing cause of conflict with Calcutta.

Major Vetch, the Deputy Commissioner of Gauhati, who held the appointment of Political Agent of Upper Assam, pointed out in a report of 24 July 1853 that some Miris and Abors also lived on the south bank of the Brahmaputra in the district of Dibrugarh. It was they who, in commutation of their claims to *posa*, were paid an allowance annually. On the northern bank of the Brahmaputra, however, the Abors remained a security problem on the frontier which needed to be 'strongly guarded both to repel' and 'to punish aggression'.[25]

The Mishmi tribe occupied the mountainous region between Assam and Tibet, north-east of Saikwah beyond Sadiya, trading with both the Tibetans and the Assamese. The tribe included in its fold the Mezho Mishmis, Tain Mishmis and Chulkatta ('cropped-haired') Mishmis. Of the three, the Chulkattas were by far the most troublesome to the British. Apart from the Chulkattas, the Mishmis did not constitute a serious danger to peace and security. They frequently visited all the stations in Upper Assam and found in trade a more profitable occupation. Even so, they did not cease to be a source of occasional trouble. Numerous armed posts had therefore to be maintained for defensive purposes.

The predatory tribes of Singphos and Khamtis were mainly inhabitants of the Sadiya Frontier Tract, with Walong as its north-eastern border close to south-eastern Tibet. The Tract included the Brahmakund valley and the district of Sadiya, a part of Assam inhabited principally by refugee Khamtis of Shan stock who were driven by the Singphos from their original seats to the south-east. Their immigration, however, continued. They obtained permission from the Assamese government of Sadiya to settle near the Theinga.[26] Taking advantage of the state of civil war in Assam and the Burmese invasion, the Khamtis of Sadiya and its neighbourhood united with the Burmese interest and took to insurgency. Their settlement was consequently dispersed.

The Singphos inhabited the country lying to the left bank of the Lohit river, opposite the district of Sadiya in the hills. It was accessible mainly by the Now Dehing and Theinga flowing into the Brahmaputra about ten miles above the parallel of Sadiya and formed part of Assam. The region was inhabited by the native subjects of the Assamese government who were however dispossessed by their hill neighbours in the course of the civil war which followed the government of Raja Gaurinath. After a spell of desolation they formed new settlements named after their chiefs, by which their areas of residence were always known, such as Bisa Gaum, Daffa Gaum and Satu Gaum. Nominally divided under twelve *Gaums* or chiefs, they rarely combined except for some temporary purpose of plunder. They invaded Sadiya in 1825 and carried their ravages as far as Jorhat in Upper Assam, taking captives who were sold mostly to the hill Singphos, Khamtis and Shans.[27] The Singphos attacked Sadiya and Saikwah in 1839 and several other posts in 1843. They were, however, reduced to submission by the British. Their population was dispersed and their chieftainship changed. The Upper Assam Political Agent reported in 1853 that the Singpho country had been quiet for several years, with the Singphos having become desparate opium-eaters.[28]

The Moamarias or Morans, who were subject to a chief called the Bar-senapati, were yet another group who inhabited the Sadiya region and preserved their independence against the Burmese as well as the neighbouring tribes. Though Hindus and worshippers of Vishnu, they had in the midst of the surrounding tribal ecology lost their Hinduism and taken to plunder. Captain Neufville,[29] who made a survey of the Sadiya Tract, pointed out that the residence of the Bar-senapati was on the Diburu in Lakhimpur (Dibrugarh) district, a central place removed from the earlier centre of Bara and Chota Sakri, which was dangerously near the Singpho border.

Territorial Reorganization for Security and Development 139

The outbreak of the first Burmese War led to the acquisition of much valuable knowledge and information about the north-east frontier and the vast, unknown region inhabited by the hill tribes to the north of the Brahmaputra.

Hill Tribes to the South of Assam Proper

The survey of the Brahmaputra valley and the frontier tribes, which was of immediate relevance to security and defence, began in 1825 when Captain Bedford and Lieutenant Wilcox were sent to explore the great river towards its source, under the instructions of Colonel Blacker, then Surveyor-General. Burton, who accompanied the team, went as far as Sadiya, while Bedford did the Dihong and Dibong. Wilcox, on the other hand, penetrated to the Irawaddy beyond the frontier. The *Memoir of the Survey of Assam, 1825–8* was a detailed account of the discoveries made by Wilcox through his exploration. The territory of Manipur and the surrounding country in the direction of Cachar and Upper Assam was surveyed around 1826–32 by R. B. Pemberton, a mere Lieutenant operating from Manipur, but emerging as the great authority on all eastern frontier matters by virtue of his standard work, *The Eastern Frontier of British India*.

These and other surveys of Assam revealed that the Nagas were by far the most warlike of the tribes on the southern border of the Assam valley. In *The Hill Tribes of the North-East Frontier of Bengal*, Alexander Mackenzie discussed in some depth the general characteristics of the tribe and its various clans in both ethnic and geographical terms.[30] It is not necessary or possible here to attempt even a summary of these features. However, it is necessary to indicate some of the influence exercised by the Nagas, and the strategic position they held on India's north-east frontier.

A reference to the Naga tribe first appeared in a letter of 6 February 1832 to the Acting Agent of North-East Frontier by Captain Francis Jenkins, who was at the time engaged on a special survey in Assam with Pemberton.[31] 'I have the honour to acquaint you', the letter said, 'that Lieutenant Pemberton and myself accompanied by Joub Raja of Munnipore and Lieutenant Gordon with 700 Munipooree sepoys have succeeded in effecting a passage through the Naga country and the forest between the hills and the low lands of Assam and Northern Cachar.'[32] It referred to 'the powerful tribe of Maram Nagas' whose hostile disposition made it difficult for the survey party to find a guide.

It was only a Cachari refugee from Raja Tula Ram's Northern Cachar[33] who enabled them to find easy communication from the Naga frontier with Northern Cachar to both Jorhat in Upper Assam and Gauhati in Lower Assam.

Lying to the east of the district of Sibsagar (Jorhat), are situated the Naga Hills, which form the boundary between Burma and Assam. They became a direct British concern immediately after the resumption of the territory of Raja Purander Sing in 1838. Bronson's mission, which was followed by Colonel Brodie's visit to these hills in 1842 to establish friendly contacts, proved to be of little or no consequence. In the winter of 1844 Colonel Brodie, then Deputy Commissioner of Sibsagar district, again made a tour of these hills and passed through a number of Naga villages to the north of Kohima without meeting any opposition. He took agreements from the different chiefs, in which they acknowledged the supremacy of the British and undertook to abstain from waging war with one another. These agreements, however, were according to Mackenzie's researches, hardly observed at any time.

After Colonel Brodie, British officers were thus discouraged from interfering with the Nagas or visiting the hills; nor did any occasion arise to send an expedition into their territory. Their history remained a series of petty raids and blood feuds. But they made no raids into the plains below after the British occupation of Assam, although they appeared to have received *posa* in former times. According to the *Report on the Province of Assam* by Mills (1854), the Nagas' peaceable relations with the British followed the execution by the British of three of their kinsmen found guilty of murder. In addition, the Nagas were encouraged to take up lands in the foothills of British territory, which tended to bring them closer to the British Government.[34]

According to Brodie, the Nagas were known as great traders. Their trading parties were continually to be met in winter, when they brought down cotton, chillies, ginger, etc., and took away salt, iron, fish and pariah pups. They appreciated the value of this trade, which made them peaceable neighbours.

The Nagas of the Sibsagar or Jorhat district claimed their land as far as the highlands or even into the foothills. But in no case did they actually claim ground under Assamese cultivation or any tea gardens. They had in fact very little communication with the tribes inhabiting the hills to the east and west of their town and the Nagas inhabiting the mountainous country around Manipur were an entirely different race. The area between the Brahmaputra valley and both Upper

Burma and the Bathang-Bhamo road, where the hills turn to the south-west is inhabited by some tribes who are called Nagas indiscriminately. Turning southward, the Indo-Burmese frontier marches with Manipur for some distance. There was no danger from Manipur itself, though the incursions of Burmese over the adjoining Patkai range into the valley of the Dehing could not be altogether ruled out, more especially when the boundary of Assam to the south of Manipur was inhabited by the hostile Lushai tribes, and further west by the people of Hill Tripura.

A survey report of 1847 by Baron Otto des Granges says that 'the occupants of the mountains round Manipur are the Nagas or Kookies. They are a free, independent and very active people' who, though poor and separated from all the cultivated territories around, 'have remained unsubdued by more powerful neighbours. They build their villages on the most inaccessible edges and mountain tops. They are of great muscular strength and indefatigable mountaineers'.[35] Alexander Mackenzie describes them as Angami Nagas and refers to a series of punitive expeditions sent against them without success.[36] It was only in 1878 that one of these expeditions succeeded in occupying Kohima, the centre of the Angami tribe. They were in fact the most formidable and aggressive of the tribes the British Government had to reckon with. According to a report of S. C. Bayley, Chief Commissioner of Assam, the Angamis inhabited 'the range of hills dividing the Valley [of the Brahmaputra], east of the Doyang, from that of the Surma, east of the Jynteah [Jaintia] hills, and may be said to be almost wholly within the territorial boundaries of the province'. Despite the occupation of Kohima, Bayley described the British position as 'a garrison in an enemy's country'.[37]

The Lushai and Kookie tribes inhabited the territory to the south, north, east and south-east of Manipur. They periodically raided Cachar and Manipur in the north and the Chittagong Hill Tract in the south. Colonel Lister led a punitive expedition into the Lushai country in 1849. But since no survey had yet been done of this tribal region, the expedition failed to achieve any permanent success. However, an expedition in 1871 ensured the defence of the southern boundary of Cachar. The eastern boundary of Hill Tripura was defined along with the line of boundary stretching up to the Chittagong Hill Tracts. But the danger of pressure from other tribes to the south-east remained, more especially from their internal dissensions or the choice of a new chief, which often became a cause of the usual tribal blood feuds.

The other tribes inhabiting the hills to the south were not a serious problem. To the west of the Nagas, for example, came the Jaintias, who broke into insurrection once, but soon settled down to a life of peace. The Khasi Hills lay to the south of the district of Kamrup (Gauhati). Prior to the establishment of British rule the Khasis had settled down gradually in the plains. Since the native government found it difficult to dislodge them, it remained satisfied with a nominal acknowledgement of supremacy. David Scott secured an agreement with the Khasis for the construction of a road running from Sylhet to Cachar through their territory. The road, however, could not be completed before the Khasis rose up in arms in 1829. Strict British controls had deprived them of all kinds of illegal cesses which they had so far been levying. The rising of 1829 was followed by an attack on British territory in 1831. The Khasi Hills were conquered in 1833, and the Jaintia Hills to the east annexed in 1835.

To the west of the Khasi and the Jaintia Hills come the Garos who, unlike the Khasis, did not border on any of the duars of Kamrup, although the Garo tribe did visit the markets held in one of the duars in the neighbourhood of Goalpara on the borders of Bengal. Though occasionally turbulent in former times, the Garos, like the Khasis and Jaintias, had learnt to live in peace under British rule.

Political Stability and Integration of Assam

Assam, which had earlier remained more or less outside the main currents of Indian history because of its ethnic variations and foreign political influence, came to be politically stabilized and integrated with British rule in India. This stability and integration proceeded from the improvement of communications, the establishment of military and police outposts in advanced strategic positions, the reorientation of tribal policy to unite security measures with those of welfare and development, and, finally, the administrative and territorial reorganization of the province to promote viability.

Communications and Security Arrangements

The problem of communications after the conquest was mainly the need for a road from Sylhet to Assam, a problem which was closely connected with defence requirements and questions of military and police arrangements. For, apart from limited contacts through routes confined to the hill tribes who carried on a minor exchange of commodities, Assam was connected with the adjacent parts of Bengal only by the tortuous courses of the Brahmaputra and the Surma rivers, the distance from Sylhet to Gauhati by water being about 400 miles. In his minute of 25 March 1833 on the north-eastern frontier, Bentinck admitted the ignorance of local geography and resources and appreciated the importance of Scott's advice to reinforce Manipur as a frontier. As the minute pointed out:

Previous to the late war with Ava we possessed no knowledge of the passes connecting Manipur with our territories, of its resources we were equally ignorant, and the panic occasioned by the simultaneous appearance of two divisions of the Burmese army, one from Manipur and the other from Assam, led to a very general flight of the inhabitants of Cachar and those occupying the northern and eastern borders of our district of Sylhet. Under such an emergency it was natural that every resource, however trifling, should be sought after, and the re-establishment of the Manipur dynasty seems to have been a scheme peculiarly favoured by our late agent Mr Scott as affording in his estimation a well-founded prospect of defence to our frontier in that direction by the interposition of a race of people known to entertain a rooted antipathy to the only enemy against whose aggressions it was necessary to guard, and of the fertility of whose country highly coloured descriptions had been given.[38]

The outbreak of the Burmese War dictated the necessity of a group of surveyors being sent into Assam to report on the state of communications in the new province, and it was as part of their special survey duty that Jenkins and Pemberton planned, for 'civil and commercial purposes' as well as for 'the effectual military defence' of the whole province, to establish 'direct intercourse' between its southern and northern parts. Their reports of the survey of the north-eastern frontier were so valuable that they drew the attention of the Governor-General who in a minute of 25 March 1833 observed: 'These reports, as condensing the scattered information already before government regarding extensive regions hitherto imperfectly described, and as containing much original matter relating to Cachar, Manipur, Assam and the Khasi Hills, and many valuable suggestions for the improvement of these countries, reflect great credit on the industry and intelligence of those officers, the result of whose labours sufficiently shows the importance of the duty entrusted to them, and their qualifications for the task.'[39]

The first attempt at road building was completion of the route which Scott had already proposed—from Sylhet to Gauhati and through Khasi territory to Cachar in the south. The construction of the road was, however, delayed because of the Khasi insurrection and the need for permission or acquiescence from the treaty states to make a permanent road through their country. The road and its intersecting bridges were nonetheless completed by convict labour from Assam and Sylhet.

Another important achievement of Jenkins and Pemberton was ascertaining the exact nature of communications between Manipur and Cachar on the one hand and the kingdom of Ava on the other. But by far their most valuable achievement was determining by a detailed survey the route and topography between Manipur and Assam; in the course of their travel report they made important suggestions not only on the choice of sites for the construction of a military cantonment, but also on the strength and disposition of military force required. It is clear from their report that, while advising the government, they were conscious not only of the danger of tribal aggression but also of 'the probability of another invasion from Ava'.[40]

Jenkins and Pemberton were both unanimous in their choice of Jorhat as a fit military cantonment. Between Sadiya and Jorhat they felt there was no area along the Brahmaputra that satisfied their requirements. Jorhat also seemed the most advanced position that could be occupied by the troops with advantage. It was argued that the 'complicated political relations' of the British Government on 'the eastern borders of Assam', the predatory and the 'intriguing character' of the tribes occupying the Naga Hills to the east, their nearness to the pass leading into Assam and its being the 'most richly cultivated and populous portion of Upper Assam' were all factors which tended 'to establish the superiority of Jorhat over any other station sufficiently near to afford protection to those distant parts of our territory which are the most liable to suffer from open assaults or intriguing treachery'.[41] The importance of Jorhat was further emphasized in terms of its 'uninterrupted cultivation', its large and numerous villages, its 'sugarcane, poppy, mustard, rice, moonga plantations' and the great 'variety of vegetables' and 'garden produce'. Both officers therefore pointed out in plain words that it would not be 'desirable to shake the confidence which the presence of the principal force in Assam has already excited by retrograding to any other less advanced position'.[42] They advised the construction of a fort at Jorhat for the preservation of military stores, public buildings and treasure which were exposed to destruction in the presence of a superior hostile force. For, it was 'only in Upper Assam that we are ever likely to meet an enemy of sufficient consequence to render a series of military operations necessary'. They reiterated the importance of the proposed fort and added: 'In a country so recently brought under our rule and in effecting which we have necessarily clashed with the interests of, and arrested the organized system of plunder practised by, the influential men of the country, it seems advisable to guard no less carefully against internal treachery than external attack'.

Another station which was considered suitable for a military position was Bishnath on either bank of the Brahmaputra between Jorhat and Gauhati, a situation 'particularly happy' in so far as the protection of Lower Assam was concerned. It provided, in addition, not only facilities for communication with Northern Cachar and Jaintia in the south, but also resistance against the the raids of the Dafla tribe in the north. The Political Agent of Asssam also agreed that Bishnath was 'an admirable position for defending Lower Assam upon the boundary line of which it is situated'. But as it was difficult to protect Upper Assam from here, Jorhat was recognized as a far better position.[43]

Gauhati, the capital of Lower Assam and general residence of the principal civil authority of the province, also claimed attention. Its importance proceeded from British influence in the Khasi Hills on the one hand and the establishment of permanent communications with Sylhet on the other. This increased with the improvement of communications through the Khasi tract to Cachar in the south-east and through Sylhet to Calcutta in the south-west. In view of the remoteness of Assam from Calcutta it was suggested that provision should be made both at Gauhati and Jorhat to meet on the spot demands of emergencies.

The military force that was available for the protection of the entire province in 1832 consisted of the following categories:[44]

Assam Light Infantry	937
Mounted troops	23
	960
Auxiliaries (Upper Assam)	
Beesa Gaum (Singphos)	80
Khamti Militia	200
Gunboats at Sadiya	20
Bara Senapati (Moamaria Chief)	300
Upper Assam Militia	500
	1100

Lower Assam *Auxiliaries*

Rangpur Sebundies	200
Assam Militia	200
Cachari Sepoys	65
Mauns	150
	615

Of the above, only the Assam Light Infantry units of about 80 men each, could be expected to provide safety 'in the event of another invasion from *Ava*'.[45]

Gauhati	3 companies
Sadiya	2 companies
Headquarters (Jorhat)	6 companies
Posts adjacent to HQ	1 company

The Auxiliaries were a body of irregulars supplied by such tribal chiefs as could be trusted and managed with care and supervision. Some, like Beesa Gaum, the Khamtis and Moamaria chief were even provided with muskets. The Moamarias under Bara Senapati, for instance, commanded the passes from Ava into Assam and held the area by feudal tenure on military service. The Khamti Auxiliary had 200 muskets, and their chief was likely to be attached to the British in the protection of the passes from the Patkai range into Assam on the Sadiya border. The Khamti men were likely to be disciplined sooner for regular military service than those of other chieftains. They were in fact drilled four months in the year because of the location of two companies of light infantry at Sadiya. The Beesa Gaum or the Chief of the Singphos on the eastern-most part of the Sadiya frontier had about 60 to 80 muskets with which he armed his followers who were 'excellent marksmen'. The Assam militia, however, was a mere rabble employed for the most part on civil duties.

Provision existed with each of the stations of the Assam Light Infantry for detachments under a subedar to keep any adjoining tribal tract under proper check and control. Under the charge of an adjutant, the headquarters at Jorhat had a detachment of regular artillery with two brigades of six pounder guns drawn by bullocks, a troop of cavalry with 30 men mounted on ponies formerly used for drawing the guns. The detachment at Bishnath consisted of 24 men under a subedar to guard supplies and protect raiyats (tenants) in the neighbourhood of the Dafla duars. Two of the three companies stationed at Gauhati were meant specifically for the protection of Lower

Assam. The detachments annexed to Gauhati were designed to secure the protection of the country in the vicinity of the Khasi duars as well as the market places and duars along the Garo frontier west of Goalpara. Assamese militia *paik* sepoys were in addition maintained for the discharge of jail and thana duties under the orders of the magistrate at Gauhati.

But the Gurkhas were rated far superior to any other group constituting the Assam Light Infantry. They came from the hills in the district of Goalpara and it was David Scott who introduced them into the service. Matthee, who had been an adjutant to the Assam Light Infantry since 1829, declared the Gurkhas to be the 'most fit class for Assam duty', a group which could preserve the efficiency of the corps. He observed: 'It is not only the goodness of their constitution and disposition that renders this class so valuable, but the circumstances of their colonizing their families in the Province makes them doubly efficient, as they never avail themselves of what every other class does, and which is of the first importance in Assam where our military force is so scanty, their furlough, as they are never beyond a year absent from Headquarters where they join all that is nearest and dearest to them, their wife and offspring'.[46]

The 'Hindustani' sepoys formed 'a considerable portion' of the Assam Light Infantry. The climate was of course inimical to their constitution. But they adapted, for, as the Political Agent himself noticed, they possessed a 'high sense of military honour and habits of fidelity to their employers', more especially as commissioned and non-commissioned soldiers, where they exhibited 'superior intelligence'.

As regards the quality of Manipuris as soldiers attached to the Assam Light Infantry, opinions were on the whole favourable. The Political Agency under which they operated described them as active and cheerful, a class which learned quickly. But they were not comparable to the Gurkhas. They were less steady and constantly quarrelled among themselves. It was therefore suggested that should they be deployed in any number, they should be placed under a European officer.

Major White, the Political Agent, did not hold a favourable opinion of the cavalry in Assam. Speaking generally, he said, 'infantry is the fittest arm for carrying on warfare in Assam where the jungly and mountainous native of the country renders it difficult for cavalry to operate advantageously; but as the Burmese were in the habit of retreating from entrenchment to entrenchment during the war, it might no doubt be found advantageous to have a small body of cavalry accompanying the corps to cut them off while retreating'.⁴⁷

Tribal Policy

British rule in Assam began not with the extermination and repression of the various hill tribes, but with a recognition of what appeared to be their long-established claims of *posa*. The British, however, viewed this collection as a form of blackmail and, as already suggested, made the payment of their dues contingent on good behaviour, which signified the supremacy of British power and its firm determination to interpose between them and the *raiyats* whom they once oppressed under the Ahom rulers. As part of policy, the hill tribes were even welcomed to settle in the plains as cultivators and peaceful tillers of the soil. The Akas and Daflas, for example, were expected to behave properly so long as their *posa* was paid and they were not unduly interfered with by forest regulations. Even with the warlike Abors, conciliation remained the basic principle of policy, a principle which, however, proceeded from a position of stength and did not omit to recognize that punishment for any outrage must be, as it had actually been, summary and severe. This was especially so when the Akas after half a century of peaceful existence rose in rebellion in 1883. Their clouds of poisoned arrows could not stand artillery fire and they had to submit absolutely, agreeing in 1884 to pay any fine imposed on them for acts of murder and rapine.

The tribes to the north of the Brahmaputra valley were doubtless sought to be kept under control and prevented from forcibly seizing men, property and goods in the plains below the hills. This was especially so with the Abors, whose frontier was not open 'to have recourse to the policy of permanent occupation and direct management'⁴⁸ in the manner in which other tribal territory could be held, occupied and managed if necessary. The object of policy to the north of the Assam valley was thus limited in the later half of the nineteenth century. While the Abor frontiers did not admit of occupation and direct management the frontiers of other tribes did not call for any interference except for the establishment of advanced security posts for a close watch. Though important in terms of India's border security and defence, the whole of the tribal country on the southern spurs of the eastern Himalayas did not come under direct political control or management. It could be secured by running roads

Territorial Reorganization for Security and Development   149

into the interior of the border country. But that involved enormous cost at the time, when a demand for communications in the settled districts was pressing hard for recognition. The necessity of such a control over the whole of the northern hill tracts from Tawang to the Mishmi country arose when Chinese arms effected a forced entry into Lhasa in 1910 and emerged as a serious threat to peace along the entire range of the Assam Himalaya. A policy of permanent occupation and direct management was, however, carried out successfully in the Naga, Garo, Khasi, Jaintia and Chittagong Hill Tracts where the British continued to act with firmness and benevolence to secure the twin ends of defence and development.

The necessity of more stringent control and firmness arose from measures to promote the economy of the province. There was, for example, a noticeable increase in the commercial relations of British subjects with the frontier tribes living on the borders of the administered territory. There was constant interference on the part of speculators with the revenue derived by the government from the India-rubber forests in the plains beyond the settled mahals in the district of Lakhimpur, involving disturbances with the hill tribes beyond British jurisdiction. There was, above all, the spread of tea-gardens outside the fiscal limits of British administration. These and other entrepreneurial developments involved the government in difficulties with the hillmen.

The spread of tea-gardens was by far the most important matter that demanded the special care and protection of the government against any tribal interference. The genuine tea-shrub grew wild in Upper Assam from Sadiya to the Chinese province of Yunnan, where it was cultivated for its leaf. David Scott, who first discovered it in Manipur in 1826, and Jenkins, reported the existence of real tea in 1834.[49] The Assam Company was organized in 1839 with the support of wealthy proprietors in England and India. The initial difficulties were numerous. There was a dearth of local labour, and, in the absence of complete security, occasional embarrassment occurred in placing funds at the plantations when the local government was short of resources. Attempts were made to bring labour from Bengal and the upper provinces, but their frequent breach of engagements and reluctance to remain left the difficulties unresolved until the establishment of steam communications between Calcutta and Gauhati in 1847. The Governor-General's visit to Assam in 1853 led to the extension of the railway terminus from Gauhati to Dibrugarh (Lakhimpur District), the seat of operations of the Assam Company.⁵⁰ It doubtless increased transport facilities, but the old problem of interference by frontier tribes remained unabated. It was necessary for the government to assume special powers and lay down special rules, and this was done in 1872–3.

The Angami Nagas between Manipur and northern Cachar constituted the major menace. As many as ten expeditions were led into their country up to 1850, but they remained unsubdued. The policy of conciliation did not seem to work, and when it had failed completely, a policy of non-interference emerged as a patient alternative. It meant a British withdrawal from every attempt to establish intimate relations with the Nagas. They could come in peace to attend markets in British territory, but British officers were not to visit their country without a pass or prior permission. This could be compared to the 'close-border' policy of the North-West Frontier. The Nagas, however, viewed non-interference as an expression of weakness, and the policy failed signally. There was no alternative left for the English but to stand forth as the governors or advisers of each tribe in the new province.

A forward policy was laid down by the Lieutenant-Governor, Sir Cecil Beadon, in 1862. It meant direct control and personal rule, supported by firmness and military strength. However, this was to be viewed not as a negation of the conciliatory element, for, it was argued,

> It would be a mistake to suppose that to inflict condign punishment for exceptionally gross outrages is any departure from a general policy of conciliation. To submit to outrage is not to conciliate, but to provoke further attack. But punishment has never, with the sanction of Government, taken the form of mere reprisal. Government has never sent out raiding parties to burn indiscriminately Naga villages. Its first aim has always been to discover the actual parties concerned in the raids on British territory, and then it has endeavoured to confine the punishment to those so offending. The policy of a Government is not to be learned from a single incident in its history. It must be viewed as a whole in the light of its acknowledged aims and motives.⁵¹

Despite the declared policy of direct control over the Angami Nagas after 1862, that policy thus remained for a number of years far too conciliatory. British officers were not even formally allowed to assert themselves as representatives of a paramount power to punish inter-tribal outrages and massacres. And when in 1870–1 the Government of India made a policy statement it stated in unmistakable terms that it would not assert a positive jurisdiction over communities not actually within the limits of the settled districts. It was only with the establishment of the Chief Commissionership of

Territorial Reorganization for Security and Development 151

Assam and the immediate subordination of the Assam administration to the Supreme Government that the views of the local officers began to carry the importance they deserved. In 1877 Lord Lytton, the Viceroy, and Lord Cranbrook, the Secretary of State, finally decided to advance into the Naga Hills to a central and dominating position in the midst of the warring Angami clans. The result of this was the occupation of Kohima in 1878 and the formation of the Naga Hills into a British district in 1881. The process of moving farther and farther east and putting more villages under direct management was almost unending, 'based on inescapable facts'. By 1942 the eastern boundary of the Naga district was more stabilized than ever before, 'though there was always left a large tract of unadministered territory running up to the Burma border, which might eventually be included'.[52]

South of the Naga Hills lay Manipur, a small dependent state under the rule of its hereditary Maharaja, with a central plain within a range of the mountains between India and Burma. Inhabited by Manipuris proper with their capital at Imphal, there was a larger surrounding tract of forest and hill where lived a mixed lot of Kukis, Nagas and some other tribesmen, whom the orthodox Hindu Manipuris treated as altogether lower creatures. Though annexed to Burma in the middle of the sixteenth century, Manipuri raiders soon became independent and a terror to Ava, carrying off loot, cattle and thousands of people under their Raja Gharib Nawaz (1714–54). Later, however, the Burmese defeated the Raja who fled to Assam. They raised their own nominee to the throne in 1770 and returned. With Cachar as his base, the rightful Raja made several attempts between 1775 and 1782 to oust their nominee. He was finally left in possession, because the country was so devastated that no real gains were possible. The question of suzerainty nonetheless remained undecided and the first Burmese War decided it in favour of the English in 1826. Raja Gambhir Sing of Manipur owed the restoration of his raj to British arms alone.

In a minute recorded on the north-east frontier in 1833[53] Bentinck observed on the basis of inquiries that, despite uninterrupted peace for seven years, Manipur remained unable to defend itself against a Burmese invasion. He thus preferred placing in it a small garrison of the Company Consistent with Bentinck's policy, a Political Agent was appointed in 1835 after Gambhir Sing's death in 1834. Manipur's subsequent history was nonetheless marked by internecine wars. In 1891, the Maharaja abdicated, a development which dictated the expediency of removing from Manipur one of his brothers, regarded as a danger to peace. J. W. Quinton, the Chief Commissioner, thus went to Imphal with an escort of Gurkha troops to secure a peaceful succession. Quinton entered the fort unarmed along with three officers to parley. But once inside they were all murdered by rebel Manipuris. The force left in the Residency retreated in despair and disorder. The punishment which soon followed, however, met with little resistance. The Manipur disaster led some to demand an immediate annexation of the state. But its strategic position along the Burmese frontier counselled restraint. Chura Chand, an obscure collateral descendant of a long-defunct Maharaja, was allowed to succeed under such terms as reduced the state to complete subordination. Even so, the Kukis and other tribes living in the surrounding hills remained a source of trouble. While the Kukis, for instance, rebelled in 1917, a semi-religious Kabui Naga movement in the hill portion of the state caused concern in 1931. Both were, however, suppressed with firmness.

The Lushai Hills formed the southernmost of Assam's hill districts. The Cachar plains in the district of Silchar lay on the west, the Chin Hills of Burma on the east, and the Chittagong Hill Tract to the south. The Lushai country had long remained an unexplored tract between the British districts of Cachar and Chittagong on one hand and the two protected states of Tripura and Manipur on the other. As with the Nagas, the first British contacts with the Lushais were unfriendly: for they, too, were constant raiders of the plains. But the knowledge gained in the course of the Lushai expedition of 1871–2 and the subsequent surveys of their country induced the Government of India to confine itself to a strictly defensive policy. It required the Deputy Commissioner of Cachar to act in concert with the Deputy Commissioner of the Chittagong Hill Tracts and the Political Agent at Manipur and Tripura, so that he might remain in constant personal communication with the Lushais who were to be induced to settle on their portion of the uninhabited tract. He was also to encourage trade at the marts chosen by Lushais and persuade their young men to spend a few years in a levy composed exclusively of hillmen belonging to the frontier. As a measure of security, however, all the outposts were to be connected initially by road with the headquarters at Silchar, and later with one another.[54] The Lushais themselves received during that expedition a salutary lesson which they remembered. And although the situation was not as stable as ideally desired by the British, the future prospects for a proper outlet for Cachar tea and Manipur products through the Chittagong tract could not be lost sight of and made the government act with moderation. The Lushai Hill area was taken directly under British administration in 1892.

Of the other tribes, the Khasis, as said before, doubtless put up initial resistance, but after the conquest of the Khasi Hills in 1833 they settled down as the most peaceable of British subjects. North Cachar came under British occupation after the death of Tula Ram in 1850. Large communities of hillmen became government raiyats, paying dues under the nominal surveillance of only a small police post. The people were later left to the control of their own headmen under the direct supervision of the Deputy Commissioner of Cachar with headquarters at Silchar. Within the British sphere of influence, the Garo Hills, too, gradually settled down to normalcy without raids or tribal feuds, as did the Jaintia Hills.

North-East Frontier under Direct Central Control

The political stability and integration of Assam doubtless proceeded from an improved communication system, a well-organized military establishment and a firm, though conciliatory, tribal policy. But by far the most effective step taken to reinforce stability was the introduction in 1874 of a system of direct central control by constituting Assam into a separate province under a Chief Commissioner immediately subordinate to the Government of India.

Before the creation of the Chief Commissionership, Assam's administration had been carried on since 1854 by the Lieutenant-Governor of Bengal. It was a huge charge extending in 1868 over an area of 240,000 square miles with a population of 40 to 50 million. In addition to Assam, it comprised the whole of undivided Bengal as well as all the territory of Bihar and Orissa containing as many as 54 big districts exclusive of a number of tributary states.[55] The increase of commerce, of European settlement and internal communication would by themselves have called for relief either by altering the machinery of executive administration or by reconstituting the territories included in it. But to these was added a considerable increase in the load of public business, evident in the number of letters received and issued increasing from 18,144 in 1844 to 62,878 in 1867.[56] The ravages caused by the Orissa Famine of 1866 established beyond doubt the failure of the Bengal Government to handle the vast area under its jurisdiction.

The creation of a separate province of Assam was, however, not without opposition. The Bengal civilians resisted any proposal that reduced their empire and the Bengali elites, too, viewed the separation of Assam from Bengal as prejudicial to their employment opportunities and commercial prospects. The remedy suggested by both the bureaucratic and popular elements of Bengal was in fact a Council form of government on the model of Madras and Bombay, with an extended secretariat to deal with the increased weight of public business. However, the Governor-General, Sir John Lawrence, was opposed to the Council form of government for Bengal: 'I cannot but doubt', he said, 'the expediency of having, on the same spot with the Government of India, a Governor-in-Council on a status similar to that of Madras and Bengal, that is, with a certain degree of independence, and with the privilege of corresponding direct with the Secretary of State. That the two Governments should be addressing the Secretary of State from the same capital seems to be *per se* incongruous and unnecessary. This would scarcely be compatible with the position due to the Government of India. In ordinary times it would cause a certain amount of friction; in times of difficulty it would produce grave inconvenience.'[57]

In another Minute of 23 March 1868, Lawrence continued his opposition to the Council form of Government for Bengal and reiterated that Assam be constituted into a separate Chief Commissionership. He went on further to suggest that, in order to give relief to the Government of Bengal, it might be possible to consider the separation even of Bihar which had 'no real affinity to Bengal proper, and which, together with some parts of the province of Benares, might be formed into a separate Lieutenant-Governorship, as was done temporarily by Lord Canning in 1857–8'.[58] Four of the seven members of the Supreme Council opposed the reproduction of the Madras and Bombay constitution in Bengal,[59] and the subject was dropped for a time. It was taken up again when Sir George Campbell took over as Lieutenant-Governor of Bengal in 1871. Lord Mayo had sought in the meantime the approval of the Secretary of State to form Assam into a separate Chief Commissionership, which the latter accordingly accepted in a letter of 15 August 1871. But since the administrative arrangement and territorial readjustment required consultation with the Lieutenant-Governor of Bengal, the decision of the Secretary of State came into effect only in February 1874.

The territories brought under the jurisdiction of the Chief Commissioner included Kamrup, Darrang, Nowgong, Sibsagar, Lakhimpur, the Garo, Khasi and Jaintia and Naga Hills, Cachar and Goalpara. The district of Sylhet was in the same year added to it. The total area amounted to 54,000 square miles with a revenue of Rs 52½ lakhs.

The original plan of Lord Mayo had sought to include in Assam the entire northern frontier of Bengal from the western boundary of Darjeeling as well as the districts of Jalpaiguri and Cooch Behar. But Bengali opposition stood in the way and these districts were left with the Government of Bengal. The military defence of Bhutan, however, dictated the necessity of the Buxa Duar being united with the Eastern Duars which were attached to Goalpara district. Though administered by its own raja, the territory of the Tripura Hills was also brought under the supervision of the Chief Commissioner, to enable him to form a united Lushai policy which called for a unity of control over such other tracts as Sylhet, Cachar and Manipur. But the Chittagong Hill Tract, which was administered by a deputy commissioner under the supervision of the Commissioner of Chittagong, remained with Bengal in keeping once more with respect for Bengali sentiment. Sylhet, though a permanently settled district, was included in Assam because of its vulnerability to Lushai invasions as well as from considerations of its tea interest, which stood on the same footing as that of Cachar.

The Assam case thus provided clear proof of the fact that when a subordinate government needed to be strengthened to deal with a state of emergency, it had to be placed immediately under the Supreme Government, for a strong centre alone could provide against weakness or breakdown of local authority. Another conclusion that emerges from the constitution of Assam into a separate province is that, apart from considerations of defence and economic development, the question of territorial reorganization now involved a number of other dimensions, such as language, race and culture, land systems, past history and popular sentiments based on power equations and employment privileges. It was these extra-administrative considerations that delayed the inclusion of Sylhet in Assam, dropped altogether the inclusion of Chittagong and the northern districts of Bengal, and maintained the jurisdiction of the Calcutta High Court in the newly constituted Province of Assam.

Administratively, the non-regulation principle did not constitute the basis of the scheme originally forwarded to the Secretary of State on 4 April 1871. The non-regulation pattern of Assam was finally adopted at a conference between Sir George Campbell and Sir John Strachey, Home Member of the Central Government.[60]

The revised plan provided neither for divisional commissioners nor for a judicial commissioner to assist the Chief Commissioner, who was to have full control of the entire Assam administration, exercising all the powers previously vested in the Board of Revenue and the Nizamat Adalat. Each district had a deputy commissioner, who also acted as the district judge and was assisted by a subordinate judge only in Assam proper, Sylhet and Cachar, but not in the Naga and Khasi Hills, where district officers exercised both executive and judicial functions. They were in fact their own superintendents of police in the tribal areas. The whole administration of Assam was thus reconstituted on a principle conducive to territorial integration for political stability, necessary both for peace and progress. It was designed to provide relief to the Lieutenant-Governor of Bengal and promote Assam's development under the direct control of the Government of India, which alone could take a broad all-India view and expedite measures to ensure effective security along the northeast frontier.

In a Note of 12 August 1879 Sir Steuart Bayley, the Chief Commissioner of Assam, indicated to the Government of India the existing strength of the military force stationed in Assam and discussed at length the changes he proposed by relieving the military of all frontier outpost duties and transferring these to the Frontier Police of 3,000 men.[61] Under the existing system the distribution of the central reserve of the headquarters consisted of four Indian regiments, each 800 strong, and governed by the double necessity of guarding and strengthening numerous posts along the frontier and at the same time being ready to undertake frequent punitive expeditions.

Of the native regiments of the central reserve, two were stationed at Shillong, centrally situated and equidistant from the northern, southern, and Naga Hills frontier, though not within easy reach of them. The headquarters of another regiment was at Dibrugarh, the highest point on the Brahmaputra to which the river steamers could ply. The fourth regiment was stationed at Cachar (Silchar), the furthest available spot on the Surma river. The next in the chain were the local reserves of detachments from these regiments, which were stationed at Gauhati, Jaipur, Golaghat and Sadiya. The number of frontier outposts, according to Bayley's Note, was 50, and these were manned partly by the frontier police and partly by the military, the former being drilled and armed with rifles and recruited mainly from the Nepalese or tribes inhabiting the hills and slopes of Assam.

Organizationally, however, the existing security arrangement was found wanting in administrative unity: the outposts were garrisoned by men under separate civil and military establishments where each post, instead of being a link in a chain, communicated with its own military or civil headquarters. From a military point of view the objections were even greater. The distribution of regiments over several small frontier outposts was viewed as highly prejudicial to the efficiency and discipline of the armed force. For small detachments of rarely more than 50 men, and in some cases as few as 15, were sent off to distant and unhealthy areas where they remained sometimes for as long as six months at a time. The commanding officer naturally complained that it was sometimes impossible to maintain proper standards under conditions of enforced idleness. Bayley, therefore, recommended that in view of the special experience and greater mobility of the frontier police in jungle warfare, the strength of the armed force in Assam be reduced from four regiments to two and of the armed police raised to 3,000.[61]

The Army Commission fully accepted the view that the frontier outposts in Assam should be manned by the police. It was also agreed that the frontier police force, which was known for its mobility in the hills and forests of Assam, should be increased. But the Commission did not support Bayley's proposal to reduce the military garrison of Assam to two regiments of 800 each. It advised the government to leave at least two and a half battalions of native infantry in Assam, each battalion being 912 strong, and to place in addition a division of mountain artillery, including two guns with 40 men. It also proposed to the Chief Commissioner to keep sufficient standing government carriage, elephant, mule, and coolie corps to move out half the force, fully equipped, at 24 hours' notice.[62]

Bayley's Note contained a perceptive, though brief, analysis of Assam's geo-political situation, more especially in terms of the hill tribes who posed problems of both security and defence, not only along India's eastern frontier but also along the northern slopes of the eastern Himalaya. Ever since the conquest of Assam those problems had been engaging the serious attention of the government as part of its north-east frontier policy. The annexation of upper Burma in 1885 provided the answer to the Burmese tangle. But the Brahmaputra remained unresolved until a solution was sought to be evolved in the course of the Simla Convention of 1914.

The task of development, however, remained for the most part in a state of neglect. Apart from the limitations imposed by physical conditions, human resources and tribal disturbances, by far the most serious problem was the absence of a regular service cadre for the newly constituted province, a cadre which might be invested with the planning and execution of development schemes. This is perfectly clear from a confidential note from J. B. Fuller, the Chief Commissioner of Assam, to the Private Secretary of the Viceroy as an enclosure to a letter of 5 April 1904.[63]

According to the note, Assam received its officers from Bengal 'on loan' and 'not as a gift'. Worse still, the terms on which the new province received 'Covenanted Civilians from Bengal offer the Bengal Government an irresistible temptation to plant upon [it] its undesirables'. The terms of the loan were one-sided and any officer could at any time after five years go back to his province without being compelled to return to Assam. The result was that Assam would always lose the services of senior and experienced officers who saw prospects of advancement before them in Bengal. Since Assam was a small Commission with limited scope for promotion, and since even this limited scope was often blocked by the unusually rapid promotion of military officers of the Commission, there was a regular return to Bengal of the more senior and able men, with the result that Assam retained those who happened to be 'inefficient' and undesirable.

The development of Assam thus necessarily involved the expansion of its Commission, for which inclusion of the adjoining Bengal districts of the Chittagong Division, was originally suggested in 1896–7. Considering the deplorable state of the civil service, Denzil Ibbetson, the Home Member of the Government of India, had already recognized the greatest need in Assam for development, a task which could not be accomplished except by an expansion of its Commission by including certain districts which formed part of Bengal. Ibbetson observed:

My conclusion is that agitation [against partition] notwithstanding, we should proceed with our proposal. The need of Bengal for relief is great, but the need of Assam is still greater. I did not realize fully, till I had visited Assam and talked over matters with Mr Fuller, how absolutely impossible the present situation is. He cannot censure an officer, or refuse him leave, or transfer him to a district which he dislikes, or refuse him an appointment which he wants without running the risk of an immediate demand to revert to Bengal, the vacancy being filled by one of the Bengal failures who knows nothing whatever of Assam. It seems to me that the administrative necessities of the [Assam] case are sufficient to justify us in overriding the opposition which our proposals have aroused.[64]

Assam's development needs had in fact been eloquently expressed

Territorial Reorganization for Security and Development 159

earlier in a letter of 3 December 1903 addressed by H. H. Risley, Secretary to the Government of India, to the Chief Secretary, Government of Bengal.

It requires territorial expansion in order to give to its officers a wide and more interesting field of work. It requires a maritime outlet in order to develop its industries in tea, oil and coal. The paying positions of the Assam-Bengal Railway are in the south, and the whole line, if it is to be utilized in the interests of the province, ought to be under a single administration. Assam moreover will continue to be handicapped, so long as it is dependent for its service upon what it may be fortunate enough to borrow from Bengal. A province that can only offer the prize of one Chief Commissionership, that is remote in locality and backward in development and organization, will not attract the highest type of civilians to its employ. The Government of India regard it indeed as incontestable that, with a service recruited as at present and confined within the present limits, Assam will find extreme difficulty in attaining the level of a really efficient administration; and it is for this reason that, in considering the question of changes, they are impressed with the paramount necessity for making them on such a scale as will remove this fundamental source of weakness, and will, if possible, give to Assam a service of its own, offering a career that will attract and retain men of ability and mark. No temporary opposition in the transferred towns or areas, no artificial agitation or interested outcry, should in their opinion be permitted to divert the efforts of Government from the main object, viz., the creation of Assam into a vigorous and self-contained administration, capable of playing the same part on the North-East Frontier of India that the Central Provinces have done in the centre, and that Punjab formerly did on the North-West.[65]

While planning the redistribution of territories between Assam and Bengal, Curzon was guided not only by the prime consideration of providing Assam a self-contained and independent service of its own, but also of improving the efficiency of the administration of Bengal by reducing the extent of its territorial responsibilities. He thus felt that the transfer to Assam of the entire Chittagong Division with the Tripura Hills would neither afford relief to Bengal from its existing excessive burden, nor adequately improve the Assam services. The Government of India therefore proceeded with a further suggestion, namely, the proposal to incorporate with Assam the neighbouring districts of Dacca and Mymensingh.[66]

In well-argued and elaborate Minutes of 19 May and 1 June 1903 Curzon sought to explain his scheme of territorial reorganization.[67] While its scope extended beyond Bengal proper and Assam to such other provinces as Bihar, Orissa, the Central Provinces, Bombay and Madras, the principles underlying his suggested reorganization extended far beyond considerations of administrative convenience.

It included such other dimensions as area, population and financial resources for development; land tenures, customs and economic development; and usages, race, language and culture. The Minutes and the proceedings with Notes[68] on territorial redistribution in India by J. P. Hewett (28 August 1902) and by Denzil Ibbetson (23 April 1903) are not only highly educative and informative, but extremely interesting and useful. However, as the expansion of the Assam Commission is of immediate relevance to our discussion here, only Curzon's reasons for the merger of Dacca and Mymensingh are quoted. He noted:

Geographically, Dacca and Mymensingh are separated by a clear line of division, viz., the main channel of the Brahmaputra river (though it is not here called the Brahmaputra) from Bengal. If they are joined to Assam, the latter will possess a definite and intelligible western boundary, whereas if the Chittagong Division goes to Assam, and Dacca and Mymensingh are left with Bengal, the two latter divisions will constitute a projection from the main body of Bengal obtruding itself into the heart of Assam, from which they will be separated by no ties of origin, language, religion, or administration. Not only would the transfer enable Assam to obtain an independent service, but that service would possess three separate commissionerships, which would be its prize appointments. These would be (1) the Brahmaputra valley or Assam proper; (2) Dacca, to which would be added Sylhet and Cachar; and (3) Chittagong.[69]

In anticipation, Chittagong had in fact already been constituted a port of Assam. Curzon naturally averred that it would equally serve as the port of the adjoining districts of Dacca and Mymensingh, of which it was a natural commercial outlet. For though their past associations had been almost exclusively with Bengal, the connexion between them and Calcutta was 'both arbitrary and unnatural' because of 'the numerous intervening rivers rendering communication difficult and slow'.[70] Curzon thus justified the inclusion of Dacca and Mymensing as part of the scheme originally envisaged in his Minute of 1 June 1903.

This proposed transfer of Dacca and Mymesingh, which was communicated to the Government of Bengal under Home Secretary Risley's letter of 3 December 1903, became the immediate cause for organized Bengali opposition to a measure designed primarily to improve the Assam services and relieve the Bengal Government of its excessive burden. The educated elites of Bengal confronted the Government of India in January 1904 with a series of representations; the agitation organized tended in effect to impart to an administrative measure a political dimension.

Once political considerations were allowed to influence the proposed administrative reconstruction, there developed a shift of emphasis from the original plan of erecting an independent service of Assam to the creation of an independent sphere of Muhammadan influence in Eastern Bengal with its capital at Dacca. Despite opposition, the Viceroy went ahead with his plans. In February 1904 he visited Chittagong, Dacca and Mymensingh and, in the course of a series of speeches, foreshadowed the willingness of the government to consider a wide scheme involving the creation of a Lieutenant-Governorship with a Legislative Council, an independent Revenue Board and the transfer of as much territory as would be required to justify the institution of a fully equipped administration suitable for the dignity and status of the projected Lieutenant-Governorship. As the Government Resolution of 19 July 1905 acknowledged, the object of the enlarged scheme involving the establishment of a Lieutenant-Governorship was to satisfy 'the feelings of those who were alarmed at the possible deprivation of privileges, which they had for long enjoyed and to which they attached a not unnatural value'.[71] This new move was thus not only a distinct departure from the original policy laid down in the Home Department letter of 3 December 1903, but also a political device to weaken the hold of Calcutta politics by creating a separate Muslim sphere of influence and power at Dacca in Eastern Bengal.

In its final form the enlarged scheme created a new province with the status of a Lieutenant-Governorship consisting of the Chittagong, Dacca and Rajshahi Divisions of Bengal, the district of Malda, the state of Hill Tripura, and the erstwhile Chief Commissionership of Assam, and which proposed, with the sanction of the Government of India, the creation of a new Commissionership out of the Surma Valley districts and Manipur. Darjeeling remained with Bengal. The new province was called Eastern Bengal and Assam, with its capital at Dacca and subsidiary headquarters at Chittagong. The Government of India had of course proposed that the new province should be called 'the North-Eastern Provinces', but in his letter of 9 June 1905 the Secretary of State rejected that proposal and preferred to call it 'Eastern Bengal and Assam' on the ground that the important commercial interests represented by the tea industry would complain if the name of Assam, now so widely known in world markets as the chief source of India tea, was to disappear from the list of Indian provinces'.[72] The new unit comprised an area of 106,540 square miles and a population of 31 million, of whom 18 million were Muslims and 12 million Hindus. It was given a Legislative Council and a Board of Revenue of two Members, with the jurisdiction of the High Court of Calcutta left undisturbed. The existing province of Bengal was left with an area of 141,580 square miles and a population of 54 million, of whom 42 million were Hindus and 9 million Muslims. The influence of Hindu Bengalis with the Government of India at Calcutta was, however, not without avail. The slogan of Bihar for the Biharis remained a voice in the wilderness and Bihar was not separated from Bengal. Bengal gained, in addition, some area to the west from the Central Provinces. Its Oriya-speaking district of Sambalpur, and the five Oriya states of Patna, Kalahandi, Sonpur, Bamra and Rairakhol were all annexed to Orissa which, like Bihar, continued to be governed from Calcutta, despite their linguistic and ethnic distinctiveness.

The original scheme of 3 December 1903, as said before, did not extend beyond the merger of the Chittagong Division with Hill Tripura and the districts of Dacca and Mymensingh. The addition to these the districts of Pabna, Bogra and Rangpur as well as the northern districts of Rajshahi, Dinajpur, Jalpaiguri, Malda and the state of Cooch Behar provided a clearly defined western boundary consistent with geographical, ethnic, social and linguistic demands. But, as the Chief Commissioner of Assam had earlier feared, it was an 'annexation of Assam by Eastern Bengal', a concentrated Muslim Bengali belt, possessing a relatively higher level of economy, education and administration. The Chief Commissioner of Assam had earlier determined to 'close the Brahmaputra valley to outsiders' and his Secretary's letter of 6 April 1904 described them as 'foreigners' said to have held up the progress of Assam.[73] He was reported even to have passed an order that 'all district appointments shall be filled up from residents of the district'. This clearly meant that 'if the people of the transferred districts could get no appointments outside of that area, they would lose heavily'.[74] All fears of such loss thus disappeared with the creation of the new province of Eastern Bengal and Assam under a Lieutenant-Governor. The 'outsiders' continued to enjoy their privileges at the cost of 'residents'.

However, the enlarged scheme furthered British interests in both political and economic terms. It concentrated in a unified province the typically local Muslim population of Bengal, with Dacca as a separate, though natural capital, and the entire tea industry, with the exception of the Darjeeling gardens, as well as the greater part of the jute tracts, under a single Lieutenant-Governorship with the long-established Divisional Commissionerships remaining undisturbed. This arrangement was especially beneficial to the tea planters, both European and Indian. It associated under a single government the tea-growing areas supplied by free labour with those worked by indentured labour. The economic resources of the new province as well as its administrative self-sufficiency and commercial outlets were even otherwise viewed as a bulwark of defence against any attack on the north-east frontier of India.

Administratively, the entire deliberations which resulted finally in the formation of the new province of East Bengal and Assam signified a new trend. They indicated a growing emphasis on the reduction of an administrative charge as a means to promote efficiency as well as security and development. The Orissa Famine sparked the idea of separating Assam from Bengal. But there were other economic dimensions which called for a basic change in the formulation of administrative policy, both functionally and structurally. These will be noticed in the chapter that follows.

Administrative Pattern for Tribal Areas

The Punjab case was recognized as a model for the administration of unsettled and tribal areas in terms of both security and development. The annexation of Punjab in 1849 was followed by the appointment of a Board of Administration, functioning directly under the Government of India, with Lieutenant-Colonel Sir Henry Lawrence as President, and his brother, John Lawrence, and Charles E. Mansel as its Members. The administrative hierarchy consisted of the President on top, followed by Commissioners at the divisional headquarters and Deputy Commissioners in the districts, equivalent to District Magistrates and Collectors in the regulation provinces. In addition, there were Assistant Commissioners, comparable to Joint Magistrates, and Extra Assistant Commissioners, a class of uncovenanted servants called *adalutees* and *kardars*, the latter being comparable to *tahsildars*, the highest grades of Indian officials in the North-Western Provinces.

The principle on which Henry Lawrence reconstituted his administration was that of a non-regulation territory and, based on the combination of powers where every civil functionary, from the Board of Administration to the *kardar*, was vested with judicial, fiscal and magisterial powers. It was consistent with local tradition and tribal practice. It ensured ready and simple justice without the observance of such forms and technicalities as fettered the regular courts of law in the regulation provinces, where prosecution was made extremely difficult. And above all, it provided for the strong administration needed for a newly-conquered and outlying frontier province inhabited by several mutually exclusive tribes posing problems of security. The administrative system so introduced was, in addition, invested with the conduct of political relations with protected Sikh states as well as the tribes of the north-west frontier.

The non-regulation form of government continued to function for four years under the able guidance of Sir Henry Lawrence whose later appointment as Resident at Hyderabad called for a change. In a Minute recorded on 26 January 1853, Dalhousie recommended that the Board be abolished and that its governing authority placed in the hands of a single officer to be styled 'Chief Commissioner'. The Chief Commissioner to be so appointed in place of the Board was in all respects to 'occupy the same position and discharge the same function as the Board of Administration'.[75] He was to be assisted by a Judicial Commissioner, comparable to the Sadar Adalat, and a Finacial Commissioner, who functioned as a Board of Revenue. Under a Resolution of 4 February 1853, the Board was abolished and John Lawrence appointed the first Chief Commissioner. The appointment so made by the Government of India was regularized in 1854 by an Act of Parliament, which empowered the Governor-General in Council, with the sanction of the Home Government, to take any territory in British India under his management, and then to provide for its administration. It was under this statutory provision that Assam was separated from Bengal in 1874. A Chief Commissioner was regarded as administering his province on behalf of the Governor-General in Council who might resume or modify the powers he had himself delegated. The Chief Commissioner therefore enjoyed a status lower than that of a Lieutenant-Governor, whose appointment proceeded from the Crown as a statutory obligation.[76] The object of a Chief Commissionership was in fact to provide for a smaller charge than that of a Lieutenant-Governor, so that a Chief Commissioner had a closer knowledge of his people and officers, and a firmer personal grip on the entire administrative machinery. There was to be no legislative council in a Chief Commissionership.

The Punjab situation, however, changed and the burden of the Governor-General in Council considerably increased with the annexation of the Kingdom of Pegu in 1852 and of Oudh in 1856, with the acquisition of Bhonsle's territory in Nagpur in 1854 and the transfer of parts of Berar to the Company in 1853 by the Nizam. Functionally, too, the burden increased. The whole direction of postal communications was vested in the Governor-General in Council himself under an arrangement made in 1854. Added to this were the control of the telegraph system in India, all the railways, the Public Works Department established in 1855, the expansion of education under Wood's plan of 1854, and the elaboration of the Legislative Council provided by an Act of Parliament just a year before. In a Minute of 25 February 1856, therefore, the Governor-General proposed that Punjab should be formed into a Lieutenant-Governorship,[77] a proposal which, on account of the outbreak of the Mutiny in 1857, remained for a time inoperative. It was by a Foreign Department Notification of 1 January 1859 that the proposed Lieutenant-Governorship was actually constituted and Sir John Lawrence appointed as the first Lieutenant-Governor.[78] The North-West Frontier Province, which formed part of the Lieutenant-Governorship, had nonetheless to be carved out as a Chief Commissionership in 1901 to provide a close grip over the Pathan tribes.

The formation of the Local Government of the Central Provinces in 1861 was a better example of a Chief Commissionership as the most suitable form of government for backward tribal tracts, administered on a non-regulation principle by both civil and military officers who from top to bottom united in their office revenue, police, judicial and magisterial powers. It had earlier formed part of the territory belonging to Raghuji Bhonsle, the Maratha Chief of Berar, who reduced the Gond king and established himself at Nagpur in 1743. The Nagpur kingdom included under Raghuji II, a lineal descendant, the whole of the Central Provinces and Berar, besides Orissa and some of the Chota Nagpur states.

In 1803 Raghuji II joined Sindia in the second Maratha War against the British. On his defeat he was obliged to accept a British Resident and cede Cuttack, Sambalpur and part of Berar. When he died in 1816 his son was murdered by Mudhoji, who assumed powers and joined the last Maratha War against the English in 1817. Defeated, he lost to the Nizam the rest of Berar and to the British his territories in the Narbada valley. 'The Saugor and Nerbudda Territories', which were included in the North-Western Provinces in 1835, were an important result of the British victory of 1817. Appa Sahib was for a time reinstated but was soon deposed. A grandchild of Raghuji II was placed on the throne to reign as Raghuji III until 1853, when he died and his territories were declared to have lapsed to the paramount power.

Geographically, the Saugor and Nerbudda Territories conjoined the Nagpur area and formed a compact area. But administratively they remained under two separate charges, one under the Lieutenant-Governor of the North-Western Provinces at Agra and the other under the Supreme Government at Calcutta. It became difficult to control the northern districts which were seriously disturbed during the Mutiny in 1857. By a Foreign Department Resolution of 2 November 1861, the Governor-General in Council therefore decided that the Nagpur area be joined with the Saugor and Nerbudda Territories and that the whole tract be constituted into a separate Chief Commissionership of the Central Provinces.[79] While the former comprised the districts of Bhandara, Chand, Chindwara, Nagpur, Raipur and Sironcha with the dependencies of Bastar and Karonda, the latter included Betul, Damoh, Hoshangabad, Jabalpur, Mandla, Sagar and Seoni. The structure of administration followed the pattern already established in Punjab. A readjustment of certain districts took place after the creation of the new province,[80] but the Punjab non-regulation pattern of district administration remained unaffected under each of the three Divisional Commissioners with their headquarters at Sagar, Jabalpur and Nagpur. A separate Division of Chattisgarh was created on the transfer of Sambalpur to the Central Provinces in 1862. Yet another Divisional Commissionership was added with that of Berar in 1903.

By 1905 the new territory, called the Berar Division, formed one Commissionership consisting of the districts of Amraoti, Ellichpur, Wun, Akola, Buldana and Basim, each under a Deputy Commissioner. As in other districts of the Central Provinces, the Deputy Commissioners in Berar were assisted by Assistant and Extra Assistant Commissioners, all of whom combined civil, magisterial and revenue functions. The tahsildars at the bottom also exercised revenue and criminal powers. There was a Superintendent of Police who functioned under the control of the Deputy Commissioner. The civil judicial function of each district vested in a district judge who also exercised the powers of a Sessions Judge. Appeals from the district judicial authorities lay with the Additional Judicial Commissioner at Nagpur, who acted as the provincial High Court. Under the arrangement of 1861 a Divisional Commissioner had the powers of a Civil and Sessions Judge in his judicial capacity and an appellate jurisdiction in revenue cases. Apart from a union of functions, the Chief Commissioner on top was also to be Agent of the Governor-General, a purely political function directed towards the proper conduct of relations with Indian states. With this unified system of government, a Chief Commissionership was considered best suited for the political stability and development of backward areas with largely tribal populations.

The annexation of Oudh was likewise followed by the establishment of a Chief Commissionership. In 1856 Major-General Outram, Resident at Lucknow, was appointed 'Chief Commissioner for the Affairs of Oudh and Agent to the Governor-General' with powers to conduct operations for the occupation of that province. As in the Punjab case, the administrative hierarchy introduced consisted of a Chief Commissioner on top, with a Judicial Commissioner and Financial Commissioner immediately under him, followed by Divisional Commissioners and then Deputy Commissioners with the usual complement of Assistant and Extra Assistant Commissioners in the districts. The tahsildari system was at the bottom renovated by appointing a tahsildar over a jurisdiction of two to three thanas and investing him with powers to collect revenue, to preserve the peace, to administer justice, and, finally, to execute all orders received from the government. In its letter of 4 February 1856 the Government of India justified the institution of a Chief Commissionership on the grounds of a state of emergency arising from the occupation of what was then a 'foreign' territory.

Rising Trend of Language Politics

The personal rule of a strong, honest and upright Chief Commissioner was doubtless ideal for the task originally contemplated. But the downward filtration of education as well as the freedom of the press and the advancement of communications had long been creating conditions where the change or progress had to meet the articulated needs of the people to be governed. In other words, there was a perceptible shift of emphasis from purely administrative considerations to popular wishes as reflected through the columns of newspapers. The agitation against the partition of Bengal marked an effective and organized beginning of this political trend which dwelt on the unity of the Bengali language and refused to look into the merits of the administrative measure that cut across linguistic barriers. The question of an appropriate administrative set-up for tribal development or the welfare of backward groups was altogether brushed away before the rising tide of language politics, which gave rise to all kinds of practical complications. For it was not a simple question of language, but involved deeper questions of educated employment, power and political influence.

The Bengali Hindus who had a virtual monopoly of professional and political influence at Calcutta, found with the partition of Bengal a sudden reduction in that influence, a considerable weakening of their hold, and a shrinkage of their employment opportunities in the new province, where by virtue of their majority the Muslims who belonged to the same language group, were bound to have a more legitimate claim to local recognition. The Bengali Hindus thus organized the *swadeshi* movement to boycott British goods, and some of their numbers took to violence to mark the beginning of revolutionary activity. Politics thus proceeded to black out administrative issues of reform for tribal welfare, or the welfare of the weaker sections of society. It is true that some Muslims in Calcutta also lent support to the anti-partition agitation. But such instances were few and far between. The boycott and *swadeshi* movement remained for all practical purposes a Bengali Hindu movement, which raised another question, the question of Hindu-Muslim communal politics, a backward-looking revivalist trend in no way concerned with the development of backward groups.

Mainly three issues were involved in the agitation against the partition of Bengal: the demand of the Bengali Hindus for the unity of the Bengali language, the emergent claims of Muslims for special recognition as a majority community, and Hindu-Muslim communal politics arising from the partition, which was supported by the East Bengal Muslims, its main beneficiaries. The voice of the Assamese who had been tacked on to Eastern Bengal was still too weak to matter.

The situation which called for an answer to the issues so involved in the partition, emerged from the enlargement of the Legislative Councils effected under the Indian Councils Act (1909). This was made clear to the Secretary of State in a letter of 25 August 1911 from the Governor-General:

In the pre-reform scheme days the non-official element in these Councils was small. The representation of the people has now been carried a long step forward, and in the Legislative Councils of both the Provinces of Bengal and Eastern Bengal, the Bengalis [Hindus] find themselves in a minority, being outnumbered in one by Biharis and Oriyas, and in the other by the Muhammadans of Eastern Bengal and the inhabitants of Assam. As matters now stand, the Bengalis can never exercise in either province that influence to which they consider themselves entitled by reason of their numbers, wealth and culture. This is a substantial reason which will be all the more keenly felt in the course of time, as the representative character of the Legislative Councils increases and with it the influence which these assemblies exercise upon the conduct of public affairs.[81]

The experience of the post-partition situation was sought to be met politically without losing sight of administrative weaknesses arising from large territorial jurisdictions. The remedy supplied not only reduced the extent of administrative charges to provide for regular supervision, security and development, but also met the demands of politics, both linguistic and communal.

Assuming the approval of the proposed transfer of India's capital from Calcutta to Delhi, the Government of India made three recommendations: (1) to reunite the Bengali-speaking districts of the Presidency—Burdwan, Dacca, Rajshahi—and administer them through a Governor in Council; (2) to establish at Patna a Lieutenant-Governorship in Council, to administer Bihar, Chota Nagpur and Orisa with a Legislative Council—an area of 113,000 square miles and a population of 35 million; and (3) to restore the Chief Commissionership of Assam with an area of 56,000 square miles and a population of only 5 million. It was thus Assam which, in the absence of a self-contained unit of provincial administration, remained neglected.

The creation of a Governorship in Council for reunited Bengal was an important feature which in the past had been considered incompatible with the presence of the Supreme Government at Calcutta. A reunion of the Bengali-speaking districts and the Governorship in Council were supposed to be adequate compensation for the withdrawal of the Government of India from Calcutta. The creation of a separate Lieutenant-Governorship Bihar, Chota Nagpur and Orissa was, on the other hand, intended not only to reduce the administrative charge of Bengal by keeping it limited to the Bengali-speaking population, but also to safeguard the interests of Muslims who in divided Bengal had commanded an overwhelming majority in the eastern districts. With the amalgamation of the Bengali-speaking Divisions the Muslims would still be numerically superior to the Hindus only if Bihar, Chota Nagpur and Orissa were formed into a sepatate administration. They had already been given special representation in the Legislative Councils and Dacca had been developed as a second capital where the Governor might reside to keep in touch with Muslim sentiments and interests.

Bihar, which was a Hindi-speaking part of Bengal, deserved on its own merits a separate administration and that, too, on the same linguistic and cultural basis as formed part of the Bengali agitation against the partition. In view of the marked political awakening in Bihar at the turn of the century a unanimous feeling arose among its educated elites against their being yoked unequally with the Bengalis. The Government of India therefore considered the ensuing Coronation Darbar to be the best opportunity to announce the uniting of Biharis with Oriyas, who had little in common with the Bengalis.[82]

All these changes concerning Bengal, Bihar and Orissa were enacted under the Government of India Act of 1912. The Act recognized only three Governorships of the Presidencies of Fort William, Fort St George and Bombay, which were referred to as the Presidencies of Bengal (not Fort William), Madras (omitting Fort St George) and Bombay. Under the Government of India Act (1919) the Presidencies of Bengal, Madras and Bombay, and the provinces known as the United Provinces, Punjab, Bihar and Orissa, the Central Provinces and Assam were declared as 'Governor's Provinces'. Orissa remained with Bihar until the Government of India Act (1935) provided for its being formed into a separate Governor's Province. The object again was better security and development.

In Orissa, as in Bengal and Bihar, the obsession caused by the politics of language not only tended to shift emphasis from the primary considerations of security and development, but also to cloud the broad all-India political perspective and make it subservient to regional interests. The downward filtration of education, which Sir Charles Wood had devised in 1854 to serve as an instrument of mass education through the medium of local vernaculars, came to be utilized as a middle-class instrument in the advancement of elitist interests through a resort to language politics at local and regional levels. Orissa was no exception. Apart from the territorial gains already secured from the Central Provinces under Curzon, Orissa acquired in 1936 such Oriya-speaking tracts from Madras as Ganjam, Berhampur, Gopalpur and Chatrapur.

The creation of large and extensive units of an administrative charge which contributed to the neglect of local or regional interests had come down as a heritage of the Company's rule. The territories acquired by it were placed under one or the other of its three Presidency Governments, and for historical reasons the Presidency of Fort William took the lead in the acquisition of territories and Calcutta became the seat of its Central Government in India. The Governor of Bengal became its Governor-General and President of the Council attached to Fort William. The Bengal Government in fact became the Company's Supreme Government in India. The other two Presidency Governments of Fort St George and Bombay were made subordinate to it.

Prior to the Charter Act of 1833, the whole of the Lower Provinces of Bengal including Bihar, Orissa and Assam as well as the entire North-Western Provinces, stretching from Banaras to Delhi, with nine large Divisional Commissionerships, were all governed from Calcutta. Added to this were the unsettled territories acquired from the Marathas in Central India and annexed for administrative purposes to the upper provinces under the Bengal Government itself. The need to establish a high official authority in the North-Western Provinces had been realized as far back as 1808. Two of the Commissioners[83] especially deputed to report on the revenue resources of these vast tracts of territory were the first to emphasize the need for a separate government to conduct political relations, to improve internal administration, to relieve Calcutta of all kinds of administrative details, to give directions to local officers on the spot, and, finally, to make the administration of the North-Western Provinces accessible to the inhabitants of these regions. But nothing came of their recommendations. The matter was taken up again in 1829 by the Civil Finance Committee. The establishment of a separate Local Government was in principle recognized as both necessary and desirable. But objections arose on financial grounds, on the extent of patronage to be delegated, and on the control of political relations. The Governor-General himself was opposed to any division of authority.

Bentinck did not want a dilution of the authority vested in the Supreme Government, but 'the removal, for a time, of the seat of [the Supreme] government to the upper provinces', an opinion in which his Council was 'pleased to concur'. The minutes he recorded on the subject on 10 February 1829 and then again on 15 September explained his reasons assigned for this. This line which divided the lower from the upper provinces, for example, was '600 miles from Calcutta'. But more than that was the difference of situation, culture and habit of men, 'a totally different order of men and things, demanding the utmost attention and care of the government'. As the September minute added,

An unsettled revenue, a warlike population, new judicial institutions, the great misgoverned kingdom of Oudh, and the many ill-regulated states in Central India, extending above eight hundred miles further. Let it be recollected that throughout this space, the communications are very imperfect; that the rate of the *dak* averages three miles per hour; that there is not a post carriage or post horse in the country; that all travellers except those going by *dak* at a heavy expense, must go by daily marches, like the caravans, or an army in the field, with tents and complete establishments of servants and cattle; that the conveyance of water is equally slow, months being required to do what in all European countries would occupy not a tenth of the time. We may, I hope, expect much improvement in the use of steam, and in the introduction of carts and horses for the conveyance of the mails, instead of runners; but this must be the work of time, and for the present, we must consider things as they are.[84]

Bentinck therefore saw 'no alternative but the removal of the Supreme Council itself' to a place in the upper provinces. He preferred Meerut as a temporary arrangement as an 'experiment'. He argued that 'the presence of the Supreme Government in the north of India will not only be useful in enabling its members to come to the most satisfactory and expeditious conclusion respecting Central India but upon many other questions regarding both the revenue and judicial administration'[85] of British territory which still remained unsettled in the North-Western Provinces. He consulted expert opinion on the constitutional and legal implications of his proposal and saw no objections to it.

But the measure so proposed did not receive the approval of the London authorities who feared that moving the government with its departments would involve considerable expense to which the Company would not agree. But there were other reasons, too, to justify a rejection of Bentinck's proposal. Doubts immediately occurred, for example, on the legality of the proceeding; for Sergeant Bosanquet, the Company's standing counsel, was of the opinion 'that the Governor-General in Council is not authorized to remove the seat of government to any of the provinces at a distance from Fort William'. The Court of Derectors therefore directed that in case the seat of government had been removed from Calcutta before the receipt of their dispatch, the members of the Council must without fail return to Fort William immediately. The Governor-General could of course remain in the upper provinces if he so deemed fit. He must in that case nominate a member of the Council to be Vice-President and Deputy-Governor of Fort William.

A still more serious consideration was the fact of India being 'governed by a distant maritime power', a peculiar circumstance which needed to be taken into account in the choice of the seat of government. And Calcutta had for that matter 'the advantage of a

Territorial Reorganization for Security and Development 173

certain communication by sea with the country'. Moreover, it had 'the further advantage of being placed at the greatest distance from all the powers whence attack can be apprehended on the side of Hindustan, and of being protected more especially towards Ava by rivers which our naval means must enable us, at all times, to command'.[86] In view of 'the experience of the past ages' the Company was in fact opposed to 'establishing the seat of the empire at Agra or at Delhi, or anywhere near the frontier of the North-Western Provinces, in a position in which it would be exposed to the sudden incursion of nations of cavalry, and to the first brunt of any hostile movement of the powers bordering upon India'.[87]

A change, however, came about under the Charter Act of 1833. It provided that the territories then subject to the jurisdiction of the Government of Bengal be divided into two separate Presidencies, namely, the Presidency of Fort William and the Presidency of Agra, with a Governor in Council for each. The Court of Directors was, however, authorized to postpone the appointment of the Council in any of the two Presidencies, or to reduce its size, or even to determine whether it should be ruled by a Governor alone or a Governor in Council. So far as the Presidency of Fort William was concerned, the Governor-General himself was to be its Governor.

The Charter Act was to take effect from 22 April 1834. In a letter of 27 December 1833 the Court of Directors communicated to the Governor-General in Council the appointments made for that purpose While Bentinck was to be the Governor-General under the new Act, his Council was to consist of William Blunt, Alexander Ross, William B. Martin and Thomas Babington Macaulay, to be respectively the first, second, third and fourth Ordinary Members. The letter further intimated that the Governor-General of India 'shall also be Governor of the Presidency of Fort William in Bengal' and that Sir Charles Metcalfe had been appointed 'to be Governor of the new Presidency of Agra'.[88] In the event of the death, resignation or coming away of Bentinck, Sir Charles Metcalfe was to succeed to the office of Governor-General.'[89]

The Act, however, did not come into operation from 22 April 1834. Bentinck moved to Ootacamund for health reasons on 15 March 1834 and remained there till October, and Sir Charles Metcalfe functioned under the old Acts as Vice-President of the Council and Deputy Governor at Calcutta. The projected Council under the new Act, too, remained circumstantially unconstituted.[90] On his return to Calcutta in the autumn of 1834 Bentinck took his seat in the new Council of India by a Notification issued on 11 November. On the same day, while he assumed his separate powers as Governor of the reconstituted Presidency of Fort William in the Lower Provinces of Bengal, Sir Charles Metcalfe took the prescribed oaths and assumed charge as Governor of the Presidency of Agra. The Notification also declared that the seat of the Agra government would for a time be at Allahabad and its function restricted to internal administration, not extending to the management of political relations.

Despite statutory provisions for the creation of a separate Presidency of Agra, the vested interest of the erstwhile Calcutta Government in fact kept on resisting its separation as a self-contained and full-fledged administration. Metcalfe doubtless regarded his appointment to Agra as an implied 'compliment' and 'intended elevation'[91] in rank. Bentinck, too, admitted that this appointment was more than ever his 'due' and that the manner in which it was done with the complete 'unanimity of the high authorities' in London was especially 'gratifying'.[92] Even so, Metcalfe was obliged to continue at Calcutta and acted on Bentinck's advice to allow the new arrangement for Agra to remain in a state of suspension during Bentinck's prolonged absence at Ootacamund, an illegal advice which had to be validated subsequently by an Act of Indemnity. And when Metcalfe took the oaths of office as Governor of the Presidency of Agra, he found the seat of his government stationed at Allahabad and limited to internal administration.

From a letter which Metcalfe addressed to Bentinck on 8 May 1834, it appears that the latter had already been anxious to exclude Metcalfe from any connexion in the management of the Company's political affairs. Bentinck and his secretariat did not wish to part with any of the powers wielded by them in the undivided Presidency of Fort William. And even when a separate Presidency of Agra was created by an Act of Parliament, their efforts were directed towards retaining its political management as well as the exercise of its patronage. Metcalfe's fear was in fact not unfounded. For he said that in case the Governor of Agra was not invested with the powers formerly exercised by the Resident at Delhi in relation to independent states, he 'would in the eyes of neighbouring states be a very inferior person to what the Resident of Delhi was in former days'.[93] He thus pressed for 'a nearer superintendence' over diplomatic relations rather than 'a distant one' from Calcutta. 'We are disposed to think', he said to Bentinck, 'that the Government of Agra might be vested with the powers formerly held by the Resident at Delhi exercising

Territorial Reorganization for Security and Development 175

authority over the diplomatic agents in that quarter and subject on its own part to the constant control of the Supreme Government. The range of intermediate authority to be exercised by the Government of Agra would necessarily be extensive. It would naturally include our relations with the Sikhs and Afghans and all states north-west of Delhi as well as Rajputana and the states in the Delhi and Agra frontier. All those above mentioned were formerly connected with the Residency of Delhi as was also the court of Indore and the province of Malwa'. Metcalfe indeed included within the range of his superintendence the states of Bundelkhand and the state of Gwalior, the state of Oudh as well as the states connected with the Saugar and Nerbudda Territories.[94] But Bentinck would not have less than the whole cake for himself.

This applied equally to the exercise of patronage and the selection of officers for an efficient transaction of business for the Agra Government. But Metcalfe was not allowed to appoint even the one secretary he had proposed. It was Bentinck who nominated Macsween to be secretary to the Government of Agra. The vested interests of a larger unit of administration thus obstructed the progress of territorial reconstruction to promote development and security, which, with the increased function of the state, called for reduction in the size of administrative units, easier contact for purposes of efficiency and development and additional investment in manpower and material requirements. There emerged in fact a dichotomy between considerations of power and patronage as an instrument of central administrative control, and the exigencies of development which required territorial redistribution as well as administrative and financial decentralization. The politics of regionalism, linguistic or otherwise, flowed from a failure to resolve this dichotomy. Such a dichotomy was not possible in the period immediately following the Charter Act of 1833, when centralization was recognized as the best means of effecting economy and enabling the Company to recoup its finances from the revenues of India. It arose from recurrent famines and the need for development which called for a decentralizing trend which the Indian middle-class elites turned into politics at the turn of the twentieth century.[95]

The dominant force before the transfer of government to the Crown was in the main centralization. When Bentinck resigned office in March 1835, Metcalfe took over as Governor-General after handing over charge of the Agra Government to William Blunt. The Bentinck-Metcalfe controversy doubtless ended. But the authorities in London did not want to have it repeated. The duties assigned to the Government of Agra had already been of a secondary nature, merely to relieve the Governor-General in Council of the administrative details involved in the management of the North-Western Provinces. Then there was the financial consideration, the need to save. An Act thus suspended in 1835 that part of the Charter Act of 1833 which had provided for the establishment of the Presidency of Agra with a Governor and Council. It was now laid down that in the period of suspension it 'shall and may be lawful for the Governor-General of India in Council to appoint from time to time any servant of the East India Company, who shall have been ten years in their service in India, to the office of Lieutenant-Governor of the North-Western Provinces under the Presidency of Fort William in Bengal', with powers to declare and limit from time to time the extent of his authority as also the powers exercisable by the Lieutenant-Governor to be appointed.[96] Agra thus became completely reduced to the status of a subordinate 'province'.

Lord Auckland, who was to take over from Metcalfe in March 1836, was advised to treat with care and politeness the outgoing Civilian Governor-General who was requested to accept the Lieutenant-Governorship of the North-Western Provinces on a salary equal to that of the Governors of Madras and Bombay. Metcalfe was further assured that if he accepted the offer, the capital of the North-Western Provinces would be shifted to Agra, that he would be vested with the exercise of patronage as well as with full powers in the conduct of political relations with independent states, subject only to the final control of the Government of India.[97] The Sadar Court and Board of Revenue would remain at Allahabad, however, and correspond directly with Calcutta to avoid delay.

Metcalfe accepted the offer and on 28 March 1836 the Governor-General in Council appointed him 'to be Lieutenant-Governor of the North-Western Provinces, to be ordinarily stationed at Agra'. The Secretaries to the Government of Agra became Secretaries to the Lieutenant-Governor in the different departments. Though formally appointed a Lieutenant-Governor, Metcalfe's rank and complimentary honours were recognized as those of a Governor.[98] Calcutta had thus to concede the demands of a distant unit of administration which needed a certain degree of local authority in the management of its civil and political affairs—a result of Metcalfe's resistance against the concentration of power and patronage at Calcutta. But the immutability of the central authority remained unaffected.

Territorial Reorganization for Security and Development 177
The United Provinces of Agra and Oudh

The territorial extent of the North-Western Provinces, however, did not decrease. On the contrary, it was augmented with the merger of Oudh under a single charge in 1877 and the formation of the United Provinces of Agra and Oudh in 1902.

In all essential matters, except the special privileges enjoyed by talukdars, Oudh differed but little from the adjacent districts of the North-Western Provinces. The population was composed of the same classes who professed the same creeds, used the same language and followed the same customs and usages as the people of the North-Western Provinces. The system of roads and communications, too, was the same. A common controlling authority for both Oudh and the North-Western Provinces was considered administratively sound for purposes of security as well as local development. A merger of the two governments was in addition calculated to stop stagnation in promotion and conduce to economy by abolishing the Chief Commissionership of Oudh. The Governor-General in Council took all these factors into consideration before he recommended the expediency of their administrative merger in 1877.[99]

Under the arrangement of 1877 the office of Chief Commissioner was not abolished, but was allowed to remain unoccupied. Sir George Couper, who was Chief Commissioner of Oudh at the time was also appointed Lieutenant-Governor for the North-Western Provinces. The Chief Commissionership was not dead, nor was Oudh absorbed by the North-Western Provinces. The Lieutenant-Governor, who was also Chief Commissioner, was required to exercise the existing functions of both offices within each province respectively. In other words, both Oudh and the North-Western Provinces became a single charge without any material interference from their functional differentiation. Although the seat of the chief executive authority in Oudh was shifted to Allahabad, he was required to reside at Lucknow for part of the year.[100] As the following table shows, the characteristic feature of the Oudh administrative system continued under the arrangement made in 1877:[101]

	Divisions	Districts	Area in sq.miles	Population (mills.)	Regulation districts	Non-Regulation districts
Oudh	4	12	23,992	11¼	—	12
N.W.P.	7	30	81,403	30¾	27	3
Total	11	42½	105,395	42	27	15

The Chief Commissionership of Oudh was, however, not an end in itself. As said before, its object was to raise standards and improve the state of administration so as to bring it to the level of the regulation provinces. Executive and revenue officers were after the merger gradually divested of judicial functions which came to be exercised by separate civil courts. The Jhansi Division, earlier a non-regulation tract, was brought under the regulation system of the North-Western Provinces, although the 'scheduled' districts of the Kumaun Division continued to be governed by the Scheduled District Act of 1874 under which the Divisional Commissioner acted as a high court in civil cases and a sessions court in criminal matters. As a result of these changes the districts of the North-Western Provinces were in 1893 regrouped together into nine Commissionerships, seven of them being in the North-Western Provinces, and only two in Oudh.[102]

Under the Indian Councils Act of 1861, provision was made in 1886 for the establishment of a Legislative Council in the North-Western Provinces and Oudh. The territories for the time being under the Lieutenant-Governor of the North-Western Provinces and Chief Commissioner of Oudh were together constituted into a province for purposes of the Legislative Council so provided from 1 December 1886.[103]

Though nominally separate, the administration of the North-Western Provinces and Oudh for all practical purposes was united, the offices of Lieutenant-Governor and Chief Commissioner being held by the same person. Moreover, the old name of the 'North-Western Provinces' had become anachronistic since the establishment of the North-West Frontier Province in 1901. By a Proclamation of 22 March 1902 it was therefore declared that the territories then under the administration of the Chief Commissioner of Oudh 'shall henceforth form part of , and be subject to, the Lieutenant-Governorship of the North-Western Provinces, and that the Lieutenant-Governorship so constituted as aforesaid shall be designated the Lieutenant-Governorship of the United Provinces of Agra and Oudh'.[104] It became a Governor's Province under the Government of India Act, 1919.

In terms of area and population it remained one of the largest provinces. The whole area constituted under the Mughals the Subah of the Nawab of Oudh, stretching from Banaras to Delhi. The Nawab had to lose the territory of Banaras and then the ceded and conquered districts which together formed part of what came to be known as the North-Western Provinces. The annexation of Oudh in 1856 and its merger in 1877 paved the way for a unified pro-

Territorial Reorganization for Security and Development 179

vince again in 1902. The politics of language which emerged as a force in the wake of the partition of Bengal as the basis of territorial reconstruction acted as a reinforcement in so far as the U.P. was concerned. It was not only Hindi-speaking, but the actual hub of 'Hindustan', and its Hindustani culture provided unity to its body politic.

This chapter brings into relief the problems of territorial reconstruction, more especially along India's land frontiers to the north-west and north-east, which were not only exposed to the dangers of foreign aggression, but also open to internal unrest arising from the unpredictable character of the hill tribes given to blood feuds and violence. Since geographical and ethnic issues were both involved in questions of frontier defence and internal security, it was not easy to choose between the advantages of a buffer territory and a forward policy of firmness to stand behind a line of strong strategic outposts. In the case of the north-west frontier, a 'close border' policy to effect penetration into tribal areas through the agency of middlemen was sought to be replaced by a forward policy. The result was the Durand Line. But since ethnic distribution under the line so drawn remained unreconciled with geographical unity or compactness, it led to uprisings and serious loss of life on the British side. Curzon took lessons from the past experience of tribal management and adopted measures conducive to both security and development. These were directed towards the disposition of the armed force and the creation of a separate province under a Chief Commissioner. While the former signified a withdrawal of troops to secure army headquarters in a 'settled' district, with detachments for purposes of security in tribal territories, the Chief Commissionership so constituted in 1901 represented an administrative arrangement best suited to tribal societies for development purposes.

As regards the policy dimensions of territorial reconstruction in the north-west, Sandeman's concept of loose political control over tribal territories had left an impression on the advocates of the forward policy. It was perhaps the influence of Sandeman's thinking that led to a recognition of the advantages of treating Afghanistan as a friendly buffer state, a view which followed the experiences of the first World War.

Consideration for the need of a buffer territory with loose political control between the Assam Himalaya and the Brahmaputra valley had engaged the attention of the British much earlier than the Simla Convention of 1914. The idea was perhaps not pushed forward earlier because of the difficulties arising from the absence of proper communications through tribal territories to the north of the Brahmaputra valley. Even so, these were dotted with military outposts which were protected by detachments of the Assam Light Infantry as well as by the tribal militia raised for police purposes and armed with muskets in some cases. Later, the survey team of Pandit Nain Singh served a major political purpose. Nain Singh and his team's report on the topography of the tribal tract between Tawang and Odalgiri provided vital data on a tract which later formed part of the agreement which Sir Henry McMahon made with the Tibetan plenipotentiary at Simla in 1914.

Assam proper and the entire north-east frontier were for security and development purposes formed into a Chief Commissionership, a form of government already recognized from experience as the best administrative arrangement to promote the welfare of backward areas. Great care was taken in the choice of a base for military headquarters in Assam which was garrisoned with an efficient body of troops, strong enough to punish any wanton aggression, though cautious enough at the same time to exercise restraint and refrain from creating unnecessary foes among hill tribes by using repressive measures. The policy followed was in fact a show of military strength rather than its actual use. Even the most warlike and formidable Abor and Naga tribes were treated with care and circumspection. Use was of course made of payments to chiefs of *posa* allowances which, though seen as blackmail, was also an aid to fight poverty. The backwardness of the Assamese, however, flowed from the virtual absence of local manpower to assist in the development of industry and commerce. They remained subject to exploitation by European entrepreneurs and Bengalis who constituted a kind of colonial infrastructure within the broad framework of imperial rule and European settlement.

A serious limitation of Assam's Chief Commissionership was, however, its dependence on Bengal for its civil service. To make it self-contained and provide a commercial outlet to the sea Chittagong and a couple of other Bngal districts were proposed by Curzon to be transferred to Assam for its proper development. That created grounds for political agitation which had to be met politically by the creation of an altogether new province of Eastern Bengal and Assam, an arrangement where Assam became annexed to the eastern Muslim majority districts of Bengal. It created as a counterweight to Calcutta new centre of Muslim political influence at Dacca, a centre which had little or no relevance to the interest of the Assamese who became all the more subservient to Bengali dominance. All that the new province did was to bring under unified control the whole of the tea industry, except Darjeeling which remained with the old province of Bengal.

The partition of Bengal indeed marked the beginning of a new era where politics, more especially of language, started superseding administrative and other essential considerations underlying territorial reorganization. This was a popular trend which emphasized the extension of the regulation system to facilitate agitational activity by making prosecution for political crimes difficult, and demanded the creation of linguistic provinces to enable even the half-educated to participate in the process of law-making and the conduct of government. While, under the pressure of politics, a general tendency was slowly growing on the part of the Government to allow the regulation and non-regulation distinctions to wear off except in name and style, the National Congress had no scruples in extending recognition to demands for linguistic provinces. Though British policy was in favour of reducing the size of larger provinces to facilitate personal contact with the people, the formation of a province on the mere basis of language was considered unsound in terms of both administrative efficiency and all-India political unity. The Montagu-Chelmsford Report (1918) therefore rejected the very idea of a linguistic division as altogethr impracticable. While recognizing common speech as a natural basis of provincial individuality, the Statutory Commission too turned down proposals for a linguistic demarcation of provinces. Considerations of compactness and national integration led it to emphasize a balanced approach to the problem of provincial reorganization by taking into account not merely a common language, but also financial resources, ethnic and religious considerations, economic and cultural interests, geographical contiguity and defence requirements. Guided solely by the emotion-bound irrationality of *swaraj* politics, the leadership of the National Congress made no effort to see any merit in the recommendations of Sir William Marris, a highly experienced administrator, who had drafted the Montford Report, or of Sir John Simon, who presided over the Statutory Commission. The Congress leaders under M.K. Gandhi would not attempt to see any problem except through their exclusive political lens. They advocated that the wishes of people alone should finally determine the question of provincial reorganization. Little did they realize that the people of India were not a homogeneous community, but a cogeries of mutually exclusive elements, with diversified loyalties varying according to region, religion or caste.

Chapter IV
INSTITUTIONAL AGENCIES AND POLITICAL STABILITY

Apart from the delimitation of frontiers and territorial reconstruction, India's political stability proceeded from the several institutional agencies constituted from time to time. They were intended to keep intact the strength and unity of the central authority vested with the formulation of policy, and yet provided for change to satisfy within their own framework the emergent demands of society resulting from new conditions.

The stabilizing agencies included social and economic institutions, which promoted in professional and other fields the emergence of the middle classes with an all-India awareness in a territorial sense.[1] But these were mainly political and administrative systems which, in the Indian situation, were by far the most important initial instruments of change without seriously prejudicing India's unity or stability. During 1858–1919 the political and administrative agencies remained for all practical purposes under the full control of the Imperial Government established under the Government of India Act, 1858, a control exercised in effect by the I.C.S., its administrative agency recruited in London on a competitive basis. Together, they represented a centralizing impulse, which, with the help of the armed forces under the civil authorities and deployed in different parts of India, preserve her territorial integrity and maintained peace and tranquillity in the country.

As already discussed in the Introduction, the concept of a strong centralized government as a bulwark of imperial unity and interest in India, however, did not completely fit the colonizing schemes of European settlers. They not only wanted freedom to exploit India's agricultural resources in the conduct of their business, but also refused to be ruled by the executive-made 'regulations', operative before the Charter Act (1833), which opened the doors of British India to a free ingress of Europeans, and provided for the appointment of a fourth ordinary Member of the Executive Council who, for legislative purposes, was to be the Law Member. Macaulay, the first to hold this office, attempted through his Standing Orders to engraft on what he called 'despotism' the fruits of 'liberty' to meet the needs of increasing European colonization. The colonial element in fact represented another force which, through its emphasis on a separate legislative authority, was potentially promoting the case of representative government, a check on executive despotism. It is true that the Indian Councils Act of 1861 attempted to put the clock back. But the growing strength of the indigenous middle-class elites induced them to collaborate in their own interest with European business and make common cause with it to reduce centralization and limit the powers of the imperial political agency which, in an attempt to establish a direct link with the rural masses, sought to bypass the European planting community as well as its Indian collaborators. Despite increasing decentralization, the imperial thrust created an organic framework which enabled both political and administrative cohesion to be maintained. This chapter will attempt a brief analysis of some of the important institutional agencies which conduced to India's political stability despite the serious obstructions arising from the application of elective principles to a highly segmented, caste-ridden society, which constituted a potential cornerstone for the emergence of an alternative Indian sovereignty. For in spite of the extension of representative principles, the British government never allowed the basic unitary character of the Indian government to be seriously impaired. The London authorities remained the final arbiters on questions of policy.

The Political System

British India

Under the Government of India Act of 1858, the political agency for the government and administration of British India consisted of the Secretary of State for India in Council and the Governor-General in Council at Calcutta functioning under the immediate control of the home government. The latter controlled in turn the subordinate governments of the Presidencies of Madras and Bombay as well as the provinces or territories appointed by the Governor-General under the Charter Act of 1853. The Chief Commissioners of some British territories, who were directly appointed by the Governor-General under the powers given to him by the Government of India Act (1854), also formed part of the same political arrangement.

The Secretary of State for India in Council

The institution of the Secretary of State for India in Council signified centralization to ensure unity in terms of both policy and action.

Institutional Agencies and Political Stability

The Secretary of State in Council was invested with such powers as jointly or separately belonged under the Company to the Court of Directors and the Board of Control, but the change brought about under the Crown was qualitatively perceptible. While the Court and the Board, for example, had functioned in committees as two separate entities representing in principle even separate interests, with their decisions taken separately through the majority vote, the Secretary of State was an individual minister of the Cabinet whose decision on all matters was final, his Council being invested with no more than the exercise of a 'moral influence' and a power to 'compel him to record his reasons in writing' where he differed from his Council and acted independently despite their disapproval of a course he might take. As a minister of cabinet rank his action was of course subject to a more direct parliamentary supervision. But since the India Office establishment was maintained out of the Indian revenues and was not subject to votes by Parliament, the Secretary's parliamentary accountability had little or no meaning, except that the weight of English public opinion acted as a restraint—not so much against centralization as the abuse of authority. The India Office was, for instance, perturbed when the Royal Commission on Expenditure sought to place the 'Home Charges' on the estimates, a move which might render the course of its functioning subject to the irrationality of party politics in Parliament. Sir Arthur Godley, the Under Secretary of State, therefore persuaded Lord Welby, the Chairman of the Commisssion, to see that the India Office remained free from the inner politics of the House of Commons on questions of Indian policy.[2] The centralization of the London authority thus remained unaffected and flowed largely from the Secretary of State being the Crown's minister without receiving any salary from the Crown's Treasury.

It is only in financial matters that the 'Council of India' was to act as a watchdog; for the Government of India Act (1858) specifically provided that no grant or appropriation of any part of the Indian revenues could be made without the concurrence of a majority in the Council. A similar majority was, of course, also required to fill by election any of the ordinary vacancies occurring from time to time among the members of the Council appointed by the Crown. Every member elected by the Court of Directors or appointed by the Crown could hold office during good behaviour, but did not have the right to sit or vote in Parliament. In this respect, the Government of India Act (1869) added to the control of the Secretary of

State: under the new Act he was authorized to fill vacancies in the Council, not during good behaviour, but only for a period of ten years, subject to re-appointment at his discretion. Besides, the majority vote of the Council was altogether dispensed with in the appointment of Executive Councillors of the various governments in India. The Secretary of State could alone make these appointments as the adviser of the Crown.

Private correspondence between the Secretaries of State and the Viceroys constituted another important source of accession to the authority of the former. It enabled them to bypass the Council and tended to divest it of the functions which it was entitled to exercise by the Act. The nature of communication, according to the Statute, came under three heads only. These were (1) official communications called 'public' which passed through the Council; (2) 'urgent' communications which the Secretary of State might on his own authority send to the Governor-General in Council, with a subsequent explanation of the causes for his so acting; and (3) 'secret' communications, in which the Secretary had the power to act on his own authority without being obliged to explain the reasons for his dispensing with the advice of the Council. But, in addition to these authorized statutory communications, there had always been the practice of the Secretary of State and the Governor-General communicating with each other privately by telegram and letter. If the private telegrams or letters happened to be explanatory of the official communications sent, they came as a matter of course before the Council of India in London or the Council of the Governor-General at Calcutta. But since they also dealt with personal matters, it was usually the practice for the Secretary of State and the Governor-General to take away their private telegrams and letters at the close of their tenure of office; though developed as a regular channel of communication, these letters had no statutory sanction. The Mesopotamia Commission of 1917 condemned such correspondence as the Council of India was deprived of its legitimate role in restraining the exercise of power.

The Governor-General of India in Council

As said before, the government and administration of India were vested in the Secretary of State for India in Council and the Governor-General of India in Council, a subordinate agency dominated by the London authorities who, with the development of quicker communications, brought India into much closer dependence on the Secretary of State. His influence and control over the Government in India, already great by virtue of his Cabinet rank and parliamentary position, grew steadily greater. As seen in Chapter II, the problem of delimiting India's frontiers began to be assessed and judged in the context of an over-all imperial situation and not in the immediate local interest, which the Governor-General of India often emphasized. The importance of the India Office acquired a new dimension in the total British imperial context.

The availability of quicker means of communication was evident immediately in the Secretary of State's power to control and direct the affairs of India. In a letter of 24 November 1870 he chose to remind the Government of India that 'serious embarrassment' would follow if a clear understanding was not maintained about the basic principle underlying the political system. 'That principle is that the final control and direction of the affairs of India rest with the Home Government, and not with the authorities appointed and established by the Crown, under Parliamentary enactment, in India itself.' The subordinate position of the Government of India was indeed reiterated by pointing out that 'the Government established in India is (from the nature of the case) subordinate to the Imperial Government at home. And no Government can be subordinate unless it is within the power of the Superior Government to order what is to be done or left undone, and to enforce on its officers, through the ordinary and constitutional means, obedience to its directions as to the use which they are to make of official position and power in furtherance of the policy which has been finally decided upon by the advisers of the Crown'.[3] This warning was preceded a year earlier by the Council of India Act, passed on 11 August 1869, which invested the Secretary of State with powers to fill vacancies taking place not only in the Council of India but also in the ordinary membership of the Governor-General's Council and the Councils of the various Presidencies.

The introduction of a legislative authority for the making of laws and regulations in India, however, led to the development of a situation potentially opposed to centralization or despotism. Macaulay was the first to attempt to indicate a distinction, albeit on a vague colonial principle, between executive and legislative functions. Independently of the Charter Act of 1853, Dalhousie had through his 'Standing Orders' pressed Macaulay's framework into a shape comparable to a mini colonial legislature, imparting to its proceedings a much more formal character than was contemplated by the Act of 1853. What Dalhousie did was in conformity with the popular view in England, which looked for something comparable to colonial self-government, or the self-government of a colony of British settlers in India.[4] But the Mutiny was a pointer to policy makers. The racist European community of Calcutta demanded ruthless repression of the mutineers and even advocated the recall of Canning, the Governor-General, who would not adopt the extreme measures against the Indian population that they advocated. A colonial pattern of government with a separate legislative authority was therefore seen by policy makers in Britain as the worst political arrangement in the Indian situation, where the government of a ruling race of Europeans could be foisted on the Indian people.

Another reason for official opposition to the development of a separate colonial legislative authority were the indigo disturbances of 1859. These showed how the European indigo planters had, on the basis of the proprietary tenures obtained from local zamindars, used their position as landlords to compel cultivators to grow indigo by a system of enforced advances and sell it at rates fixed by themselves, which did not cover the cost of production. The losses of the raiyats grew steadily as the prices of other crops kept rising until, in 1859, they combined against the planters. The cultivation of indigo consequently received a setback from which it did not recover.[5] The impression of the indigo disturbances of 1859 was perhaps still fresh when in a letter of 18 August 1861 Sir Charles Wood wrote to Sir Bartle Frere: 'Do you think a jury of indigo-planters would convict a planter, or acquit a ryot, or of how they would legislate for matters pending between them?' In fact, Wood apprehended the danger of 'partial legislation or administration in case the colonial urge for a representative government was conceded'. He, therefore, laid down that at its meetings for making laws and regulations the Council was not to be a separate and distinct body from the Council of the Governor-General. It was not to sit permanently for such purposes and was to be called together when projects of law, prepared by the proper officers under the supervision of the executive government, were ready for discussion. But, as already noted, the distinction between the Executive Council and the Council for making laws and regulations remained, both structurally and functionally. For the Indian Councils Act of 1861 provided for admission to it of non-official Europeans and Indians to participate in the important work of legislating for India. Called 'additional members', they were appointed to assist the Executive Council. But their function to 'assist' was not to signify subordination or loss of independence when they assembled in 'session' on a permanent basis, despite the statutory provisions to the contrary. This proceeded from the prescriptive effect of the separate rules of business or Standing Orders which governed the process of legislation.

Indeed, Wood was confronted with a crisis arising from increasing colonization, which demanded little or no executive interference with the economic operations of European settlers, who, in turn, sought control over legislation. But if legislation was allowed to proceed with discussion or debate according to the rules of the House of Commons, it would be 'hopeless to look for the co-operation of the native gentlemen', whose participation in the process of lawmaking was considered politically essential; for they could not 'speak or understand a word of English', nor would they have anything 'to do in the Council but to sit and listen to what they could not comprehend'.[6] This applied more especially to the Governments of the North-Western Provinces and Punjab[7] which had 'intelligent, wealthy and influential native gentlemen', attached to British rule, 'with a large stake in its stability and power', though 'ignorant of our language and not familiar with our ways'. It was realized that their participation in the Council of the Governor-General for making laws and regulations would be of great help; but they would be no more than passive spectators in the Council if it were to function on the lines of the House of Commons. The immediate post-Mutiny situation in fact presented a dilemma, where the choice between representative government and executive despotism could not be mutually exclusive. The Indian Councils Act of 1861 represented an attempt to strike a compromise, with a perceptible tilt in favour of executive control. However, it was not long before the demand for representative government came to be reinforced with the emergence of a new Indian middle class of English-educated elites who collaborated with independent European business in support of that demand. The Indian element in fact came to the fore because of the limited extent of colonization under the Company's rule, and pressed for the constitution of legislative councils based on a representative principle at the centre as well as in the provinces.

In the context of India's traditional social system and its dependent economy under British rule, the new middle-class elites emerged chiefly from legal, educational and administrative changes, not from economic diversification. Of the professions so emerging, law became by far the most important and powerful, and it was not until after the first couple of decades of the twentieth century that technical and business professions came slowly to rise to importance.

In the first phase of development, it was the Indian agents and employees of the Company who built large fortunes and bought considerable landed estates. In Bengal they largely supplanted the old aristocracy and commercial monopolists. They and their descendants were the first to receive English education and became the beneficiaries of British rule. The Tagore family illustrates the manner in which the descendants of a servant and agent of the Company rose and became a new aristocracy combining both wealth and English education. The first Indian who competed successfully in the I.C.S. examination in 1864 was Satyendranath Tagore, a grandson of Dwarkanath and elder brother of the poet Rabindranath Tagore. English education, however, started spreading after 1870 among ordinary well-to-do and even lower middle-class families. Even so, it was the new landed aristocracy and legal luminaries who dominated public life and politics before the introduction of an elective principle under the Indian Councils Act of 1892. Their interest was represented by the British Indian Association established in 1851, the first organized body of Indians. Though basically a landholders' organization, the Association included among its members 'the representatives of the most important interests, whether territorial, commercial or professional'.[8] They were invited to serve on the legislative councils. Between 1862 and 1892 as many as 35 of them became members of the Bengal Legislative Council and the Tagores were by far the most influential in this group. Maharaja Ramanath Tagore (1800–77), who became a member of the Bengal Legislative Council in 1866 and of the Central Legislative Council in 1873, was a brother of Dwarkanath Tagore. A Brahmo Samajist and member of the Calcutta Corporation, he had helped in the foundation of the Association, of which he remained President for ten years. Prasanna Kumar Tagore (1801–68) was another foundermember of the Association who became a member of the Bengal Legislative Council in 1863. Maharaja Bahadur Sir Jotindra Mohan Tagore, a member of the same family, became a member of the Legislative Council in 1870 and of the Central Legislative Council in 1877.

Ram Gopal Ghose (1815–68), the son of a shopkeeper, who had risen to the position of a big merchant, represented the business interest of the British Indian Association. Like many of his contemporaries, he was also educated at Hindu College. He became a member of the Bengal Education Council established in 1845, and in 1862 was nominated to the Bengal Legislative Council established under the Indian Councils Act (1861) Of the 35 members of the British Indian Association who had served on the Bengal Legislative Council before 1892, nearly one-third were lawyers, especially of the Calcutta High Court. But since their chief stake in society was landownership, even the lawyers in the Council fought for the preservation of zamindari rights, as did the Maharaja of Darbhanga. The European planting community, who acquired a variety of proprietary rights from the Indian zamindars, remained allied with them against the interests of cultivators. Since the indigo disturbances of 1859 an identity of interest had developed between zamindars and planters.

Representative Government and Despotism as Instruments of Stability

The growth of a separate legislative body, which followed the freedom of colonization after 1834, encouraged the demand for representative government. The former was, in fact, a prelude to the introduction of representative institutions, signifying non-official participation in official or governmental transactions. It marked the beginning of a political process which, if carried to its logical conclusion, was bound to result in the official bureaucratic hierarchy being rendered subject to control by elected representatives. Representative government was in reality a political mechanism of the socially dominant, interposed between the mass of people and the government. Bureaucratic despotism, on the contrary, admitted of no such interposition. It was not like the discretion-based Mughal autocracy, but a rule-bound executive dominance. It meant a direct approach to the people by officers of government, without the intervention of middlemen or so-called representatives claiming to give public expression to what the people felt about government measures. Which of these two alternatives could in the given circumstances secure the political stability of the country? The Indian Councils Act (1861) doubtless raised this question and even sought to provide an answer in favour of bureaucratic despotism in the Weberian sense.[9] But since the colonizing element had under parliamentary enactments come to be recognized as a modernizing agency for India's development and also for imperial strength, its demands for legislative checks on executive action could not altogether be set aside; nor could the government afford to ignore the wishes of the new aristocracy and the upper layer of middle-class professionals who, in consequence of official policy, were viewed as an indigenous social base for political support. The Indian Councils Act (1861) was thus no more than an *ad hoc* arrangement, a result of confusion in policy-making arising from a desire to protect the weaker sections of society within the framework of legislative councils comprising vested non-official elements of planters, zamindars and lawyers.

What protected the raiyats against landholders was, however, the influence and authority of the Executive Councillors as legislators, as well as the power of the Governor-General to finally approve the passage of a Bill. This constitutional arrangement contained the seeds of representative government within the framework of despotism. In the year of the indigo disturbances of 1859 the Central Council for making laws and regulations, for instance, passed a Rent Act which defined the cultivatior's right of occupancy and stated the grounds on which the rent of an occupancy raiyat might be increased, a concession to the planters who justified enhancement on their own grounds. For they claimed that the raiyats were compensated for any disadvantages connected with the cultivation of indigo by being permitted to hold their other lands at low rent. Accordingly, when the raiyats refused to supply them with indigo in the old feudal fashion, the planters retorted by demanding increased rents. The tenants refused to pay and the planters filed suits for enhancement under the new law. The litigation thus started and the lead given by the planters was taken over in Bengal by enterprising zamindars like Babu Joy Kissen Mukerji, the father of Raja Pearey Mohan Mukerji, a member of the Central Legislative Council, in 1885. An era of agrarian disputes set in everywhere, leading to serious disturbances in the district of Pabna in 1872–3.

The policy of letting things alone under the influence of *laissez-faire* was thus recognized as not being the answer to increasing agrarian troubles which not only affected law and order but also tended to weaken the moral foundations of the ruling Imperial Government. It was necessary for the Government of India to intervene and enact a measure for the relief of tenants against arbitrary eviction from their lands. The Bengal Rent Bill was accordingly introduced in 1882 on the recommendation already made by the Famine Commission appointed in 1880.

The Bengal Rent Bill proposed to concede to tenants certain rights which the zamindars considered highly prejudicial to their interests. Under the Bengal Tenancy Act of 1859, for example, an occupancy right accrued to a tenant who held his land for twelve years. But in case he was transferred from one plot to another, he ceased to be entitled to that right. This gave rise to uncertainty as tenants were often transferred without the benefit of occupancy accruing to them. The Property Bill therefore provided that, if a tenant had held land continuously for twelve years, he would become a settled raiyat with full occupancy rights transferable by sale. It proceeded further to declare that any person holding land as a tenant was to be treated as having held it for twelve years unless otherwise proved. One of its clauses declared as null and void all contracts entered into previously between zamindar and tenant during the previous twelve years. Yet another provision obliged zamindars to give compensation for any disturbance caused by eviction. Under the Bill even non-occupancy tenants were sought to be protected against arbitrary ejection. It also included provisions to restrict the powers of zamindars to raise rents. Thus, for example, all rent was limited to one-fifth of the gross produce of land, a provision which landholders opposed.

The Bengal Tenancy Bill was of great importance not merely in reflecting the government's policy of promoting agricultural development by tenurial security and thereby seeking a broad rural base for political stability, but as indicating the manner of legislation which, despite Wood's precautionary measures, had adopted the recognized norms of the House of Commons, with the press involved in the whole process. After prolonged debate in the Council, the Bill was, for instance, referred to a Select Committee which discussed it in as many as sixty-four meetings. Concerned as it was with questions of land and its control for welfare purposes to obviate the recurrence of famines, the Bill could not be summarily disposed of, as Sir Charles Wood had envisaged in the *laissez-faire* context. The representatives of vested interests were bound to resist what they considered to be an invasion of private property. The *Englishman* of 14 February 1885 thus wrote:

There has probably never been a document laid on the Council table subject to so many dissents as the Report of the Select Committee on the Bengal Tenancy Bill presented yesterday. The Select Committee was exceptionally large, its deliberations have been protracted over an almost unprecedented number of sittings, and each member seems to have been quite determined to have his own way.... The main interest which attaches to the dissents is not, however, a personal one. It consists rather in the revelation which they afford as to the diametrical opposition of the two parties in the Council on the great issues involved; and as to a conflict on general principles as strongly waged within the Select Committee as it has been waged outside in the Press.[10]

As Lord Dufferin indicated in a letter of 17 March 1885 to Lord Northbrook, the two groups involved in the Central Legislative Council debates were the zamindars and the government. The zamindar party consisted of the Maharaja of Darbhanga, Pearey Mohan Mukerji, Rao Sahib Visvanath Narayan Mandlik and Syed Ameer Ali. Except for the Maharaja of Darbhanga, the three Indian members of the Legislative Council were lawyers. No Indian representative voiced the case of tenants in the Council, a role that was left to government officials. Through the intercession of the Viceroy (Lord Dufferin), however, the Council made several concessions in favour of the zamindars, which watered down the effects of the original Bill. The Viceroy in his presidential address clearly admitted this: 'I fear', he said, 'that the enumeration I have made of these modifications which have told so largely in favour of the zamindars, will have renewed the pang felt by those of my honourable colleagues, who were opposed to their being made, and who, so far from admitting that the zamindars have been hardly dealt with, contend, on the contrary, that this Bill still falls short in giving adequate protection to the raiyat.'[11]

Still not satisfied with the concessions made, the Maharaja of Darbhanga and Pearey Mohan Mukerji proceeded first to move a postponement of the Bill; but when that fell through, they sought to oppose it. The Bill was, however, passed thanks to the votes of the official majority.

The Bengal Tenancy Act of 1885 was in fact a test case. It showed clearly that, while a representative institution or government held out the promise of advancing organized sectional group interests, the bureaucratic instruments of the executive government would alone take an overall view and protect the interests of those unable to make their voices heard. This end could perhaps not be achieved except by the power of the Imperial Government to veto Acts passed in India. The Secretary of State's letter of 24 November 1870 thus emphasized that his final sanction of important Indian Acts was as essential for welfare purposes as was his centralized power in the field of executive action. 'The Imperial Government', as the letter pointed out, 'cannot indeed insist on all the members of the Governor-General's Council, when assembled for legislative purposes, voting for any measure which may be proposed, because on such occasions some members are present who are not members of the Government, and are not official servants of the Crown. But the Act which added these members to the Council for a particular purpose made no change in the relations which subsist between the Imperial Government and its own executive officers. That Government must hold in its hand the ultimate power of requiring the Governor-General to introduce a measure, and of requiring also all the members of his Government to vote for it.'[12] The Imperial Government's control over Indian legislation on important subjects was considered essential for both stability and general welfare, particularly in the context of the Indian Councils Act of 1861, which was potentially conducive to the growth of organized interest groups with powers to propose legislative measures.

In fact, welfare measures were the antithesis of *laissez-faire,* and implied centralized control. The humanitarian, welfare approach to public administration and the need for executive control over the market were two important concepts that emerged gradually from the experience of recurrent famines in the nineteenth century. A change came about during the Bihar Famine of 1873, a change that first became noticeable with the failure of the monsoon in 1868–9 when district officers were asked to see that no preventible deaths occurred. Employment through minor works of public utility was sought to be introduced as a precautionary measure.

During the Bihar Famine of 1873 the government took other effective steps to prevent loss of life. Tests and restrictions were relaxed in respect of wages or the amount of work done. Money or grain was made available to those who were '*prima facie* in want'. Cultivators were invited to take 'loans of money or rice repayable without interest'.[13] But the most remarkable development during 1873–4 was control of the market. Grains were imported and reserves built up for public distribution. Some district officers even seized private stocks and distributed grain among the needy at rates considerably cheaper than the market rate. The power of trade was thus sought to be neutralized from considerations of public welfare.[14]

The business communities, both European and Indian, doubtless complained. But since the government could not, in the interests of its own stability, afford to ignore the sufferings of the rural masses arising either from drought or from the transfers of their lands to moneylenders or proprietors, it used its executive authority to enforce welfare measures. Sir George Campbell, who presided over the Orissa Famine Commission, realized the necessity of executive interference with the operation of free trade, developed the subdivisional system to promote direct official contact with the people, and took steps for the collection of agricultural statistics over the heads of zamindars in Bengal. The Bengal Tenancy Act of 1885 and the subsequent introduction of cadastral surveys by the government formed part of the same policy of establishing direct bureaucratic links with the rural population in all the districts settled in perpetuity with zamindars. The institution of Chief Commissionerships was also intended to function as a direct central agency for security and developmental purposes in backward tracts.

Despite the influence of the colonial element, the Imperial Government and the Government of India, in fact, went ahead with what might be called organic reforms, with improving the police and the prisons, with codes of 'law and legislative authority', a hierarchy of courts, a trained civil service and all the apparatus of a modern executive. The laws governing revenue or rent regulated relations between proprietors and cultivators; the rural population regarded officials as their representatives in the councils of government. They could bring their troubles to the notice of the government where, as the *Report of the Deccan Riots Commission* showed, they hoped to be treated with sympathy. These conditions were all conducive to political stability within a unitary political arrangement.

But this arrangement was not to last. Two main antithetical developments had started appearing as a sequel to British policy itself: the emergence of organized associations expected to advise on legislative proposals and the rising influence of the urban middle-class elites who made use of these associations to advance their interests in the name of all classes of society. The problem of British policy was how best to enable the Government of India to continue representing the interests of all classes of people without disregarding the small middle-class minority that was emerging as an organized force with backing from the Indian press and the landed gentry, a consequence of the steady fragmentation of large estates. The Indian Councils Act (1892), which recognized the principle of representative government, was a partial answer to the problem.

The formation of political associations was encouraged by the Standing Orders of Dalhousie's Council, which provided for legislation on the basis of petitions from individuals or organized bodies. The Indian Councils Act (1861) limited the scope of petitions, for under the Act the Council for making laws and regulations was to be treated as no more than the Council of the Governor-General, who alone had powers to make rules governing legislative business. But as the legislative proceedings of the Governor-General in Council showed, his legislative measures had not only to reckon with public opinion as reflected in the press, but also with individual and group which the government itself involved if consulted. The government did not necessarily accept the opinions offered, but the practice of inviting comments from private bodies on legislative proposals influenced legislation and, through legislation, the government itself. This led to the establishment of such larger associations as the Poona Sarvajanik Sabha, the Bombay Presidency Association, the Madras Mahajan Sabha, the British Indian Association, the Indian Association, the National Muhammadan Association, and, finally, the Indian National Congress itself.

At the instance of A.O. Hume, a retired civilian, Lord Dufferin himself agreed to the formation of the National Congress in 1885 as a consultative body, for he considered it desirable 'that Indian politicians should meet yearly and point out to the Government in what respects the administration was defective and how it could be improved'.[15] In his presidential speech on 27 December 1886 Dadabhai Naoroji too appreciated the freedom with which the consultative bodies spoke and made suggestions for constitutional advancement. The role assigned to the Congress in its early days was in fact one of collaboration with the government. Even George Yule, the Chairman of the Bengal Chamber of Commerce, was invited to preside over its annual session in 1888. However, the rising influence of the educated middle classes was at the same time taking the Congress towards being what Naoroji called 'a nursery for sedition and rebellion against the British Government'.[16] Interestingly, in a private and confidential letter to Northbrook, Dufferin wrote in June of the same year of 'Babu politicians' coming from the ranks of middle or small proprietors: 'In conclusion, however, I think I can safely say that however annoying may be the violence, childishness and perversity of the Bengalee press and of your Babu politicians, their influence at present is neither extensive nor dangerous. The mass meetings and all the paraphernalia of the Indian Caucus, though they make a noise and may appear effective and formidable in the telegrams which the Associations transmit to England, do not really amount to much for the present, but it does not follow that some years hence what is now in the germ may not grow into a very formidable product.'[17]

It is neither necessary nor within the scope of this work to narrate in detail the controversy that developed between the government and Indian radical agitators for political concessions. However, it is relevant to briefly note the nature of early radicalism in Indian politics and the British reaction to it in terms of India's political unity and integration. The political radicalism of the 1880s was a consequence of the wide gap between the declared educational policy of the Government of India and its implementation. In the words of Macaulay's minute of 2 February 1835, the policy was to create a middle class of educated Indians who might constitute a link between the government and the Indian masses and assist in the governance of the country according to European standards. But while Indians were severely handicapped by the British refusal later to hold simultaneous competitive examinations in India to enable Indians to qualify for superior civil service positions, the government itself claimed to stand forth directly as the best representative and protector of the millions whom it governed through its own bureaucracy. The government was loath to recognize the principle of representative government, which it regarded as an instrument of vested interests in society.

Writing to Northbrook in a private letter of 16 October 1886, Dufferin justified his opposition to the enlargement of the Legislative Council. 'There would be great difficulty', he said, 'in getting hold of the best men, and then, when we have got them, they represent after all only an infinitessimal section of the people and the interests of a minute minority in reference to a great proportion of the subjects with which legislation deals.' He contended that, while the lawyer members of the Council were connected with the 'money-lending classes' and opposed to any relief to debtors, the Brahmins on the education question did 'everything they could do to prevent the lower castes sharing in the advantages, whereas we should wish to do the very opposite'.[18]

On questions of taxation, too, Dufferin was similarly confronted with opposition from V. N. Mandlik of Bombay and Pearey Mohan Mukerji of Calcutta, two leading lawyer members of the Central Legislative Council. They opposed, for instance, the Income-Tax Bill introduced in 1886 to effect an equitable distribution of the tax burden for stability and development purposes. 'We look abroad', said Dufferin, 'and we see that the peasant pays his salt tax, which, though it has been reduced, still supplies us with a yearly net income of £6,000,000; that the landowner pays his land tax and his cesses; that the tradesman and merchant pays his licence tax; but that the lawyer and doctor, the members of the other learned professions, the officers of government, and other persons occupying an analogous status, and the gentleman at large pay little or nothing. . . . There is not one of us who pays any really serious sum from his income into the imperial exchequer.'[19] The Income-Tax Bill was passed because of the support of the Viceroy and brought the professional and service classes under direct taxation.

The lack of official support for Congress demands for representative government became a cause for the popularity of a revivalist and divisive trend in politics, which regarded religion and culture as the basis of political action, and was known both for radicalism and violence in approach and method.[20] A confidential inquiry into the Madras Congress (1887), for example, disclosed the existence of seditious literature circulated by the revivalist, radical element of the organization, with the declared intent of converting the Congress into 'an Indian Parliament', a demand which was said to have been backed by 'tens of millions of men'.[21] The demand was noted by Muslim leaders who viewed it as an attempt to impose Hindu interests. The divisive tendencies of Indian politics were further stimulated by the Report of Lord Dufferin's Committee on Provincial Councils (October 1888) which, instead of proposing direct territorial constituencies, suggested that they be divided into a couple of divisions based on the interests of religious minorities as well as landed and commercial interests, with all forming part of a panel to be elected by members of local bodies, in addition to official nominees to both the divisions.[22]

But Dufferin's proposed liberalization of the provincial councils by introducing the elective principle did not signify a radical departure from the Indian Councils Act (1861), which attempted to render the Excutive Government so strong as not to be handicapped by any expansion of the legislature. The idea was to treat legislative councils as a consultative arrangement, not as representative bodies that were potentially antithetical to imperial authority. The responsibility of preserving India's political unity and stability was viewed as wholly Imperial, not that of Indians.

The proposal of Dufferin's Committee to recognize local bodies as elective constituencies was turned down by the Secretary of State on the ground that the local bodies were too recent in origin to decide on the fitness of candidates. He rejected Dufferin's scheme for the expansion of provincial councils, although he favoured the extension of their powers to discuss the estimates, to interpellate the executive and to originate suggestions. On Dufferin's departure the several provincial governments were consulted and the Government of India then expressed its general approval of the political analysis of Sir Auckland Colvin, the Lieutenant-Governor of the North-Western Provinces and Oudh, who suggested caution in the application of the elective principle. The Government of India thus advised the Secretary of State that the proposal to reform the Indian Councils should not absolutely preclude the Governor-General in Council 'from resort to some form of election where the local conditions are such as to justify a belief that it might be safely and advantageously adopted for the appointment of additional members by nomination or otherwise'.[23]

The Indian Councils Act (1892), however, sedulously eschewed any reference to the term 'election'. It merely empowered the Governor-General in Council with the sanction of the Secretary of State in Council to make regulations as to the conditions of nomination of additional members whose strength in the case of the Supreme Council was in no case to be less than ten or more than sixteen. Even so, because Gladstone pressed for elections to be given as reasonable a trial as the circumstances in India permitted, the British Government. in transmitting the Act of 1892 explained the intentions of Parliament. It said:

Where Corporations have been established with definite powers upon a recognized administrative basis, or where associations have been formed upon a substantial community of legitimate interests, professional, commercial or territorial, the Governor-General and the local Governors might find convenience and advantage in consulting from time to time such bodies, and in entertaining at their discretion an expression of their views and recommendations with regard to the selection of [additional] members in whose qualifications they might be disposed to confide.[24]

It was understood that the discretion of the recommending bodies was to remain unfettered from political considerations. In consultation with Local Governments, the Government of India drew up regulations which Lord Kimberley, the Secretary of State, accepted. In the case of provincial councils a restricted official majority was maintained and the majority of non-official seats left to be filled by recommendation. And although the use of the term election was avoided, the acceptance of nominations by the recommending bodies as a matter of course established beyond doubt the fact of election to an appreciable extent. The Act of 1892 also followed the recommendations of Lord Dufferin's Committee so far so to give the councils the right of asking questions, and of discussing, though not of voting upon, the annual budget. This was an advance over the Act of 1861. The functions of the councils were henceforth to be more than merely advisory.

The same principle of election disguised as recommendation was also adopted in 1892 for the legislative council of the Governor-General. If an official majority was to be kept, it became impossible to admit more than ten non-officials in the case of the central legislative council. Four of these seats were thus allotted to non-official members of the four provincial councils,[25] where municipalities and district boards constituted the recommending bodies, and one to the Calcutta Chamber of Commerce. The remaining five non-official seats were left for nomination by the Governor-General to secure the representation of the vast residuary area and population.

Thus, the elective principle was in practice recognized and introduced in 1892, though not without demur or disguise. The Government could doubtless lay down general qualifications for 'recommendations', but political sense made it impracticable for it either to insist on selection from a panel of names proferred, or to reject individual nominations at its discretion. Functionally, too, there was an advance over the 1861 Act. Even so, the unitary character of the political system remained intact. Legislation formed part of the function of the Executive Government. The Governor-General could on his own responsibility issue an ordinance which had the force of law for six months.

An electoral constituency was not created under the Indian Councils Act (1892) and there was no jurisdictional demarcation between the central and local legislature. The Central Council could legislate for the whole of India. This was considered necessary for India's unity and integration, for Indian society remained composed of a large number of distinct nationalities and mutually exclusive social and ethnic entities with hardly any allegiance to a common political direction. The educated classes, who pressed for the representative principle in India's political constitution, were themselves divided— ideologically, socially and culturally. Above all, the sluggish development of the legislative councils, the potential political alternative to the Imperial agency, tended to become a forum for bickerings rather than a force for political unity.

The legislative councils were, none the less, intended to contribute politically to stability in so far as they constituted the only obvious means of knowing popular reaction to the laws enacted by the

Government. They were regarded as an alternative to what Sir Bartle Frere called the 'perilous experiment of continuing to legislate for millions of people, with few means of knowing, except by a rebellion, whether the laws suit them or not'. Considerations of policy therefore dictated the necessity of associating with the legislative councils, both supreme and subordinate, such Indians as might by their ability and experience enlighten the British in the discharge of their political functions, without divesting them of any essential portion of that Imperial authority which was necessary for their very existence as the paramount sovereign power ruling over a variety of nationalities in different stages of development. It was recognized that the public utterances and the opinions expressed by Indians in the enlarged councils would be reported by the press which, consistently with the safety and integrity of the Indian empire, could be used to explain official policies, correct wrong impressions or controvert a false statement, thus satisfying members of the council and the public at large.

The enlargement of the Central Legislative Council, as also the provincial councils, thus became recognized as a means of securing the stability of the Indian empire, more especially in a developing situation which, in the context of the Russo-Japan War (1904–5) and the Partition of Bengal, was marked by continuing unrest and a tendency towards organized violence. To provide an answer to the problem of Indian political advancement as a means of ensuring peace and stability, Lord Minto on his own initiative but with the full cognizance and approval of the Secretary of State, proceeded to take steps towards increasing the representative element in the central and provincial legislative councils. The result was the Morley-Minto Reforms, the Indian Councils Act, which was passed on 25 May 1909. It not only increased the number of non-official members, but also introduced changes to augment the powers of these councils.

Consequently, the official majority maintained in the provincial councils since 1861 was abandoned. It was decided to face the risk of a non-official majority by relying partly on the use of the veto, on the statutory restrictions attaching to provincial legislation being subject to central approval, and on the concurrent powers of legislation possessed by the Governor-General's legislative council for the enactment of necessary laws which a provincial council might refuse to enact. The Act redesignated 'the councils for making laws and regulations' as 'Legislative Councils'. It enlarged provincial legislatures to a maximum limit of 50 additional members in the case of the larger provinces and 30 for the smaller. The official and nominated non-official members together formed a small majority over elected members, except in Bengal which had a clear elected majority.

The Indian Legislative Council was enlarged to the extent of 60 additional members, of whom 28 were officials, 27 elected members and 5 nominated. The object of this nominated official majority was to ensure the stability of the government in the event of an adverse vote as well as to see that it did not have 'to enter into negotiation with rival factions to save its measures from rejection'.[26]

Functionally, too, the change introduced was not imperceptible. The regulations drawn up under the Act recognized the principle of election, which the regulations of 1893 had admitted in practice, though not legally. The legal recognition of the elective principle therefore necessarily involved legal qualifications being imposed for election. The deliberative function of the councils was considerably enlarged, for the new Act empowered councils to discuss the budget thoroughly before it was finally settled, to propose resolutions on it, and, finally, to divide the councils on the issues raised during the discussions, not only on budgetary provisions, but on all matters of general public importance. The resolutions adopted were to operate as recommendations to the Executive Government. The right to ask questions, too, was enlarged by supplementaries being permitted to the original question already raised. The regulations governing interpellation, in fact, served as an inquest into official performance.

It is unnecessary to review here the official or private correspondence that finally resulted in the constitutional reforms of 1909 to meet the nationalist demand for greater political advance. However, it is important to recognize the fact that the slow but steady political advance was unwittingly creating conditions highly prejudicial to the maintenance of Imperial supremacy. The enlarged legislative councils, with their growing powers, were developing as an embryo for the emergence of an alternative Indian sovereignty, despite Lord Morley's disclaimer: 'If it could be said that this chapter of reforms led directly or indirectly to the establishment of a parliamentary system in India, I, for one, would have nothing at all to do with it.' His emphasis on 'the sovereignty of the British Parliament', which was reinforced by Minto's view of the reforms as no more than a 'constitutional autocracy', was doubtless intended to preclude any approximation to an alternative Indian sovereignty. In the context of very limited constituencies and indirect franchise provided under the Act of 1909, such a possibility was considered 'unthinkable' in India. But this was to be only 'until an Indian electorate which was not then in sight, had arisen to take the burden from its shoulders'. Nonetheless, features of the Morley-Minto reforms were admittedly recognized as constituting 'a decided step forward on a road leading at no distant period to a stage at which the question of responsible government was bound to present itself'.[27] That stage came in 1917. The announcement made by the British Government on 20 August 1917 defined the goal of British policy as 'that of increasing association of Indians in every branch of the administration and the general development of self-government institutions with a view to the progressive realization of responsible government in India as an integral part of the British Empire'. Clearly, it envisaged a gradual transfer from Imperial to Indian hands of the responsibility for the defence, unity and stability of India. The change so envisaged was to be within the framework of recognized law and the constitution, a principle of policy which was briefly suggested by Gokhale in his political testament of 1914 and later elaborated in a Memorandum on Post-war Reforms submitted by nineteen members of the Indian Legislative Council in 1916.[28] The *Montford Report* (paras 291 and 292), the *Report of the Joint Select Committee* (1919) and Sections 19A and 45A(d) of the Government of India Act (1919) were, as shown below, all unanimous in recommending the transfer of authority from London to New Delhi without a fresh and formal enactment.

It is important to notice the electoral system sought to be introduced, both to satisfy the requirements of political stability and to meet the demands of the educated classes for representative government. In 1892 the Secretary of State and the Government of India had agreed that, while introducing the machinery of election, stress must be laid on representation by classes, races, communities and interests,[29] an electoral principle that was considered to be in keeping with the requirements of India's heterogeneous and status-bound society continuing in a state of mass illiteracy. In view of the seeming middle-class bias against the interests of the rural masses, it was feared that a regular territorial system of election on Western lines ran the risk of getting only such men elected as had nothing in common with the people whose support was considered vital for Imperial stability. In the words of C. H. T. Crosthwaite, the Home Member, 'Instead of men representing the upright and intelligent opinion of India, we shall have men of the political agitator class, notoriety-hunters working for their own ends regarded by most of their fellow countrymen with suspicion and dislike.'[30] As Morley observed, this was so because 'Indian gentlemen of position ordinarily refuse to offer themselves as candidates to a wide electorate, partly because they dislike canvassing, and partly by reason of their reluctance to risk the indignity of being defeated by a rival candidate of inferior social status'.[31] Since the European form of representative government was not considered suitable for Indian conditions, the existing electoral arrangement was therefore sought to be reformed by the Indian Councils Act of 1909.

The whole body of correspondence on the Indian question was dominated by the concern of Lord Minto and his bureaucracy to devise an electoral system which might secure the representation of a wide range of interests. The regulations of 1893 had failed to accomplish this, more especially in respect of Muslims. As many as 45 per cent of the non-official members elected to the Central Legislative Council since 1892 belonged, for example, to the professional Hindu middle class; the landholders obtained 27 per cent of the seats, while Muslims, who were of major significance in terms of all-India political stability, commanded only 12 per cent against 23 per cent their total population in the country.

The seeming decline of Muslims with the rise of British rule was brought to the notice of Lord Ripon's government in 1882 by Syed Ameer Ali of the National Muhammadan Association, Calcutta, through a memorial.[33] Their plight in terms of employment was reviewed by the Indian Public Service Commission (1886-7) and the Census of India (1901).[34] For instance, even in 1903 the total number of Muslims employed in Bengal, on a monthly salary of Rs 75 and above was only 283, as against 2,263 Hindus.[35] Lord Minto first took a political view of the state of Muslim decline; without prior reference to the Secretary of State or consultation with the provincial governments, he conceded the principle of a special, separate electorate when a deputation of Muslims waited on him under the leadership of the Agha Khan on 1 October 1906. In these circumstances the All-India Muslim League was established at Dacca on 30 December 1906 to forward as an organized body the claims presented to the Viceroy by the deputationists. The principle so conceded formed part of the electoral system under the new Act, an arrangement which was regarded by the British as the 'soundest solution of the [political] problem', and one that supplied 'the requisite counterpoise' to the 'excessive influence' of the 'educated and professional classes', a counterpoise which was further reinforced by creating an additional electorate recruited from the landed and monied classes.

Apart from the Muslims, landholders and the Planters' Association, who elected their representatives directly, the professional classes were to be represented indirectly through a general or mixed electorate comprising groups of district boards and municipalities, corporations and universities, returning members in the first instance to a provincial legislative council which, in turn, elected representatives to the Legislative Council of the Governor-General. The regulations established under the Morley-Minto Reforms of 1909 doubtless followed the principle of representation by classes, communities and interests, a principle already adopted in 1893. But to make that principle an effective instrument against the risk of imbalance in the distribution of seats, care was taken to provide a special Muslim 'counterpoise' against middle-class professional elements who had thus far dominated. Besides the five seats reserved for Muslims in the Indian Council through separate electorates, Muslims could secure further seats through the general as well as the landholders' constituencies, in addition to a resort to nomination in case these openings were still found insufficient. Their strength in the Indian Legislative Council thus rose to eleven out of thirty-two non-officials. The Muslim community no doubt had a special and genuine case, but it was handled against a background of official prejudice against agitational politics, which distorted and created an imbalance in the political behaviour of revivalist and radical nationalists.

The principle of indirect election, which had been continuing since 1893, was not without merits as a more representative and stabilizing arrangement than had existed earlier. It had the support even of Gokhale, the eminent national leader. His 1914 draft scheme for post-War reforms recommended indirect elections by different constituencies and interests, with emphasis on four-fifths of the legislators being elected. In Bombay Presidency, for example, he wanted each district to return two members, one representing municipalities and the other district and taluk boards. Apart from Muslims, the other bodies, communities or interests he recommended for representation were the Bombay Corporation, the Karachi Chamber of Commerce, the Ahmedabad millowners and even the Deccan Sardars and Lingayat Communities.[36] The Congress-League Scheme of December 1916 was likewise weighted on the side of indirect constituencies. It recommended that the franchise for the central legislature should be widened as far as possible on the lines indicated by the Muslim constituencies for provincial legislative councils, and that elected members of the provincial councils should also form an electorate for the return of members to the Indian Legislative Council.[37]

However, the 'Memorandum of the Nineteen' members of the Central Legislative Council, presented to the Viceroy in November 1916, favoured a broadening of the franchise and its direct extension to the people on a territorial basis. The 'Nineteen included such eminent members of the bar as Tej Bahadur Sapru, Madan Mohan Malviya, Mahammad Ali Jinnah, Bhupendra Nath Basu, Mazharul Haq, V. S. Srinivasa Sastri and others.

The Southborough Committee, which visited India in 1919, recommended the indirect system, but influenced by Edwin Montagu, a protagonist of the Congress, the Joint Select Committee held the contrary view, which was accepted by Parliament. The Government of India Act (1919) thus introduced direct election by territorial constituencies, even though care was taken to ensure that the enfranchised population of British India did not exceed 3 per cent of the total. The consensus of opinion, even amongst non-official bodies, did not at this period oppose educational and property qualifications for voters, and as late as 1928 the Nehru Report did not omit to impose some qualifications and, further, emphasize that election must be on the basis of proportional representation with a single transferable vote to enforce the principles of representative government in the full sense.

The White Paper issued in May 1933 recommended the continuance of the existing system based on direct territorial constituencies, as well as the restrictions suggested by the Franchise Committee concerning the age, education and property of voters. The Joint Select Committee (1934), which made a deeper examination of the electoral problem in terms of its relevance to representative principles as an integrating force, differed from the Statutory Commission's Report as well as the White Paper, both of which were unanimously in favour of direct elections. The Joint Committee stressed the peculiar conditions of Indian society where, in a state of mass rural illiteracy, direct election could accentuate social divisions. It felt that the apprehended risk could become palpable with the extension of franchise from the existing 3 per cent to a minimum of 10 per cent of the population which the Franchise Committee had already suggested. The Joint Committee feared that the franchise so extended might in due course lead to adult suffrage and a consequent threat to peace and tranquillity arising from political exploitation of social divisions from motives of personal or group interest:

> Indeed, any considerable extension of the franchise under a system of direct election would cause an inevitable breakdown. We do not believe that constituencies both of large size and containing an electorate of between 2,00,000 and 3,00,000 people can be made the basis of a healthy parliamentary system. We think that Parliament and Indian opinion should face these facts and should recognize that direct election, apart from its immediate merits or demerits at the present time, cannot provide a sound basis for Indian constitutional development in the future. We cannot believe that it would be wise to commit India at the outset of her constitutional development to a line [of direct election] which must prove to be a blind alley.[38]

The Joint Committee therefore recommended a system of indirect election by the provincial legislatures, which, in the given circumstances of the country, was considered the most practical arrangement.

The proposed indirect system, which was accepted by Parliament, was the result of 'a careful and prolonged examination of the matter in all its aspects'. The Committee was conscious of the arguments brought against it by protagonists of the direct system, and agreed that the indirect system be subject to review in future. But it did not omit to say:

> that the ultimate solution [of the electoral problem] may well be found in some variant, either of the system whereby groups of primary voters elect secondary electors who vote directly for members of the federal assembly, or of the system whereby those already elected to local bodies, such as village panchayats, are the voters who vote directly for members of that assembly. Systems of this kind apparently work with considerable success in many countries where conditions are not dissimilar to those in India. But the discovery of the best method of adopting those ideas to India's needs ... is clearly one which should be made by Indians themselves in the light of their experience of the practical working of the [indirect] representative institutions under the new (1935) constitution.[39]

The Indian bureaucracy and national leadership, with only a few Gandhian exceptions, remained none the less united in their support of a centralized parliamentary constitution. They did not accept the idea of village panchayats being established as primary units of government as an efficient instrument of stability and integration. The Draft Constitution prepared between October 1947 and mid-February 1948 made no mention whatever of village panchayats. It consisted primarily of borrowed and modified provisions from the British and American constitutions and the Government of India Act (1935), an arrangement based on direct territorial constituencies, with their ever increasing populations. Rajendra Prasad and some others admittedly wanted the Constitution of India to begin with the village and go up to the Centre through the provinces, an indirect system based on a principle which he considered the most representative, broad-based and politically stabilizing. In a letter to B. N. Rau, the constitutional adviser, Prasad pointedly said: 'I strongly advocate the idea of utilizing the adult franchise only for the village panchayat and making the village panchayats the electoral college for electing representatives to the provinces and the centre.'[40] Interestingly, the idea he so commended for acceptance as part of the Draft Constitution, was in keeping with the suggestion made by the Joint Committee of Parliament in 1934 in the interests of stable constitutional development.

However, Rajendra Prasad's suggestion was not accepted. In his reply Rau said that the Constituent Assembly had already decided in favour of direct election for the lower houses at the Centre as well as in the provinces and that there was no going back on that decision. The Congress leaders doubtless paid lip service to the Gandhian concept of a broad-based and decentralized political constitution, but their commitment to direct election based on adult suffrage meant clear acceptance of a centralized parliamentary constitution and the rejection of the village panchayat as a basic unit of India's political system.

The Indian States

Though governed by treaties, grants and *sanads*, the 'Native States' formed an integral part of a single geographical whole called India. The British slowly and gradually devised a constitutional machinery to make that geographical whole a unified political entity. Passing from the original policy of non-intervention in all matters beyond its own 'ring-fence' to that of 'subordinate isolation' initiated by Lord Hastings, Britain moved on to a kind of relationship which bordered on union and co-operation within the broad framework of a 'two Indias' concept, which recognized the personal rule of the Princes and their control over internal administration as the basis of overall unity.

The development of an integral connexion with the States in the post-Mutiny period consisted not only of their relations with the British Crown but also of their growing interest in many matters common to the land to which they and the British provinces alike belonged. There were basically three conditions which acted as inducements to integration, conditions which flowed from history, from institutional interactions and changing patterns of political development in British India and adjoining areas.

Historically, the term 'Native States' was applied to a collection of more than 600 rulerships, ranging from States with full autonomy concerning their internal affairs, to States where the Government of India exercised through its agents large powers of internal control, and down to the owners of petty estates. The general clause in many of the treaties with the major states which stipulated that the chief would remain the absolute ruler of his kingdom, did not preclude interference. In the earlier days, British agents at the princely courts assumed responsibility for maintaining order, for preventing supposedly 'inhuman' practices, and to guide the hands of an allegedly weak or incompetent chief as the only alternative to the termination of his rule. The Government of India, in fact, continued to acknowledge as trustee the responsibility for the proper administration of a State during a minority, and also an obligation to check 'misgovernment'.

But the right to interfere in practice apart, the treaties themselves were a formal guarantee of security from without. The paramount power conducted the relations of all 'Native States' with foreign powers as well as other princely States. It intervened as soon as the internal peace of their territories appeared to be threatened. The States, on the other hand, shared with the Government of India an obligation for the common defence of the country. These features developed in course of time an identity of interests between the States and the paramount power, which brought both sides into closer relations, a kind of relationship which encouraged the princely States to fight for the allied imperial cause when World War I was declared in 1914.

However, this cooperation did not signify a denial of the 'two Indias' concept which called for mutual abstention from interference in the internal affairs of each other. The independence of the 'Native States, in matters of internal administration carried with it a counter-obligation of non-interference in British Indian affairs. Care was sought to be taken on either side to see that no attempt was made to cross this frontier.[41] It was on this basis that a perceptible process of infiltration went on and which contributed to unity. The Government of India, for example, helped the States in times of famine; it provided for the training of their officers to supervise their financial administration, improve agriculture and irrigation, and induced many of them to adopt the civil and criminal codes of British India along with the acceptance in several cases of British India's educational system. Such co-operation was extended likewise to matters of police and judicial administration as well as to the use of the railways and telegraph systems carried through many of the States. It is true that there was no uniformity in the level of the various States' development, but the characteristic feature of all was the dominance of the chief and his personal rule, which remained for most purposes unimpaired under the terms of treaties with the major States.

Common interests tended increasingly to bring the British provinces and the States closer together. Matters common to both were, more especially, those which touched on defence, tariffs, exchange, railways, posts and telegraphs, salt and opium. The evolution of the Chiefs' Conference, first initiated by Lytton's proposal to constitute an Imperial Privy Council, was intended to create an awareness of the reality of these shared interests. Its result was ephemeral, but Dufferin's institution of the Imperial Service Troops was more effective in promoting the feeling of a community of interests. Curzon planned to constitute a Council of Ruling Princes with the same object in view, and Minto first tried an Imperial Advisory Council and then an Imperial Council of Ruling Princes. It was Lord Hardinge, however, who sought the collective opinion of the Princes as trusted colleagues; and, while responding to his invitation, the Maharajas of Gwalior and Indore laid equal stress on the 'identity of interest between the two halves of India'.[42]

Hardinge's government was, in fact, credited with the perceptible evolution of two main trends: one towards the promotion of self-governing institutions leading to provincial autonomy; the other towards the formation of a Council of Princes for the discussion of questions affecting the States as a whole. The former symbolized a break from the continuing unitary character of government and sought to substitute without prejudice to India's unity a popular political mechanism. The latter, on the other hand, represented a concrete step in the direction of a one-India concept, which was not to be imposed but established with the co-operation of the Princes.

What immediately contributed to the growth of provincial autonomy was, of course, the creation of an Executive Council which the Indian Councils Act (1909) provided even for a Lieutenant-Governorship, an Executive Council which partially included Indian members functioning in the context of a non-official majority in the local legislature. The presence of local leaders in the executive councils with the added strength of the non-official legislative members constituted a popular political forum for the advocacy of provincial autonomy as an effective means to promote provincial interests. But the Government of India's dispatch of 25 August 1911 indicated for the first time the kind of political arrangement where unity could be preserved despite the provinces being autonomous. For the main consideration, as the dispatch pointed out, was 'gradually to give the Provinces a large measure of self-government, until at last India would consist of a number of administrations, autonomous in all provincial affairs, with the Government of India above them all, and possessing power to interfere in cases of misgovernment but ordinarily restricting their functions to matters of imperial [all-India] concern'.[43] In a confidential memorandum (October 1915)[44] addressed to the London authorities, Hardinge reiterated what he called 'the surest and safest policy' of 'provincial self-government' which he considered 'necessary to maintain the faith of the people in British justice'.[45] The line of future policy which Hardinge suggested bore no immediate fruit, but the principle of policy he stressed was not without avail. The conditions which led to the announcement of its goal on 20 August 1917 and the establishment of Dyarchy in the provinces under the Government of India Act (1919), justified his calculations.[46]

Noticeable changes also occurred in the other half of India, namely, the States. Lord Chelmsford, who succeeded Hardinge, utilized the system of conferences to discuss general questions affecting all the States. The Gaekwar of Baroda even expressed the hope that the system of conferences, which had by that time become annual, might develop into a permanent council or Assembly of princes. Looking ahead to the future, the joint Montford Report of 1918 in fact visualized India as presenting the 'semblance of some form of federation', with the provinces becoming in time 'self-governing units' and the princely States involved in all matters of common concern touching, for example, on defence, security and development. A forward-looking Report, it thus recommended the establishment of a permanent Council of Princes annually invited to appoint a Standing Committee and provide means for joint deliberation between the Government of India and the Princes on matters of common interest to both. The Report in addition suggested measures to improve relations with the Crown and remove occasional misunderstandings arising either from the exercise of powers by the Crown to determine succession or to deprive a prince of his rights and dignities. Emphasis was in this regard laid on the appointment of Commissions of Inquiry prior to any action being taken for the settlement of disputes.[47]

After due consultation with the provincial governments, the projected Chamber of Princes was set up on 8 February 1921. It is true that the Chamber and its Standing Committee did not for some time fulfil all the expectations formed of them. As the Butler Committee reported in 1929, their decisions did not bind the Princes either as a body or individually. Some of the leading Princes even refused to attend meetings of the Chamber. The Nizam of Hyderabad, for instance, always 'adopted an attitude of entire detachment from it'. Even so, as the Committee acknowledged, 'the constitution of the Chamber and its Standing Committee was a great and far-reaching event. It meant that the Paramount Power had once and for all abandoned the old policy of isolating the States and that it welcomed their co-operation'.[48] As recommended by the Montford Report, the political practices of the paramount power were duly codified. The policy of secrecy was given up and the old process of reaching decisions without discussion was substituted by open conference and consultation, which opened a new chapter of harmony between the paramount power and the Indian States.

Within the framework of a unified system of administration under the supreme authority of the Government of India, matters had thus started moving towards the concept of federation, an all-India constitutional arrangement designed to forge unity between the British India provinces and the Indian States. It was to be a new kind of central organism or federal polity for which there was no precedent. For federations had in the past resulted from the union of independent, or at least autonomous, states which agreed to come together and create a federal authority by the surrender of a certain defined part of their sovereignty or autonomy. The Indian States doubtless possessed a semblance of sovereignty or autonomy, but the British provinces were not even autonomous. Both in administrative and legislative matters they were subject to the control of the Government of India, and such authority as they exercised had only been devolved upon them under a statutory rule-making power by the Governor-General in Council. No federal polity could thus come into being unless the Indian States agreed as a body or individually to join it by a surrender of such part of their independence as they themselves determined, and the British provinces came to be invested with an autonomy that only an Act of Parliament could determine.

The legal basis of the projected federal constitution as an instrument of all-India unity was explained by the Joint Select Committee of Parliament as follows:[49]

It is clear that in any new constitution in which autonomous Provinces and Indian States are to be federally united under the Crown, not only can the Provinces no longer derive their powers and authority from devolution by the Central Government, but the Central Government cannot continue to be an agent of the Secretary of State. Both must derive their powers and authority from a direct grant by the Crown. We apprehend, therefore, that the legal basis of a reconstituted Government of India must be, first, the resumption into the hands of the Crown of all rights, authority and jurisdiction in and over the territories of British India, whether they are at present vested in the Secretary of State, the Governor-General in Council, or in the provincial Governments and Administrations; and second, their redistribution in such manner as the Act may prescribe between the Central Government on the one hand and the Provinces on the other. A federation of which the British India Provinces are the constituent units will thereby be brought into existence; but since the rights, authority and jurisdiction which will be exercised on behalf of the Crown by the Central Government do not extend to any Indian State, unless the Ruler has agreed to their exercise for federal purposes in relation to the State, it follows that the accession of an Indian State to the Federation cannot take place otherwise than by the voluntary act of the Ruler. The Constitution Act cannot itself make any Indian State a member of the Federation; it will only prescribe a method whereby the State may accede and the legal consequences which will flow from the accession. There can be no question of compulsion so far as the States are concerned. Their Rulers can enter or stand aside from the Federation as they think fit. They have announced their willingness to consider federation with the Provinces of British India on certain terms; but whereas the powers of the new Central Government in relation to the Provinces will cover a wide field and will be identical in the case of each Province, the Princes have intimated that they are not prepared to agree to the exercise by a federal Government for the purpose of the federation of a similar range of powers in relation to themselves. This is a further aspect of the matter which differentiates the proposed Federation from any other; for not only will some of the constituent units be States whose subjects will continue to owe allegiance to their own Rulers, modified only within the federal sphere, but the powers and authority of the Central Government will differ as between one constituent unit and another.

A closer association of British India with the States had, in fact, been stressed by the Statutory (Simon) Commission, although the question of an all-India federation was deemed to be the ultimate aim. The announcement of the proposed Round Table Conference in September 1930, which followed the Report of the Commission, was intended to discuss the question as part of the main issues involved in India's constitutional development. The Congress had resolved to boycott the Conference, but all other shades of opinion and interests were duly represented. The Indian States sent a strong delegation, including some of the prominent rulers. The delegation arrived in London towards the end of October 1930, and the actual work began formally on 17 November. The interval was utilized for informal discussions between the British India delegates and representatives of the States, the whole object being to evolve a suitable scheme for all-India unity in constitutional terms, not a mere 'Council for Greater India'. Contrary to general expectations, the idea of federation was unanimously accepted as the best constitutional device to maintain India's unity and integration. At the instance of Sir Tej Bahadur Sapru, the Liberal leader, the Maharaja of Bikaner and the Nawab of Bhopal declared on behalf of the princes that they would come into the proposed federation provided their autonomy in matters of internal administration remained inviolable. The White Paper issued at the conclusion of the Third Round Table Conference[50] contained the proposals that emerged during the deliberations of the Round Table Conference. The legal observation which the Joint Committee of Parliament made was, in fact, a result of what had already transpired on the issue of federation, including the Communal Award which was modified later by the Poona Pact in 1932.

Consistently with its legal interpretation of the proposed federation, the Joint Committee recommended that the Ruler of a State be required to execute an Instrument of Accession signifying to the Crown his willingness to accede to the federation in respect of those matters which he had agreed to recognize as federal; such matters were to be handled by the federal authority to be established by a Constitution Act. Outside the limits to be so defined by the Instrument of Accession, the autonomy of the States and their relations with the Crown were to remain unaffected. In order to ensure that the Instruments of Accession followed a standard form, the White Paper even set out in a separate list the number of subjects which were to be exclusively federal. But as the proposed federation was not to be brought into existence except by a Royal Proclamation, and as no such Proclamation was to be issued until at least half the number of States in the Federal Upper Chamber had signified their desire to join the Federation, its formation remained subject to acceptance by the major Indian political parties.

The Joint Committee of Parliament recognized that the essential conception of British parliamentary government did not exist in India. The principle of majority rule was, of course, formally accepted, but constituted as India society was on a non-territorial, communal principle, there was, for example, little or no willingness on the part of the minority to accept decisions of the majority without demur. Despite this, the Joint Committee felt that such handicaps 'must be faced, not only by Parliament,' but by Indians themselves'—even if it was 'impossible to predict whether, or how soon, a new sense of provincial citizenship [on an all-India territorial principle], combined with the growth of parties representing divergent economic and social policies [on secular lines], may prove strong enough to absorb and obliterate the religious and racial cleavages which thus far dominate Indian political life'.[51] A successful working of parliamentary government under provincial autonomy was in these circumstances not without serious hazards unless the new Constitution Act provided for statutory 'safeguards' as a precondition for the introduction of parliamentary practices based on the British constitution, although it was recognized unequivocally that the new Indian constitution must contain the seeds of that growth in the provinces. The Joint Committee thus emphasized that the 'Federal Legislature must be constituted on different lines from the [existing] Central Legislature of a unitary state'. 'The Statutory Commission, the Committee added, recognized this truth and had proposed a new form of legislature at the Centre, specifically 'designed to secure the essential unity of British India'.[52]

The question of the 'essential unity' of the country did not arise so long as the unitary constitution created under the Government of India Act (1858) remained for all practical purposes operative even under the Government of India Act (1919). It arose from the difficulties of a proposed federation composed of disparate units, namely, the British India provinces and the Indian States. These difficulties were sought to be resolved by a new form of legislature which, for purposes of unity and integration, combined both into a federation through the same Act. This was considered necessary, for, apart from the uncertainty involved in the accession of the Indian States or their individual princes, the creation of autonomous provincial units without a corresponding adaptation of the existing legislative and executive arrangements could give full play to the centrifugal forces of provincial autonomy without any provision 'to counteract them and to ensure the continued unity of India'.[53] The creation of a new Central Legislature on federal principles, with provisions for a strong political executive with powers to exercise 'special responsibilities' in the provinces on behalf of the Centre was recognized as

Institutional Agencies and Political Stability 217

the best path to constitutional advance without threatening India's unity. This statutory recognition provided regional freedom without prejudicing the national edifice.

Interestingly, the Statutory Commission emphasized full provincial autonomy as the only form of constitutional advance over the 1919 Act. But since the projected inclusion of the Indian States envisaged a federal constitution, which clashed with the unitary principles of earlier developments, the Commission held that it would be impossible to attain that federal objective except by a constitutional evolution of several stages, the first stage being 'centrifugal', with the provinces constituting themselves as individual states in thier own right before being asked to surrender part of those rights voluntarily to form a federal government, representing the second, centripetal stage. According to Simon, the Central Government was to be a 'stop-gap' arrangement, and the constitutional advance in the provinces treated as experimental.

There was no quarrel with the concept of full democracy in the provinces. There had been a sustained demand for it ever since 1921. But the new political situation created by the Civil Disobedience Movement called for a definite constitutional advance at the Centre, which Simon wanted to leave more or less unchanged, or vaguely expressed in respect of both the Central Executive and Legislatures, with a view to maintaining the facade of a strong Centre. The principle on which the Government proceeded to act was, however, to strike a balance between standing fast at the Centre on the erstwhile unitary tradition and the existing political pressure involving the transfer of power from the British Parliament in London to the Legislative Assembly in New Delhi. This search for a political balance gave rise to the idea of 'reservations' and 'safeguards'. The principle of reservation led to dyarchy at the Centre, while that of safeguards not only ensured full ministerial responsibility in the provinces for all departments, but eventually also a 'transfer of power' to the Indian Legislature, to be there exercised by responsible ministers, subject, in specific cases, to the 'overriding powers' of the Governor-General and the provincial Governors functioning as the agents of Parliament, the sovereign authority.[54]

This new approach of the Government of India was perhaps the result of the welcome given to the idea of federation by almost all the Princes present at the first Round Table Conference which met in November 1930. It was reinforced by the announcement of the British Prime Minister who, on 19 January 1931, said: 'With a Legislature constitued on a federal basis, His Majesty's Government will be prepared to recognize the principle of the responsibility of the Executive to the Legislature.' The acceptance of the federal principle by the Princes and of the principle of responsibility at the Centre by the Prime Minister thus destroyed the relevance of Simon's proposals regarding the 'centrifugal stage', in the provinces, which was completely given up. But this did not signify any weakening of the Centre. The principles of reservation and safeguards, which constituted the core of the constitutional advance under the 1935 Act, were most important instruments of unity and stability in India's plural society.

While stressing the need for strong executives in the provinces the Joint Committee did not intend to devalue the legislature. 'We have no wish', it said, 'to underrate the legislative function; but in India the executive function is, in our judgement, of overriding importance. In the absence of disciplined political parties, the sense of responsibility may well be of slower growth in the legislature, and the threat of a dissolution can scarcely be the same potent instrument in a country where, by the operation of a system of communal representation, a newly elected legislature will often have the same complexion as the old'. Consistent with the views of the Statutory Commission, the Joint Select Committee, too, emphasized that 'there must be an executive power in each Province which can step in and save the situation before it is too late. This power must be vested in the Governor, and so strongly have we been impressed by the need for this power, and by the importance of ensuring that the Governor shall be able to exercise it promptly and effectively, that, among other alterations in the White Papers, we have felt obliged to make a number of additional recommendations in regard to the Governor's sources of information, the protection of the police, and the enforcement of law and order'.[55] These, as well as provisions relating to the accession of the Indian States, were incorporated in the Government of India Act (1935), a legal and constitutional scheme which provided full scope for regional development as part of a strong and unified all-India framework. It is not intended here to attempt even a brief analysis of this constitutional arrangement in its totality, an arrangement which contained the seeds of sustained political unity and ordered progress. It was not given a fair trial. The federal portion of the Act contained special provisions for the administration of defence, ecclesiastical affairs, foreign relations and the tribal areas. These subjects were to be administered by the Governor-General 'in his discretion', with the assistance of counsellors. Other federal subjects were to be administered by the Governor-General with the assistance of ministers responsible to the Federal Assembly. The Governor-General was, in addition, invested with special powers to prevent the disruption of peace and tranquillity, to safeguard the financial stability of the country, to protect the minorities and their legitimate rights, to ensure the necessary enactment of legislation, and to promulgate necessary ordinances. This portion of the Act, however, did not come into operation, thanks to the opposition of the nationalist leadership to dyarchy at the Centre, and the outbreak of war which resulted in the suspension of negotiations with the Princes.

Pending the establishment of a federation, the erstwhile Central Government and Legislature continued to operate, with certain necessary modifications, under the transitional provisions contained in Part XIII of the 1935 Act. These provisions, in fact, remained in force till the transfer of power to Indians, subject mainly to two changes. The relations of the Crown with the Indian States were, for example, no longer the concern of the Government of India. These were transferred to a new functionary called the Crown Representative who was, in fact, the Governor-General himself, exercising that function through Political Agents or local Residents. In the case of the provinces, however, accession to the Centre was automatic. The provincial governments were made autonomous without an all-India federation, with separate lists of subjects in addition to a concurrent list. The administration of a province was to be carried on normally by the Governor with the assistance of a council of ministers responsible to the legislature, except in cases where the Governor had special responsibility with powers to override his ministers. Besides, the Act had also provided under Section 93 that, if at any time the Governor of a province was satisfied that its administration could not be carried on in accordance with the provisions in it, he could by proclamation take upon himself the administration of the province.

With its Faizpur Resolution of December 1936 the Congress started to combat the safeguards and end the whole constitution, but finally it decided to contest the elections to the provincial legislatures; and when the results of the elections held early in 1937 turned out in its favour, it also resolved to accept office, leading to the formation of homogeneous ministries of its own in such provinces as Madras, Bombay, the Central Provinces, Bihar, Orissa, and the United Provinces. Congress coalitions were later formed in the North-West

Frontier Province and Assam. These eight ministries held power till October 1939 when, instead of deciding to co-operate in the conduct of the war against Fascism, they resigned office. The Governors of these provinces had to take over under Section 93 of the Act.

The projected all-India federation was, in fact, intended to preserve India's unity without prejudicing the country's constitutional advancement, a development which envisaged replacement of the unitary imperial apparatus by a national alternative as an instrument of political integration in a plural society. V. P. Menon, who possessed an intimate knowledge of the constitutional machinery from inside and held a nationalist view of the whole development, has acknowledged the unifying spirit of the Act, which the leadership of the Congress failed to appreciate adequately. 'The underlying concept of an all-India federation', he says, 'was to preserve the essential unity of the country. But it is sad to reflect that in the clash of politics, the struggle for power, the wrangle for ascendancy, and the scramble far gains on the part of the political organizations, politicians and the Princes, the scheme of federation became a tragic casualty. The Congress condemned it for reasons mostly divorced from facts and realities.'[56]

The resignation of the Congress ministries was necessarily followed by a period of constitutional deadlock during the War; none the less there were efforts by the British government and some Indian leaders to resolve the deadlock as a means to win the War without affecting the unity of the country or the prospects of its constitutional progress. The efforts so made included, for example, the 'August Offer' of 1940, the proposals of the Cripps Mission and the 'Johnson Formula' of 1942, the Bhulabhai Desai-Liaqat Ali Pact of 1944, the Sapru Committee recommendations and the Wavell Plan of 1945. A new chapter opened with the victory of the Allied Powers and the success of the Labour Party in England in the elections of 1945. On 19 February 1946, Lord Pethick-Lawrence in the House of Lords and Prime Minister Attlee in the House of Commons simultaneously announced the decision of the British Government to send a special mission to India. It was to consist of three Cabinet Ministers, and in association with the Viceroy, it was to seek an agreement with Indian leaders to resolve the constitutional impasse. The Cabinet Mission arrived in India on 24 March 1946 and stayed for nearly three months. Only when the Congress and the Muslim League, the two major contestants for power, failed to reach an agreement, did the Cabinet Mission announce its proposals—on 16 May 1946. Its proposals re-

Institutional Agencies and Political Stability 221

jected the League's demand for a separate, sovereign Pakistan, which included districts in Punjab, Bengal and Assam where the population was predominantly non-Muslim. The basic principle of the Mission's recommendations was to provide for an all-India constitution, a Union of India, embracing both British India and the Indian States, and dealing with such subjects as defence, foreign affairs and communications. The recommendations relating to the States were for the first time made specific in order to give stability to the proposed all-India constitution. The States were to retain all powers other than those ceded to the Union. They were to be represented in the final Constituent Assembly, subject to a maximum of 93 members. In the preliminary stage they were to be represented by a Negotiating Committee. They were to be associated even with the settlement of the Union Constitution in the same way as the provinces or their groups proposed by the Cabinet Mission.

Consistent with the proposals of the Cabinet Mission, the Chamber of Princes set up a Committee to negotiate the entry of the States into the Constituent Assembly, its personnel being announced by the Chamber of Princes on 21 November 1946. The Constituent Assembly appointed in December its own corresponding committee to confer with the Negotiating Committee of the Princes. N. Gopalaswami Ayyangar had in the meantime prepared some preliminary notes concerning the representation of the princely states. These were adopted by the Constituent Assembly on 22 January 1947. The participation of the States in drafting an all-India Constitution along with the provinces was thus assured.[57] The Interim Government set up at the Centre was composed of the political leaders of the major parties, including the Muslim League which joined in October 1946. They exercised wide powers within the existing 1935 Constitution. As a result of elections held to the provincial legislatures early in 1946, Indian governments responsible to legislatures in the provinces had also come to power, with Muslim seats captured by the Muslim League throughout India. But the Constituent Assembly which met on 9 December 1946 got bogged down.[58] The Muslim League, which included the majority of representatives from Bengal, Punjab and Sind, as also the representative of British Baluchistan, decided not to participate in the Constituent Assembly. The Congress and the Muslim League thus fell apart. The result was India's partition, the recognition by the British government of two separate and sovereign Constituent Assemblies, to which power was transferred in accordance with a plan laid down in the policy statement of 3 June 1947.[59] The British had tried to unite India, but division resulted from cumulative developments.

The Administrative System

The East India Company started off in Bengal as a revenue collector or zamindar, and the English occupation of that province was, in one way or another, based on its rights as revenue collectors under the Mughal Emperors. The exigencies of the revenue service induced the English to elaborate within the indigenous framework a system of government and administration that, in course of time, assumed distinct European attributes, a process which the author has already analysed in earlier works.[60] It was based on the principle of *contract*, a legal principle enforceable by courts of law, not a principle of executive discretion based on local custom or usage and exercisable capriciously either by men in authority or by powerful individuals of status enjoying social sanction on the basis of caste, affluence or political influence. Born of the nascent principle of rising capitalism, the new contract-based system called for uniformity in the administration of law, which, in turn, signified a centralization of authority invested with the enforcement of the new principle. It was not only a progressive machinery, but also one with a vertical depth in the penetration of official or bureaucratic influence, an influence which under the Mughals remained, in the absence of improved communications, by and large confined to urban administrative headquarters.

Executive Control over Rural Economy

The land economy was thus the basic source from which the administrative system evolved in the early years of British rule in India. There were in the main three categories into which this agrarian economy was divided, namely, the Zamindari system of settlement in perpetuity, the Raiyatwari tenure, and a virtual combination of the two called the Mahalwari system of periodical settlement. Doubtless, there were in the beginning variations in administrative arrangements. There were, for instance, Boards of Revenue functioning as controlling authorities in all provinces except in the Presidency of Bombay which, unlike Madras, had Divisional Commissioners but no Board of Revenue. But the district was common as the unit of administration. In spite of tenurial variations, the provincial government's general authority descended through the Divisional Commissioner in a direct chain to the district officer, functioning as Collector at the head of the revenue organization, and as magistrate at the head of the machinery of law and order, both being equally important and related to the promotion of the country's rural economy. The district officer came to form a vital administrative link for purposes of India's political unity and stability. He continued to be the direct representative of the Government in the district, responsible for the execution of its policy; and, although answerable immediately for a due discharge of his revenue and magisterial duties, he, as part of an overall responsibility, kept in touch with the activities of all 'special departments', established from time to time for development purposes. In a memorandum submitted to the Indian Statutory (Simon) Commission, the Government of India recognized the key role of the Collector in the district administration of British India in terms of his direct contact with the people as an instrument of political stability. As an administrative unit, the memorandum acknowledged, he had grown 'to be part of the people', not merely as an 'administrative convenience', but 'as an essential part of the organization of the community' in the sense that he 'corresponds to certain fundamental characteristics of the people'.

As the memorandum of the Government of India rightly pointed out, the position which district administration attained in course of time was a result of growth, of an interaction between the Cornwallis system of the judge-magistrate as its presiding officer, and the Munro system of a collector-magistrate as an alternative better-suited to the conditions of an authoritarian and plural society as well as to the exigencies of its agrarian economy. While the former was in concept dominated by the principle of *contract* and its resultant rule of law as the best guarantee for the security of private property, the latter was influenced by considerations of justice for the weaker sections of society. Experience established the validity of Munro's contention. But Cornwallis's emphasis on respect for the rule of law stayed. It rendered the Collector's judicial decisions subject to review by civil courts.

While enacting his Regulation for the introduction of the Permanent Settlement in the diwani provinces of Bengal, Bihar and Orissa in 1793, Cornwallis acted on the assumption that the zamindars were the actual proprietors of the lands included within the limits of their zamindaris. The judge-magistrate system, which divested the Collector of his judicial authority in the settlement of disputed cases, was intended to reinforce that assumption in legal terms. It laid down elaborate regulations which had already been growing in complexity since the days of Warren Hastings.

The complex conditions arising from the introduction of the Cornwallis Code were sought to be remedied partially by arrangements under which peoples of the newly annexed territories inhabited by aboriginal tribes came to be governed by a 'non-regulation' system, its main object being to bring officers of superior rank into direct contact with the tribal people. It was with this end in view that certain particular districts in the older provinces were likewise withdrawn from the operation of the general regulations in order to have them administered on less elaborate principles. The north-east frontier of the Bengal district of Rangpur, Assam, Arakan, and Tenasserim were all made non-regulation territories; and so were the south-west frontier tracts of Orissa and the tributary mahals. The districts of Jalpaiguri, Darjeeling and the Hill Tracts of Chittagong later came under the non-regulation system, a system which was based on the concentration of power.

The exigencies of direct and deeper official contact with the people in rural districts called for change in two more directions in Bengal, namely, reduction in the size of districts as territorial units of administration and the introduction of the subdivisional system under the Company's government itself. Districts were earlier created by a division of large estates of individuals. But conditions changed when commerce increased and the reference of disputes to law courts grew rapidly with the ownership of land passing from the hands of big zamindars to those of new families and proprietary communities of coparceners. It became necessary to subdivide districts into police-circles and subdivisions, not into large estates of individuals who earlier had some police functions to discharge. Even so, by 1856 there were not more than thirty-three subdivisional magistrates in the whole of Bengal.[61]

A subordinate cadre of Indian deputies in the revenue and executive lines of administration had been introduced by Bentinck as a measure of policy to relieve pressure on the administration as well as to please Indian public opinion. But they were in general ill-trained and prone to corruption. As realized by Sir Frederick Halliday, the Lieutenant-Governor of Bengal, the basic issue was to make the Collector the head of district administration and invest him with both executive and judicial powers in the same way as the North-West Provinces had already done on the lines of Madras and Bombay. In a minute of 18 February 1857 Canning lent full support to the views of his Lieutenant-Governor who favoured a combination of powers in the district officer as a prinicple of administration best suited to a patriarchal society.⁶² Canning's recommendation received the sanction of Lord Stanley, the Secretary of State, who in a dispatch of 14 April 1859 recognized the soundness of the Collector-Magistrate system as the best form of district administration and one which would bypass the zamindars and enable the people to directly approach the government, an approach which the Collector-Magistrate sought to establish by extensive touring.

In Bengal, however, the difficulties of establishing mass contact posed major problems. The Permanent Settlement had denied to the Executive that revenue machinery by which the rest of India was governed in detail. Its agricultural statistics were collected by the police and all orders from district headquarters to outlying parts travelled through the corrupt and oppressive police. The Collector-Magistrate operated under serious constraints; for in all revenue and executive matters he was subordinate to his Commissioner, while his judicial decisions in criminal cases were upset by the sessions judge. The indigo planters were another constant source of trouble. They had acquired from the zamindars various forms of proprietary tenures and compelled the cultivators by a system of enforced advances to grow indigo and sell it at rates fixed by the planters themselves. With the rise in the prices of other crops, the raiyats' loss from the cultivation of indigo became increasingly heavy. Their combination against the planters dealt a blow to its cultivation in Bengal. But Act X of 1859, which the Government passed to regulate cultivation, became an instrument to enhance rents. It led to agrarian disturbances and called for the intervention of the Government, resulting in the Bengal Tenancy Act of 1885, which was passed against the united opposition of the planters and zamindars. This expression of executive concern for the protection of cultivators had become perceptible since the administration of Sir George Campbell who, in the course of his Lieutenant-Governorship during 1871-4, effected a complete departure from *laissez-faire* by introducing sweeping changes, including the collection of statistics of all kinds, the development of the provincial civil service, and the extension of the subdivisional system which had been greatly developed by Grant after the indigo troubles of 1859.

The effect of the Permanent Settlement, which obstructed a free and beneficial intercourse between district officers and the rural population, proved especially ruinous to the hill tribes. From a minute recorded by Denzil Ibbetson on 23 April 1903 on the subject of territorial reconstruction, it appears that the rents which the zamindars collected from the tribal population of Chota Nagpur, for example, were nominal. But these rents, as Ibbetson pointed out, 'are supplemented by a number of praedial servitudes, and of those extra cesses so dear to the Indian landlord, both of which lend themselves to abuse and oppression'. These conditions had over the previous fifty years 'afforded a favourite field for missionary enterprise', a development which encouraged 'the aborigines' to resist landlords. Thus, the tribal territory became a seat of 'almost continuous agrarian trouble. It began with what is known as the Kol rebellion of 1831, and since then the fire has never gone out'.[63]

In a letter of 4 March 1832 Metcalfe wrote to Bentinck of the 'peaceable' and 'industrious' character of the Kols of Chota Nagpur, including Palamau and Ramgarh, where an insurrection had spread on account of rising land prices and the exactions of the several layers of alien middlemen interposed between the cultivators and the zamindars. 'The Kols of Chota Nagpur', the letter said, 'seem to have acted with thorough unanimity and concert in their insurrection. They thoroughly cleared their own country of foreigners and interlopers, and then made incursions into Palamau and Ramgarh. We hear of exactions made on the part of the raja, the jagirdars and zamindars, and farmers in land revenue; and on the part of our nazirs and mamadars in abkari and opium. The truth no doubt will come out at last. I am not incredulous as to the exactions, which have, I dare say, taken place, as they do everywhere, but the effect seems to be too decisive without some previous effort to obtain redress'.[64]

Tenancy legislation was doubtless enacted from time to time to provide redress. But the men who administered it were trained in the general traditions of the province, a legal labyrinth. In the words of Ibbetson, 'The jungle folk dread the machinery of administration; what they understand is the personal rule of a strong man; and they prefer even a good deal of injustice which they can understand to a great seal of justice which they can't.' The main trouble arising from the Cornwallis Code was to tack on to an advanced province territories inhabited by aboriginal tribes, so that the laws in force in advanced tracts were misused to the detriment of backward peoples and tribes. The uniformity of the legal and administrative system which the British sought to establish could not provide justice without a corresponding change in their social and economic counterparts. The weaker sections of society could not perhaps engage the attention of Government except by acts of violence. The Munda rising of 1899–1900 was another case in point.[65]

Institutional Agencies and Political Stability 227

The violent outbreak which occurred in the Santhal Parganas in 1855 was a confirmation of the same principle of policy, a policy which treated violence against socio-economic ills as a simple law-and-order issue to be resolved by a resort to suppressive measures. The 'agrarian agitation' which produced a relapse 'into what might be called a barbaric insurgency' and resulted in the Santhal rebellion of 1855, 'arose from the resentment of a tribe of primitive cultivators at their impotence to resist the exactions of Bengali and Bihari landlords. About 30,000 Santhals overran a large expanse of country, roasting Bengalis, ripping up their women and torturing their children. The rising was quelled by strong military force and afterwards the Santhal Parganas were constituted into a separate district and ruled on a simpler system designed to secure closer personal contact between British officers and the people.'[66]

The establishment of closer personal contact with the people, though not a permanent solution, undoubtedly restored confidence in the administration. Institutionally, it partook of the character of the non-regulation system administered by a mixed cadre of officers belonging to both civil and military services. This class of officers was responsible for the transformation of primitive tribes into generally peaceable and law-abiding cultivators. This happened not only with the Santhals, but also with the Bhumij of Manbhum, the Larka Kols of Chota Nagpur, the Khonds of the Orissa hills, and the hill tribes of the North-East Frontier which had over the years formed part of Bengal.

A basic trouble of tenants in Bengal was the absence of registers which recorded their status, their rents, and the particulars of their holdings. In the absence of records of their rights they found themselves at the mercy of rapacious landlords, pleaders and court underlings. Surveys of districts had of course begun in 1834–5. But these were not cadastral, field to field surveys, comparable to those in the neighbouring North-Western Provinces. In the Lower Provinces they were intended merely to demarcate village boundaries and collect some useful stastistics. The emphasis laid by the Famine Commission of 1880 on the security of occupancy rights as a means to promote agriculture led to the Bengal Tenancy Act of 1885 and the subsequent introduction of cadastral surveys on a general plan to secure the rights of tenancy. Under the Act of 1885, a survey and record of right was first completed for the greater part of Bihar and for a number of scattered estates in various parts of the province. These operations were eventually extended to the whole of Bengal. This added enormously to the work of the Government, and was a new trend in the administration which began immediately with the transfer of the jurisdiction in rent suits from the civil to the revenue courts.

Growing Power of the Secretariat

The basic cause of the increasing load of administrative business was the government's concern to promote agriculture by land reforms and save the cultivating community not only form the ravages of recurrent famines but also the exactions of landlords and their allied underlings. This welfare concept of administration extended the responsibility of the government and called for executive interposition in the operations of a *laissez-faire* economy. As recorded by the Home Secretary, H. H. Risley on 6 December 1904, 'the main cause of the increased pressure on the Lieutenant-Governor' of Bengal arose from this basic economic premise. Risley attributed this to

closer intervention of Government in disputes between landlords and tenants. The necessity for such intervention dates from the rent movement arising out of the breakdown of the Bengal indigo system; it has become more frequent and more minute since the passing of the Tenancy Act; and there is every prospect of its being very largely extended. Its effect is by no means confined to the revenue officers of Government but reacts upon every branch of the administration. Disputes about rent produce a crop of criminal cases followed by a series of civil suits, and it is difficult to say where their influence stops. Apart from this the general progress of the province, the growth of the city and port of Calcutta, the increased attention given to education, the large number of municipalities, the wider activity of District Boards, the demand for improved sanitation and the reforms contemplated in the Police, all contribute to increase the number of references which must come before the Lieutenant-Governor.

These were, of course, administrative problems which kept on growing. The difficulties increased, in addition, thanks to 'the intensely litigious spirit of the people' which grew with the spread of education, 'the incessant criticism of the Press', and the importance which the people attached to 'the right of interpellation' in the Legislative Council which was exercised freely in Bengal.[67]

An important result of the growing pressure of work at the political level was a corresponding increase in the initiative and power of Secretaries to the Government. For, so long as the work continued to be as heavy as it was bound to be under modern conditions of administrative and functional expansion, the Lieutenant-Governor, who was the political head of the Government of Bengal, was in practice 'driven to adopt his Secretaries' suggestions in a large number of cases without going thoroughly into the papers himself'. But, apart from the increase in the load of public business, as Risley himself acknowledged, the influence and power of the Secretariat grew enormously because of the frequent changes of political bosses. 'When Governors and Deputy Governors were constantly changing', he said, 'the Secretary's influence must always have been great, and it was maintained later on by the natural increase of work'.[68] With the Imperial stress on the welfare content of relief measures for the agricultural classes, the Secretaries tended to assume in practice the position of 'Ministers of Departments', informally exercising discretion without responsibility.

Interestingly enough, the position was not dissimilar in so far as the Government of India also was concerned. A very important subject like territorial redistribution, involving the transfer of Orissa from Bengal to the Central Provinces, remained, for instance, in a prolonged state of discussion at the Secretariat level without even a Governor-General of Curzon's stature being officially consulted. The matter came to the notice of the Governor-General only when Denzil Ibbetson made a noting on the file that the subject could not be 'taken into consideration' unless 'His Excellency approves of the matters'. Expressing his surprise over the manner in which the Secretariat of the Government of India functioned, Curzon noted on the file on 24 May 1902:[69]

It seems to me a most extraordinary thing that this discussion should have been going on for more than a year without any mention of the matter ever being made to the head of the Government.... And yet, during this period Secretaries and Deputy Secretaries have been calmly carving about and rearranging provinces on paper, colouring and recolouring the map of India according to geographical, historical, political, or linguistic considerations—in the manner that appealed most to their fancy: and finally on January 29th, 1902, Sir C. Riyaz recorded that the idea of transferring Orissa from Bengal to the Central Provinces *must be dropped* and that the idea of forming Orissa into a separate Chief Commissionership *cannot be entertained*.

I really feel disposed to ask—is there no such thing as a head of the Government, and what are Secretaries for but to keep him acquainted with the administration? Would it be considered credible, outside the Departments, that these really very important issues, affecting the constitution or dismemberment of provinces, should have been under discussion for more than a year without the file ever being sent, or the subject even being mentioned, to the Vicerory?

The fact remained that the Secretariat had over the years acquired an entity recognized as separate from the political head of the Government, and possessed a power and authority of its own as an Imperial administrative agency. It preserved the unitary character of India's political system and contributed to the country's stability independently of changes at the political level.

Collector as the Mainstay of Stability in Rural Districts

The agent of the Secretariat in the rural districts was the Collector who, in course of time, came to be recognized as District Officer by virtue of his direct functional contact with the people; he was invested not only with law and order and the management of most branches of revenue, but also with the handling of the manifold relations where the people and government were both involved. He was, in fact, required to remain concerned with the general circumstances of his district, touching not only on peace and fiscal administration, but also on such other matters as were related to its economy, development and general welfare. Despite local variations in the pattern of provincial administration, the District Officer remained the mainstay of stability in the body politic of India.

This overall importance of the Collector in district administration was a concept which was effectively recognized and enforced by Sir Thomas Munro, the well-known advocate of the raiyatwari system of Madras. In a minute which he recorded as Governor of Madras on 22 January 1822, Munro made it clear that in the context of Indian society the union of revenue and magisterial functions was an absolute necessity. We should form a very erroneous judgement of the important influence of the office of Collector,' he reminded, 'if we supposed that it was limited merely to revenue matters instead of extending to everything affecting the welfare of the people. In India whoever regulates the assessment of the land, holds in his hand the mainspring of the country.'[70] However, it was Holt Mackenzie, the Territorial Secretary to the Government of the North-Western Provinces, who, with the backing of the London authorities and the immediate support of Lord Hastings, the Governor-General, set the current against the Judge-Magistrate system of Cornwallis. Despite a declared policy to establish the Cornwallis system in the North-Western Provinces, Holt Mackenzie managed to have Regulation VII of 1822 duly enacted. Under this Regulation the Collector or his settlement officer was required to conduct inquiries on the spot, village by village, proceeding upwards from the persons who tilled the ground to the Government itself, noting distinctly all the classes who shared in the produce or rent of land, the extent of the interest of each, and the nature of the title by which it was held. It was the beginning of a cadastral survey for revenue purposes and a record of rights to ensure the protection of tenants as the holders of coparcenary tenures in villages held jointly by certain communities. Obviously, the new task could not be accomplished except by reviewing the Cornwallis system and re-investing the Collector with a measure of executive and judicial authority that enabled him to reach on-the-spot decisions in disputes arising from settlement proceedings. His decision was of course to be subject to review by a regular civil court. But it remained operative unless determined otherwise on appeal to a civil judge.

The combination of powers in the office of the Collector as the basic principle underlying the reorganization of district administration had, in fact, already been recognized by a regulation of 19 January 1821. This declared the Governor-General in Council competent to authorize a Collector or other revenue officer to exercise the power of a Magistrate or Joint Magistrate, or to invest magistrates with the powers of a Collector. But the arrangement so provided in 1821 or even 1822 had not yet become a firm principle of policy. Holt Mackenzie viewed the combination of powers in the office of the Collector from a much wider perspective, which treated that office not only as a vital link with the Government of India for purposes of political stability, but also as a direct link with the people to ensure that stability on the basis of popular acceptance without executive interference with the security of personal or property rights. In other words, he viewed district administration in its totality and regarded the Collector as representing that totality in relation both to the Government and the people.

In his famous minute of 1 October 1830,[71] Mackenzie sought to clarify his concept of administration for the rural districts, 'an administration' which 'shall satisfy the just expectations of the people and ... shall [at the same time] contain within itself those seeds of improvement which must be cultivated by their co-operation'. Such an administration, he pointed out, 'is to be sought in the completest possible unity of purpose throughout all the departments, and in all their grades, for it is only thus we can hope to maintain that moral control which will enable us safely to delegate to natives those duties which must be confided to them, if we would really do justice. Without this, Government must, I fear, fail either to maintain its own rights, or to protect those of its subjects'. In the absence of self-governing institutions, he said, 'the only alternative was to unite powers in the office of the Collector, a sufficiently strong connecting link between the several departments'. For, as Mackenzie argued, 'to disjoin the several parts of government, in a country which is not self-governed, is like placing the different members of the body in charge of different physicians, severally acting with their respective limbs according to individual theory, without reference to the treatment of other parts, and each holding in his hand the power of destroying life, but helpless to save, from the blunders of his brethren. It is to animate the lifeless frame with a plurality of souls'.[72]

But the union of powers in the revenue authorities was not to be without safeguards. It was recognized that the acts of revenue officers affected the interests of society very widely; they touched at every point the properties and institutions for the protection of which the courts were instituted and invested with separate judicial functions, necessary for the maintenance of people's confidence in men so trusted. The legal safeguard sought to be provided thus made the decisions of revenue officers subject to review by civil judges who had no hand in determining the quantum of assessment to be demanded. This arrangement had the tendency to encourage litigation. But the volume of litigation was considered 'insignificant' when contrasted with the mass of property to be decided about by revenue officers in the unsettled districts; for, as Mackenzie argued, 'to the revenue authorities it belongs to say whether the estate of every landowner in those districts shall be valuable or worthless, and whether millions shall cultivate their paternal fields in comfort and independence, or shall toil in poverty, or suffer exile; and in the few cases in which our Collectors are sued, it may well be doubted whether, as matters are now arranged, there be any solid reason to anticipate a better judgment from the judge, excepting on grounds that would imply on the part of the revenue functionaries an entire unfitness for the trust actually confided to them.'[73]

The other, and still more important, safeguard was the quality of selection to ensure restraint and balance in the exercise of power. Mackenzie assumed 'that all the civil servants of the Company are or will be men of fair capacity and character, fully qualified to discharge, after a sufficient course of service, the duties which attach to the Collectors and Magistrates of districts, and that a large number of them will be found equal to the more difficult and important duties that belong to the administration of civil and criminal justice.'[74]

As already noted this principle, of district administration was accepted by Bentinck's Government in 1831. Regulation VIII of that year divested civil judges of their powers to receive any claims of rent or revenue unless the complaint was preferred as a regular civil suit in their court. The Governor-General in Council, on the other hand, transferred on a general plan the office of Magistrate to Collector in most districts. Under Regulation VIII the Collector became the first officer in the district, invested not only with executive powers of arrest and punishment of offences, but also with powers for the summary trial and determination of rent or revenue cases arising from disputed claims to landed property. This became a feature of Indian administration which provided a regular institutional link between the secretariat headquarters and rural districts. The exigencies of estates' resumption proceedings doubtless dictated the necessity of separating, for a time after 1837, the magistracy from the office of Collector. But by 1859 Bentinck's arrangement of 1831 was recognized as a settled norm of district administration in the context of Indian society.

Mackenzie's stress on the Collector-Magistrate as the only suitable link for purposes of political and administrative cohesion proceeded from yet another consideration. He believed that 'the protection afforded by tribunals bound by strict laws, and authorized to interfere only upon full legal proof and with specific acts distinctly prohibited, must be very imperfect'.[75] The poor and weaker sections of society could, rather, look forward to justice only through the executive agency of the Government. And since the protection of the people against public officers in a despotism was by itself a problem, more especially in respect of acts done under the cover of law, Mackenzie emphasized the necessity of executive intervention by official superiors acting with influence and discretionary authority as the best safeguard.

The official arrangement enacted in 1831 on Mackenzie's recommendations was also considered best fitted to accomplish the political objectives of unity and efficiency. But Mackenzie did not fail to realize the dangers of pure bureaucratic involvement in the transaction of business at the grass-root level of society, an involvement which could be misused by officials acting against the interest of village communities, the only traditional check on despotism and recognized as barriers to local penetration for political and social change:

Nothing can be more striking than the scorn with which the people have been practically treated at the hands of even those who are actuated by the most benevolent motives; for, since the world began, there is probably no example of a government carrying the principle of absolutism so completely through the civil administration of a country, if that can be called civil which is in its spirit so purely military; nay, which sets the people aside in the management of their own concerns much more than the *sepahee* in the government of the army.

He called this an error that affected 'the village associations [panchayats], which form the great bond of society throughout so large a part of India, but which have been greatly misunderstood and disturbed. These institutions seem to afford one of the most important of all the instruments we could use to insure the good government of the country and the comfort of individuals'.[76]

Indeed, Mackenzie viewed village panchayats as symbolizing institutional unity at local levels, forming a poupular link with the district under the Collector-Magistrate. For he said:

Without them, or some substitute similarly resting on popular principle, we must, I fear, have a miserable and disunited people, whom it is scarcely possible to govern otherwise than as the slaves of our native servants, whereas with them, these men might, I conceive, be made really servants of the public; at the same time our judges and magistrates would assume their proper station as governors, according to their several degrees, in all they ought to govern, and would leave the people to transact, with their protection, support and control, the innumerable concerns of civil life, which they alone can tolerably administer. One cannot too often refer to the principles which ought to restrain us from the error into which we constantly fall of doing too much, both legislatively and executively, under a government which excluded the idea of political freedom.

Thus, a basic principle underlying Mackenzie's advice was that, in the absence of self-government, the scope for functional interference should be restricted in the interests of promoting civil liberty. He argued:

absolute power may thus be made consistent with much civil liberty; and this large and divided nation with no risk of political disturbance but laws arbitrarily imposed by a despotic government can have comparatively little effect in checking the abuse of power, except as they restrict the occasions of its being exercised; and it seems to be vain to think that we can by any legislative provision secure the community from extortion and vexation, if we once allow, or require the government officers to interfere perpetually in the minute details of the people's business. We have unfortunately acted on an opposite principle, interfering in almost everything, neglecting popular institutions where they exist, and never attempting to create them where wanting.

He therefore urged the government to 'use every endeavour to revive and maintain the system of village or parish government'.[77]

The whole trend of administrative reorganization was, in fact, towards a reinforcement of British paramountcy through a union of liberal and unitary principles, which, in contrast to a feudal polity,

Institutional Agencies and Political Stability 235

called for a deeper and wider interference by public officers with the affairs of the people. This interference was, however, sought to be contained within reasonable limits by a resort to local bodies later created for political reasons to obviate the dangers of minute official interference with matters of local interest. These local bodies were not panchayats in a traditional sense. They were created under acts of the legislature. Their operations were subject to law and open to inspection. The local bodies were provided a popular base without prejudicing the unity of district administration.

Administrative Personnel to Secure Unity and Stability

The personnel of civil administration, which contributed to India's political unity and stability, was organized into disciplined and compact services and kept the country territorially integrated. Though functioning under the well-regulated machinery of its London authorities, the Company's Indian administration was over the years carried on by 'covenanted' servants appointed separately to different presidencies under a system of patronage exercisable by its directors. There was nothing like an all-India service, nor was its civil service free from acts of indiscipline or neglect of duty.[78]

Earlier works have analysed the bureaucracy in India in some depth. All that is needed here is to emphasize that the personnel of the British Government in India served in course of time as a link between the districts and the central secretariat, combining its field experience with the experience of policy-making at the secretariat headquarters, a development which helped in keeping together under a unified political control the distant parts of Indian territories despite their great regional variations in terms of language and culture.

Doubtless, the all-India character of the civil service, which was essential for India's political unity, took time to mature. It did not occur under the Company's rule. For, despite the introduction of competition as the basis of recruitment in 1855, the Bengal civilians continued to dominate. Organizationally, however, the introduction of open competition under the 1853 Act heralded a new phase, highly significant in terms of a wider all-India perspective applicable not merely to individual civil servants, but to the civil service as a class by itself, a class which became in general known for both intellectual and moral excellence. It put an end to the system of patronage, and recognized merit and character as the only qualification for admission to the civil service, open to all 'natural born subjects of Her Majesty'. The

Act did not exclude Indians from the open competition so provided. In a debate on the subject Lord Grenville made it clear that Indians would 'gain the appointments' if they went to England and pass the required examination. Macaulay, who headed the committee appointed to frame rules for the purpose, held the same view. He added that any Indian of distinction could enter the service as a matter of right. In other words, every civil servant, European or Indian, was expected broadly to subserve the all-India imperial interest, an interest which was particularly emphasized after the Mutiny. The old system of recruitment primarily to provincial cadres doubtless continued. But it was left to the Government of India to borrow any civil servant of a provincial cadre for its own purpose. The appointment to the Governor-General's Council in 1859 of Sir Bartle Frere of the Bombay cadre marked the beginning of a new trend which broke the erstwhile monopoly of the Bengal cadre in the services of the Government of India.

However, the recommendations of the Aitchison Commission imparted to the old 'covenanted' service a proper all-India character in accordance with what might be called a service concept in organizational terms. Based on these recommendations, the Government of India under its Resolution of 21 April 1892 divided the great majority of its officials into classes corresponding to the responsibility of their work and the qualifications required to perform it. These classes were mostly organized as 'Services'—the highest, or the 'Superior' Services, the 'Provincial' and then the 'Subordinate'. The Superior Services were themselves divided into the 'Central Services' dealing *inter alia* with the Indian States, frontier affairs, the railways, posts and telegraphs, customs, audit and accounts as well as such technical departments as the Survey of India, the Geological Survey and the Archeological Department. In the same class were the 'All-India Services' which the Aitchison Commission preferred to call 'Imperial', but which at the instance of the Secretary of State became known as 'Indian' services. These included at the time the Indian Civil Service and the Indian Police, but later, on the same principle, covered the Indian Educational Service, the Indian Forest Service, the Indian Agricultural Service, the Indian Veterinary Service, the Indian Medical Service (Civil), and the Indian Service of Engineers.

The All-India Services operating in both regulatory and developmental fields were by far the most important of the administrative agencies which not only kept the country together, but also maintained a regular flow of higher talent and character necessary for its modernization. In terms of India's political integration and stability, the organization of the All-India Services, with provision for promotion to 'listed posts' from the provincial civil services under the arrangement of 1892, was, in fact, based on sound principles. Though their field of operation was a province, members of the Services could be called to work with the Government of India. But they were even otherwise intended to act as the eyes and ears of the Central Government as its executive agents, spread throughout the provinces in India and subject only to the disciplinary control of the Secretary of State from whom they received their appointment. Combined with this unity of control was the rigid test of candidates before recruitment. There was no racial barrier to recruitment by open competition, but since recruitment was done in England, it no doubt prevented some deserving Indians from sitting for the necessary examinations. This became a political issue amongst Indians and led to an agitation to secure increasing employment for Indians in the Superior Services. It remained for a time limited to Indians *versus* Europeans, and then extended to Hindus *versus* Muslims, a political trend which filtered downwards with the spread of education and representative institutions. The politics of employment, racial or communal, found full support with nationalist agitations, a development which dictated the necessity of providing constitutional safeguards for the protection of the Services against political pressure or communal prejudices. The Government of India Act (1919), for example, provided that no officer could be dismissed from a Service by any authority subordinate to that by which he was appointed. It clearly meant that the officers of the All-India Services could not be dismissed by any authority in India; nor were their pays, pensions or other emoluments subject to vote in the legislatures. The Act was not to affect any of the existing or accruing rights of a civil servant who remained entitled to compensation for the loss of any of them.[79] The safeguards so provided for the protection of the rights of civil servants enabled them to exercise in the performance of their duties an independence of judgment which in no small measure accounted not only for administrative efficiency with social change, but also for political progress without prejudice to structural stability.

According to the *Report of the Joint Committee on Indian Constitutional Reform* (1933-4), the total European population of British India, including some 60,000 British troops, was at the time only 1,35,000, while the total British element in the Superior Services was about 3,150, of whom approximately 800 were in the Indian Civil

Service and 500 in the Indian Police.[80] Together, the Superior Services provided the personnel both for territorial security and development. It was through their agency that the unitary pattern of India's political system actually operated, a system which depended for its success on the efficiency and soundness of institutional arrangements rather than the personal or discretionary character of rule, as in Mughal times. The transformation of British India into a single, unitary state engendered among Indians 'a sense of political unity which was reinforced by an impartial and enlightened administrative agency, powerful enough to control the disruptive forces generated by religious, racial and linguistic divisions'. The emergence of a sense of nationality was a result of both. They created conditions of peace where the enforcement of law and order through an upright administration came to be easily accepted as a matter of course, a development which led Indians to turn their mind to broader issues of nationalism and economic interest transcending separatist divisions. The institutional agencies at both political and administrative levels contained in fact the seeds of structural change from within. Constituted as these agencies were on familiar British conceptions, they encouraged the growth of a body of public opinion 'that good government is not an acceptable substitute for self-government, and that the only form of self-government worthy of the name is government through ministers responsible to an elected legislature'.[81]

Military Policy and Organization

After the establishment of British paramountcy, the Army in India was mainly required to put down internal rebellion and overcome foreign attack, both being equally important in terms of India's political unity and stability. But who was to preserve the peace of the country, immediately essential for success against foreign aggression? Was this to be done by native soldiery or by the very limited European force? If the native soldiery were to constitute the bulk of the agency employed for the task, what was the indispensable supplement of Europeans to be? These were organizational questions which called for a settled answer, more especially in a situation where there was no unity of command on account of the continuing presidency army system.

The question of internal security as an important condition of political stability was interestingly raised by Bentinck who presided over the Indian administration in a period of comparative peace. In a minute of 13 March 1835, he acknowledged that in Madras, as in Bengal, 'there no longer exists a single chief, or a combination of chiefs, who possess even a semblance of a military force. Nor are there any large masses of the population who have the least disposition to rebel against our authority'. Even so, he prophetically confirmed the 'state of uncertainty' earlier observed by Sir John Malcolm who felt that in India, 'we are always endangered, and it is impossible to conjecture the form in which it may approach'. Bentinck shared Malcolm's apprehension, when, as he reasoned, 'one hundred millions of people are under the control of a government which has no hold whatever on their affections; when, out of this population is formed that army upon the fidelity of which we rely principally for our preservation; when, our European troops, of whose support under all circumstances we are alone sure, are so exceedingly limited in number and efficiency as to be of little avail against any extensive plan of insurrection'. Bentinck was perhaps conscious that a policy of peace and Liberalism was leading the Indian people towards a new awakening, a development which added to the state of political uncertainty and called for a remedy that lay beyond the comprehension of an army. 'The state of uncertainty', as his minute emphasized, 'is greatly aggravated by our condition of peace, by the spread of knowledge, and by the operation of the press, all of which are tending rapidly, as well to weaken the respect entertained for the European character and prestige of British superiority, as to elevate the native character, to make them more alive to their own rights, and more sensible of their power'. Bentinck was not so concerned about the possibility of losing territory already brought under British dominion, as of 'changes incidental to our *new* position of peace, a more enlightened state of mind, a higher elevation of character, knowledge, improved morality, courage, all concurring causes that must produce effects to be dealt with by a very different philosophy from that which has hitherto obtained'.[82]

The rise of a renascent India was doubtless a threat to British sovereignty. But this was a potential dimension, not an immediate problem of military policy which, as Bentinck recognized, called for 'enforcing the prudence and expediency of bettering the condition of the native army and of preventing discontent by timely concession and precaution'. In a letter written in 1817 Munro expressed a similar idea: 'But even if all India could be brought under the British dominion', he observed, 'it is very questionable whether such a change, either as regards the natives or ourselves, ought to be desired. One effect of such a conquest would be that the Indian army, having no longer any warlike neighbour to control, would gradually lose its military habits and discipline, and that the native troops would have leisure to feel their own strength, and, for want of other employment, to turn it against their European masters.'[83] Precautionary measures and vigilance were necessary, for, as Holt Mackenzie later advised, even though the sepoys had 'a great deal of attachment to their officers' this depended upon 'personal character', not upon 'attachment to the nation generally'. He felt that 'the sepoys, as long as they are well paid, will have a strong sense of the duty of being faithful to those who so pay them, to be only overcome by some powerful cause of discontent or excitement'. He thought that 'a large native army is quite essential for maintaining the tranquillity of the country' but would be very sorry 'to see its defence entrusted to them without a large European force'.[84]

Organizationally, the Presidency armies had their own weaknesses. They were in general ignorant, custom-bound and given to discontent on religious or other grounds. This was especially so with the Bengal army, which consisted of battalions composed of men of high caste from areas of Oudh. They were known as 'Hindustanis' and, in Bentinck's time, constituted the whole of the Bengal army and half of that of Bombay. They were believed by the English to be more robust than Indians from south of the Narbada. Even the Nizam's contingent and the infantry of the Raja of Nagpur consisted entirely of Hindustanis. The units of the Madras and Bombay armies were, on the other hand, composed of a mixture of men from the lower castes. In the Bengal army no provision was made in the regimental lines for the men's families, but the sepoys visited their families at regular intervals. In the Bombay army accommodation was provided for a certain proportion of families. In Madras, on the other hand, all the men's families accompanied the battalion wherever it moved.

A serious weakness of the Presidency system was that, except during hostilities, no sepoy was willing or ready to serve in territory outside his Presidency. But when a territory was pacified and annexed by the Company, the sepoy insisted on the field service allowances being continued as his due should he be called upon to serve outside his Presidency area, for he considered the latter foreign territory. There was nothing like an all-India territorial concept in the native army under the Presidency system. The result was that service in the newly acquired territories was not merely unpopular but also caused discontent if war-time field concessions were withdrawn on the cessation of hostilities. The question as to what troops were to occupy the new territories became an embarrassing one for any Local Government. Obviously, such territories had to be garrisoned by the troops of one or other of the Presidency armies, but the administration of a mixed Presidency force was beset with problems arising from social and cultural heterogeneity. In the final Burmese War, for instance, the high caste men of the Bengal army objected to being transferred across the sea and Burma thus eventually became one of the stations of the Madras army.

In the conquest of Sind, on the other hand, Bengal troops had taken a share in the operations under Sir Charles Napier. When the province was annexed in 1843, the Bengal troops declined to remain there without the compensatory allowances given during operations. Serious mutinies broke out in Ferozpur among the Bengal units, and a Madras battalion, which was ordered in relief, also broke out in open mutiny in Bombay when it learnt that the Sind allowances were to be discontinued. The Government was thus left with no alternative but to send Bombay troops to garrison Sind, which became part of Bombay Presidency. Even service in the Punjab was unacceptable to the Bengal sepoys who would not remain there after its annexation without special allowances. Their conduct amounted, in not a few cases, to open revolt.[85]

Mutinying was thus a familiar experience with Indian sepoys under the Presidency system. The upheaval of 1857 was partly a culmination of this tendency, but further nourished by religious, agrarian and political factors.

Towards Unity of Command and Military Integration

The Royal Proclamation by which Queen Victoria assumed control of the Government of India on 1 November 1858 created a new situation which, in view of the earlier experience, dictated the necessity of a reorganization of the three Presidency armies consisting of both the European and Indian forces. The task was not merely organizational, but also involved mattters of policy, more especially in respect of the caste constituents of the Indian units which had endangered military discipline and undermined political stability.

As regards the 'Native Army', the Army Organization Commission of 1859 sought to provide a remedy.[86] According to its recommendations the Indian units were to be composed of different ethnic groups and castes, with all their men being enlisted for general service regardless of caste distinctions. Equally important was the Commission's emphasis regarding the pay and allowances of the officers and men. The pay codes were sought to be simplified and adopted on the principle of fixed scales of allowances for the troops in garrisons or cantonments and in the field. A principle of efficiency and rationalization in matters of promotion was recognized as a means to tighten discipline. It was recommended that the promotion of Indian commissioned and non-commissioned officers should be governed by their efficiency rather than seniority, and that the commanding officers of regiments should have the same power to promote non-commissioned officers as was vested in officers commanding regiments of the line. This was done to increase the powers of commanding officers and involved revising the 'Articles of War' governing the 'Native Army'. As part of a marginal modification of the Presidency system, the Commission recommended that the Commander-in-Chief in Bengal be styled the Commander-in-Chief of India, and that the General Officers commanding armies of the minor Presidencies be commanders of the forces, with the power and advantages which they had enjoyed in the past.

However, the main emphasis was laid on the removal of caste prejudices by a reliance on checks and balance, as well as a resort to territorial or local designation for regiments. The earlier dominance of high castes in certain regiments was to be avoided. Each regiment was therefore sought to be composed of all sects and classes to make it reasonably balanced in secular terms. Moreover, under the reorganization of the Indian forces which took place in 1861, some cavalry and infantry units were disbanded, others were amalgamated, and the Indian artillery was, with certain exceptions, abolished altogether. The cavalry, on the other hand, was reorganized on the *siladar* system, except in three regiments of the Madras army which had in the past remained consistently loyal and disciplined.

No less important was the question of maintaining an adequate amount of European force in relation to the Indian soldiery. Before the Mutiny the 3,00,000 strong army in India consisted of 40,000 British troops and 2,60,000 Indian. By January 1867, the strength of the troops was not only reduced to 1,81,747, but the number of European soldiers was augmented to 61,747 and that of Indians reduced to 1,20,000. The introduction of measures to subdue caste predilections and increase the number of Europeans in the army was

Institutional Agencies and Political Stability 243

in accordance with thinking on military policy since the days of Clive, and, more especially, since Bentinck's time.

As regards the organization of the European wing in the army, mainly two alternatives were considered—whether the British forces in India should form a portion of the Imperial British Army, the units of which would in turn do the garrisoning in India; or whether they should be localized and maintained solely for service in India. Opinion was divided on this issue, but considering the importance of India's defence in a global context, it was finally decided that the 'British Army serving in India' should form part of the Imperial British Army stationed overseas. This decision necessitated the transfer of the late Company's European troops to the service of the Crown. Consequently, the old distinction between 'Royal Troops' and the 'Company's European Troops' disappeared altogether. The Company's European infantry became British regiments of the line, and the Bengal, Madras and Bombay European artillery were amalgamated with the Royal Artillery. This reorganization was completed in 1860, with a decision that their establishment in India should not exceed 80,000 men. The recognition of the British Army serving in India as part of the Imperial British Army was an important step that was intended not only to undo the parochialism of the Presidency system, but also to bring a wider Imperial responsibility to bear upon the defence of India.

A serious defect in the organization of the Army in India was the shortage of British officers serving in the regiments to which they belonged. This proceeded partly from a number of British officers being withdrawn frequently for staff and other duties as well as for civil employment. This was so persistent that the Indian Army Organization Commission (1859) stressed the need to strengthen the position of British officers in order to enhance the efficiency of the army. Thus, three Presidency Staff Corps were created in 1861. Inquiries revealed that, apart from the shortage of officers who could shoulder burdens caused by the absence on extra-regimental duties, problems flowed from promotional disparities also. British officers had a right to promotion in regimental succession. But officers of one regiment could not claim equality of promotion with officers of another regiment. Thus, promotion in one regiment might be faster than in another, with the result that officers in one case might, on no merit of their own, become senior to those of another.

The institution of the three Staff Corps was designed to remove both defects. In the first place, it established a corps of officers in each Presidency with a strength which ensured that regimental establishments would be maintained despite the drain made on them by the departure of officers on extra-regimental tasks. Secondly, promotion to a higher rank was henceforth to be governed by length of service. The officers of the several Staff Corps were placed on separate seniority lists. Equality of promotion was thus assured, although it did not remove inequality in the rank of officers holding appointments in different corps. In a fiction handed on by tradition since the time Clive organized regimental staffs for his newly formed Indian battalions, officers held 'staff' appointments in the combatant units with which they served. The three Staff Corps consequently consisted of a body of military officers serving on the staff of the army or of units, and in military departments or in civil employ. However, the principle of unity recognized for purposes of promotion in the army was a step in the direction of unity in command.

Certain other important changes had, in fact, already been taking place in the direction of the eventual union of the armies, which culminated later in 1895. The 'irregular' system, which was finally settled in 1863, applied to all the three armies, including of Madras and Bombay. Under this system seven British officers were attached to each Indian regiment instead of twenty-five under the system of line regiments, and instead of three as under the irregular regiment before the Mutiny. The military accounts departments of the three Presidencies were consolidated in 1864 under the Military Department of the Government of India. In 1874-5 the strength of European officers with Indian regiments was uniformly increased by the addition of a couple of probationers to each corps, who were to take the place of officers absent on furlough or other duty.

When serious defects surfaced in the conduct of the Afghan War (1878-80), an Army Organization Commission was assembled by Lord Lytton in 1879 to explore means to reduce military expenditure and suggest measures to improve the efficiency of the army in India. The Army Organization Commission reported that on 1 July 1878 the Indian armies consisted of 6,002 British officers, and 60,341 British and 123,254 Indian soldiers, all maintained at the cost of about 17 million sterling (or Rs 17 crores) out of a total Indian revenue of 63¼ million sterling. This strength of the armed forces was intended to prevent and repel foreign attacks, to suppress internal rebellion, and watch the movements of the armies of feudatory Indian states.

As the best Indian soldiers alone could hope to cope with the

Institutional Agencies and Political Stability

Afghans in their mountainous country, the Commission considered that in a state of impending complications with Kabul no reduction should be made in the strength of the total effective Indian forces. It observed:

> We fully recognize that the great extension of communications, and the improved armament of the troops, make a Native regiment a far more powerful unit than it used to be. But, looking to the extent of the Empire; to the long land frontier of India, infested by wild or hostile tribes; to the military monarchies of Nepal, Cabul, Burma, aggression from which must be prevented; to the strength of the great feudatory chiefships, such as Hyderabad and Gwalior, whose troops must be overawed; to the existence of half-civilized tribes, who have hardly forgotten their habits of rapine, within our borders; and to the other causes of disturbance which may arise among 250 million people of many nations, tribes and languages, looking to all these considerations, we feel unable to recommend the reduction of the Native army below its present strength.[87]

By far the most important recommendation of the Commission, however, was virtually to abolish the Presidency armies, a recommendation which none the less took sixteen years to bear fruit. Under the existing arrangement the Bengal army garrisoned territory from the Bay of Bengal to Afghanistan. The Madras army, on the other hand, did the garrisoning of Madras, Burma, Hyderabad, the Central Provinces, and certain stations in Bengal; while the Bombay army handled Bombay, Sind, Rajputana and Aden. The existence of three separate armies had long been considered anomalous in that they all served the same Central Government in the same country. Their amalgamation had in the past been discussed several times; but the step was condemned before the Mutiny. According to the Commission, even the Duke of Wellington objected to a thorough fusion of the three armies. For it was argued that the three armies were composed of different nationalities and that it was politically expedient to maintain their separation on the principle of segregating national divisions. The Commission did not attempt to counter this argument. 'The question', it said, 'is no longer one of amalgamating the armies of India into one large unmanageable homogeneous body, in which all national and religious differences shall be obliterated and ignored, in supersession of a system under which three armies of different nationalities are maintained. No one would, we imagine, in the present day, advocate a change of such a nature'. But the basic question it raised was: 'Is it really true that the great principle of segregation of national divisions is preserved under the existing system, and is it necessary for the maintenance of such a principle that we should have three distinct armies, three separate systems of military government, three sets of staff, and army departments?' Improved communications as well as the social and political mobilities that flowed from the freedom of the press were all operating against the preservation of the principle of segregation. The Commission naturally felt that, although it was politically desirable 'to maintain the great national divisions of the army', it was necessary 'to systematize and re-adjust them on an intelligible principle'. It recommended that 'instead of three Presidential armies of Bengal, Madras and Bombay, the armies of India should be divided into four completely separate and distinct bodies, to be called army corps [as part of an all-India organization], so distributed that they shall be deprived, as far as possible, of community, of national sentiment and interest, and so organized, recruited, and constituted, as to act in time of excitement and disturbance, as checks each upon the other'. The Commission's recommendation attempted to evolve an all-India army under a unified Imperial command.

Although the recommendations of the Commission did not immediately result in the abolition of the Presidency armies, they led to changes in that direction. In 1884, for example, the three ordinance departments were united under the Government of India, a partial reorganization of the transport services was effected, and commissariat regulations, applicable to the whole of India, were duly compiled. The Punjab Frontier Force was transferred in 1886 from the Government of Punjab to the control of the Commander-in-Chief of India. In 1891, the three Presidency Staff Corps were united in one Indian Staff Corps. The unity of military command was further accelerated under the Madras and Bombay Armies Act (1893), which abolished the offices of the commander-in-chief of the troops in those Presidencies as well as of their military secretaries, and authorized the Commander-in-Chief of India to function as that in each Presidency. This was followed by a General Order of the Government of India in the Army Department, No. 981 of 26 October 1894, which abolished the Presidency armies with effect from 1 April 1895.

The General Order divided the 'Army of India' into four commands, each under a Lieutenant-General who was subject to the direct control of the Commander-in-Chief of India. These commands were Punjab (including the North-West Frontier and the Punjab Frontier Force), Bengal, Madras (including Burma), and Bombay (including Sind, Quetta and Aden). There were, in addition, local corps under the control of the Government of India, which did not belong to any command. There were the Hyderabad Contingent, the Central India Horse and its two regiments, the Malwa Bhil Corps, the Bhopal Battalion, the Deoli Irregular Force, the Erinpura Irregular Force, the Mewar Bhil Corps and the Merwara Battalion.

The new organization doubtless brought the army under the direct control of the Commander-in-Chief of India. But the commands created were as separate as the old Presidency armies; for they included within their areas districts which had little or no connexion with each other, either territorial, sentimental, racial or strategic. Drastic measures were therefore taken to secure a complete unification of the army in India. This was done by Lord Kitchener, who took up his appointment as Commander-in-Chief on 28 November 1902.

On his arrival in India, Kitchener introduced a number of changes. One of these, which led to an effective unification of the armed forces, was the abolition of the Indian Staff Corps. The officers belonging to that Corps were redesignated, and henceforth called 'Officers of the Indian Army'. It was realized that the former title was a misnomer, for most of the appointments held earlier by officers of the Corps were regimental, not staff, appointments. From 1 January 1903, the British officers as well as the rank and file of Indian units came to belong to one corps, 'the Indian Army'. In his reorganization and redistribution of the armed forces Kitchener was guided by four principles contained in his proposals communicated to the Government of India in November 1903. These were: (i) that the main function of the army was to defend the North-West Frontier; (ii) that in peace time the army should be organized, distributed and trained in units of command similar to those in which it would take the field in times of war; (iii) that the maintenance of internal security was a means to an end, namely, to set free the field army to carry out its function; (iv) that all fighting units, in their several spheres, should be equally capable of carrying out all the roles of an army in the field, and that they should be given equal chances, in experience and training, of bearing these roles.[88]

The Command and district areas in which the army was distributed under the arrangement of 1895 could not fulfil the roles which Kitchener proposed, more especially in respect of adapting peace formations to the requirements of war. For the 'Commands' were mere geographical divisions of varied extent, with troops scattered through a number of small military stations, making it impossible to collect them for training in tactical formations in which they were expected to be employed in war. They had been localized in the newly created commands precisely as they had been in the old Presidency areas. The central feature of Kitchener's scheme was to divide the army into three corps, each comprising three divisions, and each such division to be complete in field army troops as well as in troops for internal defence. The distribution of the three corps was to be as follows: Northern Army Corps—1st (Peshawar) Division; 2nd (Rawalpindi) Division; 3rd (Lahore) Division; Western Army Corps—4th (Quetta) Division; 5th (Mhow) Division; 6th (Poona) Division; Eastern Army Corps—7th (Meerut) Division; 8th (Lucknow) Division; 9th (Secunderabad) Division.

The plan was designed to correlate peace formations with the needs of war, the object being to make sure that the army retained through vigorous and uninterrupted drilling its martial discipline and qualities. It also put an end to all traces of the old Presidency armies. The concentration of troops towards the explosive north-west frontier to meet aggression, and elsewhere in India to provide for training in tactical formations, necessitated the abandonment of several old stations, the move of a large number of troops and the building of new quarters for them elsewhere. These implications of Kitchener's scheme raised serious objections on financial grounds and dictated the necessity of phased implementation. The reorganization had not been completely carried out when the First World War broke out in 1914. But the complete unification of the army in India on the principles laid down by Kitchener enabled India's armed forces to promptly take up responsibilities in the War. They were principles under which all constituent elements of the army were sought to be trained and organized on a uniform basis to admit of interchangeability. On 1 August 1914, when the War broke out, the total strength of the fighting services of the Indian army in all ranks was 1,55,423. By the time of the armistice this had risen to approximately 5,73,484, a figure which included Indian troops serving in Palestine, Egypt, Iraq, Salonika and elsewhere in the world outside India.

The War of 1914–18 in fact created conditions for the emergence of two main trends in the military history of India. These were organizational and political. They were mutually related trends, giving rise to a military reorganization that was considered best-fitted to war experience as well as the nationalist demand for increasing Indianization. While the first emphasized reforms to suit the purposes of external defence and internal security, the second, regardless of efficiency, remained concerned with the rapid Indianization of the army as an instrument to force the pace of India's political progress, a concern shown equally in respect of the superior civil services. Both trends had merits, but time and patience were required to accomplish a balance between efficiency and Indians' political urges for power. No such balance was ultimately attained, for the revolt of a section of the Indianized army and the silent support of a section of the civil services were recognized catalysts of the final transfer of power.

The source of political discord underlying military policy was non-recognition by the British of the right of Indians in general to bear arms in defence of their country, and a general demand made by Indian political leaders during the War itself for extended opportunities of military service. Their pressure increased when the principle of providing for increased Indian representation in the civil services had already been recognized on the lines recommended by the Islington Commission, which stirred hopes for some change in British military policy too. The Indian demand for immediate consideration arose, however, from a 'recognition of the brilliant and devoted services of the Indian army in the various theatres of war'.[89] A good deal had been done by the British to acknowledge these services, but the grant of British commissions to Indian officers was the most important development.

During the War, British and Indian ranks intermingled in the same unit. There had, of course, been a certain proportion of British ranks in Indian units from early times. The process was now reversed, for Indian combatants were instead introduced into British units. The only types of combatant units in the army which did not include both groups in their ranks were the British and Indian Cavalry. Indeed, under the impact of War the places of British officers were being filled by Indians who held the King's Commission and were in every respect equals with their British comrades. This union of British and Indian elements at the levels of both men and officers was leading, though yet imperceptibly, to the evolution of an alternative agency of defence and security, an Indian military agency necessary for the support of self-government.

The First War, however, revealed some serious defects in respect of organization and equipment. The ancillary services of the Indian army were, for instance, either non-existent or undeveloped. Internal security units had to be largely depleted for war purposes because of the inadequacey of peace establishments. Technical and administrative personnel required on mobilization, on the other hand, had to be found from the establishment of combatant units. The equipment in the Indian army was also found to be outmoded for modern warfare—especially in respect of machine guns, artillery, hospital equipment and medical establishments. Neither an air force nor a mechanical transport service existed. The reserves and machinery for reinforcements were equally inadequate in comparison to European armies. India's indigenous resources had not, in fact, been sufficiently developed.

A Committee was appointed in 1919, with Lord Esher as its president, to further reorganize the army. The Committee presented its report towards the close of 1920.[90] Improvements in the terms of service of the Indian ranks of the army proceeded from the recommendations of the Esher Committee. Indian soldiers and officers with the Viceroy's Commission were specially benefited, not only in respect of pay and pensions, but also as regards other facilities like rations, clothing and housing. To strengthen the country's defence and internal security, however, the Esher Committee recommended the formation of a territorial or second line force in India modelled on the combatant recruits raised during the War. The Committee provided province-wise figures of the numbers of recruits raised as follows:

	Population in millions	Number of recruits raised
Punjab	20	326,000
United Provinces	47	142,000
Madras	40	46,000
Bombay	20	36,000
N.W. Frontier and Baluchistan	3	33,000
Burma	12	13,000
Bihar and Orissa	33	8,000
Bengal	45	7,000
Central Provinces	13	5,000
Assam	6	1,000
Ajmer	0.5	7,000

Consistent with the recommendations of the Esher Committee, in 1917 the Government of India proceeded to form the Indian section of the Indian Defence Force. But the scheme proved a failure, for the protagonists of the territorial movement had probably overrated the wish and capacity of India's urban classes to share in the burden of defence, except in the auxiliary service. The difficulties involved in arousing support for combatant service and creating a disciplined force from among the urban classes had not been anticipated.

Institutional Agencies and Political Stability 251

Although military expenditure was not subject to control by the Indian Legislature under the Montford Reforms, the Legislative Assembly was permitted in March 1921 to express its views on the report of the Esher Committee, with special reference to improvement in quality of the Indian armed forces. Its resolution, which bore directly on the reorganization of the army then in progress, pointedly said: 'That the purpose of the army in India must be held to be the defence of India against external aggression and the maintenance of internal peace and tranquillity. To the extent that it is necessary for India to maintain any army for these purposes, its organization, equipment and administration should be thoroughly up to date and, with due regard to Indian conditions, in accordance with present-day standards of efficiency in the British army, so that when the army in India has to co-operate with the British army on any occasion there may be no dissimilarities of organization, etc., which would render such co-operation difficult.'[91] The resolution clearly implied that India's military obligation was limited to the defence requirements of India only, but that its efficiency must in no wise be inferior to the standards of the British army. It was therefore emphasized that Indians must be freely admitted to all arms of His Majesty's military, naval and air forces in India, including the ancillary services and the auxiliary forces, the object being to encourage Indians, more especially 'the educated middle classes', to attain the necessary level to enter the commissioned ranks of the army.[92]

Indeed, the First War marked the perceptible rise of an alternative sovereignty. It provided conditions for Indians to utilize institutional agencies which had, over the years, attained a fair degree of maturity in political and legislative fields as well as in civil and military administration. While the growth of a separate and distinct legislative authority came to be used as an organized and legitimate forum to press for the recognition of political freedom, the increasing strength of Indians in the civil service and the army in India reinforced the nationalist struggle for self-government. The War created the compulsions, but British policy did not fail to recognize them in its quest for political stability.

Although Lord Inchcape's Retrenchment Committee (1922) recommended considerable reductions in military expenditure as a whole, it left the military advisers of the Government of India with discretion in carrying out its proposals so that no vital service of the army be unduly weakened. Moreover, the Royal Air Force had already been added to India's defence services, which, in addition, had also the benefit of the Mechanical Transport and Signal services. The artillery had likewise been reorganized, along with the reconstitution of the staff and commands of the army on the basis of modern military ideas. The resources of the country and its *post-bellum* economic conditions were, of course, not strong enough to meet the cost of equipment and reorganization, but the British government helped meet the expenses in recognition of its responsibility to maintain India's unity and stability. The problem of India's defence taken by itself had acquired an urgency as 'the frontier tribes after the War were more formidable than they had ever been before, owing to better armaments, more plentiful supplies of ammunition and a great advance in tactical skill'. Added to the problem of protecting the north-west frontiers, was the spread of lawlessness proceeding from the Non-cooperation Movement of the Congress, which occurred simultaneously with 'attempts from alien sources to propagate in India the principles of Bolshevism'.[93]

The pressure which political agitations started exerting on the resources of the army became a regular feature in the wake of the Montford Reforms: the Moplah rebellion in 1921, and the numerous occasions on which troops were called out during 1920–2 in support of civil administration highlighted this trend. The field army increasingly had to be made available in the maintenance of internal security with the advancement of constitutional reforms. All the elements of the army in the Indian situation had thus necessarily to be organized and trained on a uniform basis to render 'interchangeability' possible to suit the exigencies of external defence or internal security, as the case might be.

The solution of the political problem of self-government was directly dependent upon the military problem. Indian politicians realized that it was impossible to envisage a self-governing India without an Indianized army. Under the advice of General Rawlinson, the Commander-in-Chief and War Member of the Council in India, the Government started as early as 1923 to give Indians a fair opportunity to see that the units officered by them became efficient in every way. But, while the process of Indianizing the army was bound to be slow, many politicians, including moderates, had started proclaiming the goal of swaraj in eight or ten years' time and forcing the pace of political advance after the visit of the Statutory Commission in 1929. General Rawlinson had realized in 1924 'the supreme difficulty' of keeping the military and political processes of development 'at anything like an even rate of advance'.[94] For at the time there were only eight Indian officers with King's Commissions, against two thousand needed for complete Indianization. The utmost that could be done within five or six years, he argued, would be no more than raising the strength of Indian commissioned officers to about three hundred, a figure far below the requisite number.

A certain degree of decline in the efficiency of the civil services could perhaps be risked to meet the demands of local politics. The Lee Commission accepted this by provincializing the All-India Services in the technical fields and increasing promotions to 'listed posts' from the provincial services. But a similar risk could not be taken in the defence services. The Indian Sandhurst Committee's Report cautioned in 1927 that the number of Indian King's Commissioned officers should not be increased 'without reference to considerations of efficiency'. The reasons were obvious. As the Committee argued:

We recognize that in the Army there can only be one standard of efficiency, namely, the highest. We hold strongly, therefore, that the severity of the existing tests should not be relaxed in any way, and, if Indians capable of satisfying these tests are not forthcoming, then the pace of Indianization must for the time lag behind the number of vacancies offered. But at the same time we contend that, in order to induce the best material to accept the admittedly arduous preparation for a military career and in order to induce the educational authorities in India to lend their active co-operation, it is necessary to widen the field of opportunity.

The Committee thus proposed the establishment of what it called an 'Indian Sandhurst',[95] and for an increasing number of Indian cadets to be trained at Sandhurst in England. Its recommendations extended to Indianization in such other fields as the Artillery, Engineers, Signals, Tanks and the Air Force. An Indian 'Sandhurst' was instituted at Dehra Dun in 1933, to secure a considerable increase in the number of Indian commissioned officers and thus facilitate the complete Indianization of the army without prejudice to its efficiency.

The army authorities were guided in their efforts by a sincere desire to promote the growth of Indian unity in sympathy with many Indian leaders who were anxious to develop a more general sense of citizenship on an all-India territorial principle. But the 'potent influence of tradition and race' needed to be overcome. The steps that the army authorities had already started taking in the direction of that unity included the formation of cadet corps in the various universities out of army funds. In 1923, the Territorial Forces Act was passed, and 23 Territorial Units (including four urban battalions) formed in all parts of India. But the change was bound to be slow, given the ethnic differences of the country and the varying attainments of the different groups.

No division of loyalty in the army on grounds of India's races or religions was apprehended in the presence of British troops and the leadership of British officers. The main source of division was Indian politics, guided more by motives of power than considerations of unity or even ideology. The Quit India Movement of 1942 represented a climax which made a return to normalcy well-nigh impossible. The army remained unrelenting in its determination to preserve India's territorial integrity throughout the Second War, and on 22 October, soon after the War ended, General Auchinleck, the Commander-in-Chief, even announced a plan for the complete Indianization of the armed forces of the future. The plan, which had been under consideration for some time, aimed to provide for the future officering of the Royal Indian Navy, the Indian Army and the Royal Indian Air Force. It restricted the grant of permanent commissions to non-Indians and to others domiciled in India, but it was generally recognized that the three services would still require a quota of British officers until such time as there was an adequate supply of qualified Indian officers to fill all grades in the officer cadre.

The political imbalance which the Quit India Movement had created could not but influence the future course of development. It accelerated the pace of political advance at a time when a corresponding progress in terms of Indianization of the services could not occur for security and defence reasons during the Second War. The gap which so developed not only affected the standard of military efficiency and discipline, but created a sudden shortage of superior personnel in the civil service too, which left its own mark of inefficiency and corruption at most levels of civil administration, more especially in the provinces. The alternative sovereign authority which had over the years been emerging on the legislative front had thus to contend with serious handicaps of its own making, although the bureaucratic infrastructure and the Indian army built up during British rule continued as the mainstay of the new successor sovereignty.

Social and Economic Policies

The Rural Scene

The rural population under the Mughals was dominated by the landed aristocracy consisting of such superior officers of government as were paid by the grant of jagirs. The nobility also included another category, rajas or big zamindars, and recipients of rent-free grants of land for religious and charitable purposes. Whatever the purpose, official or otherwise, a regular feature of the Indian landed aristocracy was that it was a creature of the ruling kings. The several types of 'assignees' of a superior order constituted what might be called a feudal structure where the rajas or zamindars received from the king their arms and fiefs with knightly ceremonies. Their seignorial rights, however, arose from tax farming and the military and fiscal fiefs of a despotic state for purposes of local security and financial stability, rather than from any legal recognition of the same in respect of village lands.

It is true that in between the big zamindars and the cultivating tenants there existed a class of landed gentry or under-tenure holders known variously in the different parts of the country. But on these was superimposed a higher order of landed interest called *zamindars* in Bengal, *taluqdars* and *jagirdars* in Oudh, *inamdars* in Bombay and *poligars* in Madras. The system of revenue farming that had come to be recognized by later Mughals as a regular mode of collection was ruinous to the cultivating community, for it encouraged extortion and led to the expropriation of the hereditary and lawful owners of land. In a letter of 3 November 1772 the Bengal Government wrote to the Court of Directors that Mughal governors 'exacted what they could from the zamindars and the great Farmers of the Revenue, whom they left at liberty to plunder all below, reserving to themselves the Prerogative of plundering them in their Turn when they were supposed to have enriched themselves with the spoils of the country',[96] a development which became even more serious as those who were invested with the collection maintained a body of militia to keep the peace and used them to curb the independence of subordinate proprietors.

In *The Indian Middle Classes* I have shown how the general effect of British rule weakened the landed superstructure which prospered under the country governments, Mughal or Maratha. This was sought to be accomplished in two ways: first, by erecting a strong central political authority with a judicial and executive apparatus which left little or no scope for a zamindar to exercise any public duty or function: and secondly, by defining and recognizing the rights not only of under-tenures, but also of tenancy. It reduced the powers of a zamindar to subject them to his observance, a development which conduced to the growth of a landed middle class, more especially with the flow of commercial capital into agricultural investment.

As already noted, political considerations led Cornwallis to introduce his Permanent Settlement with the zamindars in 1793. While Munro set the current against the Cornwallis system in Madras, Holt Mackenzie in the North-Western Provinces took the cue from Munro and set at nought a scheme of permanent settlement which Wellesley had announced at the turn of the century for the ceded and conquered districts.

The idea of a permanent settlement was revived again after the suppression of the Mutiny in 1858. But in its revival the British were now guided by a land policy much wider and deeper in the extent of its appeal for political stability, an appeal which was reinforced by the inclusion of economic objectives to promote an identity of interest between European settlers and the Indian 'agricultural community generally' [97] in other words, land policy was linked with *laissez-faire* to stimulate 'the employment of British capital, skill and enterprise in the development of the material resources of India' with the collaboration of Indian counterparts. The Revenue Dispatch of the Secretary of State, Lord Stanley, which laid down this policy, applied to waste or unoccupied lands. The land revenue of any such lands granted for the cultivation of exportable products was subject to being redeemed. When applied to the grant of waste lands, it gave encouragement to agricultural development. But when extended to the estates of Indian land-holders, the principle of redemption involved a clear loss of public finances. Stanley was not unaware of this: 'But the political results of such a change cannot be overlooked. The fortunes of the Zemindar who has been allowed to extinguish his fixed annual liabilities by a single payment are from thenceforth still more intimately connected than they are at present with those of the British Government. The immunity from taxation ... renders his loyalty a matter of prudence and self-interest'.[98]

Sir Charles Wood, who succeeded Stanley, treated the proposed redemption of the land revenue as in effect a settlement in perpetuity. But he preferred a general permanent settlement. In his dispatch of 9 July 1862 he clearly indicated his preference and affirmed that 'Her Majesty's Government entertain no doubt of the political advantages which would attend permanent settlement'. He observed: 'The security, and it may almost be said, the absolute creation of property in the soil which will flow from the limitation in perpetuity of the demands of the State on the owners of land, cannot fail to stimulate or confirm their sentiments of attachment and loyalty to the Government by whom so great a boon has been conceded, and upon whose existence its permanency will depend.'[99]

While advocating the case for a permanent settlement, Wood was not thinking in terms of the old concept of a landed superstructure, which did not fit in with the requirements of the rising capitalist economy calling for an extended entrepreneurial base for economic operation. As said before, the need was for a wider social base to lend political support to the colonial economy. 'It is most desirable', he thus observed, 'that facilities should be given for the gradual growth of a middle class connected with land',[100] an entrepreneurial class which might collaborate with European settlers in the promotion of India's agricultural economy.

The Urban Scene

In pre-British days there were doubtless elements which formed the basis of the later middle-class growth in modern, bourgeois terms. They comprised a number of functional groups which, in point of income as well as respectability, came next to the aristocracy—official, landed or mercantile. In the commercial field as well as on the financial side there were, apart from big monopolists and bankers, average and well-to-do shopkeepers and money changers. Foreign travellers testify that considerable quantities of bullion flowed into India but did not find its way into productive channels. Bernier attributed the lack of productive investment to the despotic political system of the country. From a sense of insecurity, most of the wealthy buried their gold and silver 'at a great depth', and, instead of seeming independent and living in comfort, preferred to appear 'indigent'. According to Bernier, the general practice among all merchants, whether Hindus or Muslims, but especially among the former who mostly controlled the trade and wealth of the country, was to keep money concealed lest it excite the cupidity of a local governor or officer.[101]

Unlike English medieval towns, Indian cities possessed no charter of liberties or any form of popular municipal government which might otherwise have operated as checks on the arbitrary proceedings of officials. Even the upper layers of business communities assumed a posture of servility in their quest for security. It was here that the coming of the British marked a radical difference. The Hindu business communities found political backing with the servants of the East India Company, who were themselves primarily interested in the promotion of trade on a free and extended basis. They emerged as collaborators against the autocracy of the Mughal system and were precursors of the new middle class in modern times. Their collaboration with the Company was based on an identity of commercial and financial interests, which helped the latter gain a firm footing in Bengal.

The modern middle-class growth was, however, a product of a liberal education that was antithetical to monopoly but conducive to social mobility. The artisans, who had the monopoly of craftsmanship, received no education. The shrewdness of business classes was proverbial, but their education did not go beyond a hereditary knowledge of commercial accounts. Occupational specialization proceeded from hereditary callings, not from higher education which remained literary in content and restricted to the priestly classes, without any emphasis on the applied skills and sciences. Despite the potential for middle-class urban development, therefore, the immobility arising from caste organization and the despotism of the political system precluded the development of a bourgeois elite. The rule-based Company's committee form of government, and Bentinck's educational policy, heralded an era of capitalist enterprise: this was referred to by Sir Charles Wood in connexion with his 'middle class' land policy laid down in the revenue dispatch of 9 July 1862.

Bentinck's educational policy, which aimed at 'the promotion of European literature and science among the natives of India', contained the directive that all funds set apart by Government for education 'ought to be employed on English education alone',[102] was based on Macaulay's well-known minute of 2 February 1835.[103] 'In one point', Macaulay argued with the Committee on Public Instruction, 'I fully agree with the gentlemen to whose general views [in favour of Oriental learning] I am opposed. I feel with them that it is impossible for us, with our limited means, to attempt to educate the body of the people. We must at present do our best to form a [middle] class who may be interpreters between us and the millions whom we govern—a class of persons Indian in blood and colour, but English in tastes, in opinions, in morals and in intellect. To that we may leave it to refine the vernacular dialects of the country, to enrich those dialects with terms of science borrowed form the western nomenclature, and to render them by degrees fit vehicles for conveying knowledge to the great mass of the populaiton.'[104]

The object of this educational policy was extension of modern science and literature to the masses through their own vernaculars, and not through the classical oriental languages which were monopolized by the priestly classes, an exclusive minority. The responsibility of educating the masses was sought to be devolved on the English-educated elites, a responsibility which, as Wood's Dispatch of 1854 admitted, had remained neglected. However, it was increasingly recognized that, in view of the limited resources of the government, successful rural development had to depend on the education and character of the urban elites, more especially those in the growing Presidency towns, the chief centres of modern business and education. It was recognized that the advancement of Western knowledge in these urban centres, as Sir Charles Wood envisaged in his famous dispatch of 19 July 1854, could alone help Indians in 'the development of India's resources' by a resort to 'the employment of labour and capital' in rural areas on a scale that might supply the many agricultural products required for the manufacturing industries in England. This typically colonial approach to rural welfare was to be implemented through the European settlers and their emerging collaborators, the Indian middle class that was connected with land. The promotion of an English-educated middle class was thus considered not only politically expedient but also an economic necessity for Britain and India alike. The plantation industry marked the beginning of India's rural development.

Pattern of Economic Developmet

Modern capitalist enterprise in India also began with the plantation industry. The export of India's agricultural produce created a trade surplus which paid not only for the construction of railways and other reproductive public works, but also for the import of capital goods and machinery which began to manufacture locally the raw materials developed by planters.[105] The surplus of merchandise exported during 1842–82, for example, averaged nearly £16,170,000 annually. The value of treasure imported into India in the same period exceeded the export of treasure by an annual average of more than £8 million. This pattern of economic development continued until after the First World War, when, under nationalist pressure, the Indian Government was obliged to introduce protection and direct participation in the industrial development of the country.

The principle of state participation in the development of India's economy symbolized a clear opposition to *laissez-faire*, and is traceable to the experiences of recurrent famines calling for executive interference with the operation of free trade. In the course of the 1873 famines certain District Officers proceeded on their own, without any provisions of law, to seize stocks of foodgrains and distribute them on humanitarian grounds among starving villagers.[106] A series of measures followed the recommendations of the Famine Commission. Some of the important measures included not only tenancy legislation and cadastral surveys to curb the arbitrary proceedings of zamindars and planters, but also other changes calculated to promote agricultural development.

The middle class connected with land clamoured for the permanent settlement of land revenue as a remedy to recurrent famines, and quoted Sir Charles Wood's dispatch of 9 July 1862 to justify its claims. But Curzon closed the controversy by an official Resolution of 18 January 1902 which declared that 'a Permanent Settlement, whether in Bengal or elsewhere, is not protection against the incidence and consequences of famine'. He was also the first to recognize in principle the necessity of state participation to promote India's industrial development, a principle directed against the free play of *laissez-faire*. Apart from the importance of its protective effect recognized in the post-War period after 1918, the principle was reinforced because of the ideological overtones of Bolshevism, which had begun permeating in the wake of the Russian Revolution of 1917. The Government of India actually adopted a protectionist policy in 1924, which contributed to a rapid growth of Indian industrial enterprise and the managing agency system, its organizational infrastructure. From the experience of the two World Wars and the success of the Soviet Union's five-year plans emerged the concept of national economic planning initiated by the Government of India during 1944–5; this formed part of the Grow More Food Campaign started in 1942, the year of the Congress Quit India movement. On accession to power, the Congress government appointed its own Grow More Food Inquiry Committee and followed up on the principle of five-year development plans formulated by the Government of India in 1945.[107]

Problems of Social Policy

The basic problem of social policy was the imbalance between rural and urban development. It occurred under a political system which was sought to be evolved on contractual principles without matching attempts being made to provide for social and economic ingredients to make these principles evenly operative. Its effect was an unintended destabilization and distortion.

Under the Mughals status conduced to social stability, which meant a negation of change. Both the despotism of the sovereign and the caste system were hostile to change or social mobility. There was an ideological identity between the political and social systems. While Hindus provided a kind of religious legitimacy to Muslim rulers on the basis of a divine will, the latter recognized the supremacy of the caste order, the cornerstone of a status-bound society. In a statement before a parliamentary committee on petitions in 1781, Verelst, the Governor of Bengal in 1769, affirmed that the Muslims, who had usually carried their conquests 'by the Edge of the Sword', chose 'wisely' to become 'the Guardians and Protectors of the Hindoo Religion',[108] based on caste. Formal differences doubtless continued on both religious and political grounds. But Hindus held on by virtue of their traditional superiority in the field of commerce, industry and finance. In fact, the number of Hindu officers increased even under Aurangzeb, the most orthodox of Muslim kings. Hindu education, which was considered a religious exercise, remained a veritable Brahmanic monopoly and even influenced Muslims, especially in the profession of astrologers who were regarded by all 'as so many infallible oracles'.[109] The several occupational castes were likewise recognized as part of a divinely ordained social system, a system which seriously impeded the indigenous growth of capitalism or social mobility based on humanist principles. Weber truly remarks that 'a ritual law in which every change of occupation, every change in work technique, may result in ritual degradation, is certainly not capable of giving birth to economic or technical revolutions from within itself, or even facilitating the first germination of capitalism in its midst'.[110] Instances were, of course, not wanting of popular movements which challenged the validity of hereditary privileges. But while their followers were finally assimilated into the caste order, political expediency did not permit any interference with the punishments imposed by a village community for the violation of caste practices.

The contract-based legal and judicial systems which came to be established under British rule were in principle opposed to caste being accepted as divinely ordained. But political compulsions could not be ignored. The Vellore Mutiny was an early case in point. In the interests of uniformity, the Madras Government required Indian troops to wear a new kind of turban and remove their distinguishing caste marks and earrings when on parade. The sepoys feared that this was an attempt to convert them to Christianity. They revolted and killed about 200 Europeans at Vellore in 1806, which led the Directors of the Company to issue a dispatch to the Governor of Madras on 29 May 1807, which said: 'In the whole course of our administration of Indian Territories it has been our known and declared principle to maintain a perfect toleration of the various Religious systems which prevailed in it, to protect the followers of each in the undisturbed enjoyment of their several opinions and usages; and neither to interfere with them ourselves nor to suffer them to be molested by others.'[111]

This dispatch was in accord with the Royal Proclamation which Queen Victoria later issued in 1858 on the subject of respecting religious and caste prejudices. The problems of social policy, in fact, involved considerations of political stability. On the conquest of Cuttack from the Marathas, for instance, the Company's government had to take over the entire management of the celebrated temple of Jagannath and undertake, like its Maratha predecessor, to supply the deficit, if any, from its own treasury. It took care to employ 'every possible precaution to preserve the respect due to the pagoda and to the religious prejudices of the Brahmins and pilgrims'.[112] Despite objections from Christian missionaries the Government and its officers continued to be associated with the management and financing of both Hindu and Muslim shrines in Bombay and Madras also.[113]

Indeed, it was easier to ensure political stability, as the Mughals did, within the framework of a status society or *status quo ante*. It could not be so in a developing society where social relationships were sought to be changed through a productive, legal or political process on a principle antithetical to *status*. This precisely was the case when the caste-ridden society of India was rendered politically subject to the discipline of the legal and judicial institutions established under British rule on the principle of *contract*, a capitalist principle, territorial and humanist in concept, which contained seeds for the growth of citizenship, the antithesis of a 'divinely ordained' caste order. It is true that the 'Regulations' of the Company's government tended to preserve the country's customs and usages, with the result that even a practice like sati continued unchecked until 1829. But the English law which was administered in the Presidency town of Calcutta, observed no such formality. Maharaja Nandkumar, a Brahmin zamindar of great influence, was sentenced to death for forgery. The Indian Penal Code doubtless represented a compromise. But it was a compromise weighted under Macaulay's influence on the side of Liberalism rather than the iniquity of an authoritarian status society. It set in motion a process of social change which tended to alter the character and complexion of the political system. To ensure political stability even in the midst of a changing society was, in fact, an ingenious, though difficult, task. It needed not only a sound knowledge of society but also the prudence, capacity and firmness of statesmen who recognized the inter-relatedness of social and political change. Before he came out to India as Law Member in 1834, Macaulay, for instance, had foreseen the consequences of a social policy that created an English-educated Indian middle class. Having become instructed in European knowledge, he said in a debate in the House of Commons, India might in future 'demand European institutions', an eventuality, which he would treat as 'the proudest day in English history'.[114] India's political unity and stability became in course of time a function of applied science and technology, the manifestation of that European knowledge which, in nationalist terms, contributed through a separate legislative process to the emergence of an alternative sovereignty, that of the people of India.

To ensure a balance in the pursuit of social policy Macaulay had recommended mass education through the local vernaculars. In the words of C. E. Trevelyan, the object was 'to promote the extension, not the monopoly, of learning; to rouse the mind and elevate the character of the whole people, not to keep them in a state of slavish submission to a particular sect'.[115] But the agency recommend for the extension of education 'from town to country, from the few to the many' was that of 'the rich, the learned, the men of business' who alone received the benefit of Western education.[116] The plan of rural education remained inoperative. Its failure was admitted in Wood's Dispatch of 1854. The importance of elementary education for villages was revived in 1870 by the government of Lord Mayo. But its responsibility was relegated to local bodies with poor financial resources. Things dragged on until secondary education became in 1882–3 a free enterprise, which the urban rich used to their exclusive advantage for higher education and employment in the services of the State.

Even the rural population tended to move towards urban centres. For those who rose in the social scale wished their children to be educated in English. The affluent among the landed gentry encouraged this tendency, while the district and local boards did little to raise the standard and quality of education under their control. Their basic difficulty was financial, and hardly one-fifth of the total number of boys of school-going age received primary education even in 1901.[117]

The situation continued to remain in a state of neglect, even though Curzon made a beginning in the field of technical and industrial education.[118] His contribution to the promotion of modern agricultural knowledge was especially noteworthy.

Curzon regarded the Indian peasant as 'the bone and sinew of the country' who contributed 'one-fourth of the national income'. In a speech at the Byculla Club, Bombay, he suggested that the peasant should be 'the first and final object' of attention:

It is for him in the main that we have twice reduced the salt-tax, that we remitted land revenue in two years amounting to nearly 2½ millions sterling; for him that we are assessing the land value at a progressively lower pitch and making its collection elastic. It is to improve his credit that we have created co-operative credit societies, so that he may acquire capital at easy rates, and be saved from the usury of the moneylender. He is the man whom we desire to lift in the world, to whose children we want to give education, to rescue whom from tyranny and oppression we have reformed the Indian police, and from whose cabin we want to ward off penury and famine. Above all let us keep him on the soil and rescue him from bondage or expropriation.[119]

As regards the educated middle class who pressed for a permanent zamindari settlement of land revenue and representative government as a panacea for all ills, Curzon questioned their claim to speak for the Indian people, although he consistently sought to recognize their right to appointments in the services of the state.

The dichotomy between rural and urban interests also raised issues relevant to political stability. One section of the National Congress was revivalist in its approach. It was led by Tilak, a supporter of the hereditary privileges which were being shaken under British rule. The pro-tenant policy of the Government which had flowed from the experiences of recurrent famines, was already affecting the interests of zamindars, the protagonists of the Permanent Settlement, which the revivalist group in the Congress commended to the Government as the only answer to famines. The Collector, through his cadastral surveys, sought to bypass the zamindars and establish direct contact with the people, but the Congress proceeded to agitate for the separation of executive and judicial functions, a move intended to protect zamindars against executive action. Any official measure to strike a balance between urban and rural interests, more especially in the field of educated employment, was viewed by Tilak and his group as prejudicial to the Brahmanic monopoly. Measures of social reform intended to remove inequities in the Hindu tradition were opposed on the grounds of interference with religion. Steps to introduce through local vernaculars the benefits of education among the lower orders of society, or action against infanticide, the immolation of widows, the immunities enjoyed by Brahmins, or, even the famous Age of Consent Bill (1891) were, for example, used to provoke agitations among the masses who were susceptible to propaganda and rumours that were frequently based on falsehoods.

British social policy thus obviously involved great political risks. Even so, the risks were taken and the policy established the superiority of the state over religion in the governance of social relationships, a modern concept which recognized the universality of law as a cohesive political entity signifying the will of the state in its cumulative sovereign capacity. Political stability nevertheless dictated the need for a compromise based on the principle of checks and balances. Despite its distrust of the educated middle-class elites who claimed to speak for the Indian people, the Government conceded, even though grudgingly, their demand for representative government under the Indian Councils Act (1892), which followed soon after the Age of Consent Act (1891). Plot-wise revenue surveys of land were however introduced on a general plan in 1892 itself, to strike a balance between rural and urban interests by an attempt to discourage absentee landlordism and secure the rights of the cultivating community. The cadastral survey of land was completed in 1901, when the Punjab Land Alienation Act was passed in the general interest of the peasant community.

The balance was tilted in favour of the middle classes and their urban interests mainly by the crisis of imperialism arising from the compulsions of the First War, which, besides raising the strength of Indians in the armed forces, contributed to a considerable increase in indigenous industrial capacity. The weakening of the imperialist hold in terms of its military and economic position in India was a major factor for compromise with the Indian middle classes, the protagonists of urban interests and indigenous industrial enterprise. But a more important reason for compromise was the rise of Bolshevism and the Russian Revolution of 1917, which encouraged not only a spirit of political freedom, but also a revolutionary concept of socialism which had been growing during the War as an anti-imperialist force. In a letter to the Viceroy (Chelmsford), the Secretary of State (Sir Austen Chamberlain), described the Indian situation as being understandably influenced by the revolutionary ferment elsewhere. The ferment, he said, 'is working everywhere, and in India as much as anywhere. Opinion cannot but be excited by the Russian Revolution, by congratulations showered upon the revolutionaries from England and elsewhere, and by the constant appeals to the spirit of liberty and nationality which are the groundwork of most of the public declarations of the time'.[120]

The British declaration of 20 August 1917 came in this context. It led to the introduction of partial democracy through Dyarchy, a political compromise between Indian and British urban interests. And as the establishment of the Communist International in Russia in 1919 provided institutional support to Indian revolutionaries abroad and in India,[121] the emphasis on collaboration with the middle-class Congress leadership was reinforced rather than relaxed. For the Congress was to the British a lesser evil than any party or group, either in alliance with, or favourably inclined towards, the Communist International, a counterforce against imperialism on a world scale. That perhaps accounted for a virtual neglect of India's rural economy thereafter. Though critical, and frequently strongly opposed to the policies of the nationalists, in the overall imperial context, the British had no alternative but to accept the Congress as perhaps the only possible successor to British sovereignty in India.

The several agitations which took place after 1920 to expedite the transfer of full power, were, however, not without significance. It is doubtful whether the leadership of the Congress had an overall appreciation of the international situation which led to the policy announcement of 20 August 1917, which signified the British intention to withdraw in course of time. The Congress's analysis of that situation perhaps did not go beyond the limits of pan-Islamic religious sentiments arising from British action against Turkey, an ally of Germany during the War. The nationalist and khilafat leaders reveal little or no awareness of how, in the wake of the Russian Revolution of 1917, both Muslim and Hindu revivalist revolutionaries abroad had shaken off their religious beliefs and turned to socialism under the influence of the Comintern.[122] The new spirit demanded political independence as a means of promoting social revolution in the interests of peasants and workers. The Hindu-Muslim alliance in the course of the Non-Cooperation Movement was, however, a mere political contrivance to quickly get into the portals of power. While the Hindus, who dominated the Congress, manoeuvred to grasp power as soon as possible, the British endeavoured, from overall considerations of imperial interest, to regulate the pace of political progress, more especially in keeping with the demands of a plural society and national stability. But the agitational approach to power politics created, in the peculiar Indian situation, divisions on communal and caste lines, which ultimately led to the partition of the country.[123]

Chapter V

DIVISIVE TENDENCIES IN A PLURAL SOCIETY: PRE-1936 DEVELOPMENTS

The famous policy announcement of 20 August 1917, which recognized the gradual development of self-governing institutions on the principles of democracy and responsible government as the goal for India, marked, in the words of the Montford Report, 'the end of one epoch and the beginning of a new one'.[1] For the extension of franchise on a direct electoral principle and the constitutional change effected in consequence of that declaration were not a natural expansion of the Morley-Minto system, where, apart from members representing some special class or community, there was absolutely no connexion between the supposed primary voter and the man who sat as his representative in a legislative council. The 1917 announcement heralded a qualitative change, one conducive to the development of an alternative Indian sovereignty. It embodied a new principle of policy, a principle consistent with nationalism and self-determination, which formed no part of the earlier schemes of representative government.

The Montford Report, which followed the announcement, led to the establishment not only of a limited form of responsible government called Dyarchy in the provinces, but also of a bicameral legislature with elected majorities at the Centre. With a system of direct election on a parliamentary principle the change so effected afforded opportunities for the penetration of politics through political parties and groups representing the various shades of interest characteristic of a plural society like India's. In view of the limited franchise, the degree of that penetration was, of course, bound to be limited. Even so, the change marked the beginning of what might be called mass involvement in politics.[2] Democracy, even though partial, had begun to leave its imprint not only on the political behaviour of the constituents, but also on others related to them. Broadly, they consisted of the two major political communities, the Hindus and Muslims, who, despite the cultural interactions which had taken place over the centuries, maintained their distance in terms of religious faith, historical antecedents, social organization and view of life.

Divisive Tendencies in a Plural Society

The parliamentary system, which established a direct connexion between the primary voter and his representative in the legislature, afforded opportunities for the constituents to have their views and interests reflected in government policy through the proceedings of the legislative council, not on territorial principles on which constituencies were in law demarcated, but on religious and caste differentiations, on which politics had to operate in India's plural society. This dichotomy between law and politics, a characteristic feature of the Indian situation, was the cause of division on a communal basis.

Historical Background up to 1919

The Mughals did not interfere with the economic power and influence which the Hindus wielded, more especially in trade and commerce, industry and finance. The Hindus acted as monopolists and financiers for the government. James Grant's *Political Survey of the Northern Circars* makes it clear that, while the village communities had virtual control of the rural economy, the feudatory status of Hindu zamindars remained unaffected despite periodic dislocations caused by the movements of armies. The *Ninth Report of the Parliamentary Select Committee* of 1783 also confirmed that the Muslims had not generally engaged in trade and commerce, or disappropriated the ancient Hindu states or proprietors whom they conquered. Their main, if not sole, support was a share in the civil and military offices of the government, especially in its police and judicial departments.[3]

Even under the Mughals, the department of revenue was manned almost exclusively by Hindus, who continued to dominate the district staff of Warren Hastings' Amini Commission.[4] Those who lost heavily with the establishment of British rule in Bengal, for example, were mostly Muslims whose main occupations were in government service. They were hard hit by the abolition of the office of *faujdar* in 1781 and the discharge of the judges of criminal courts in 1790. In both cases they were replaced by Europeans. On the acquisition of the Diwani in 1765 the Nawab of Bengal was paid an annual sum of £420,000 to enable him to meet his own expenses and those of his government which he was allowed to retain for a time. This kept well-placed Muslims in their jobs. But, while the Nawab's annual grant was later reduced, lucrative posts were taken over by the British under Cornwallis. The only senior Indian officers whom Cornwallis retained were the high-ranking law officers, both qazi and pandit, whose posts were abolished with the establishment of High Courts in 1862.

In the North-Western Provinces, however, Muslims held higher posts than Hindus, especially in the judicial branch of the Company's service; for the Nawab of Oudh continued to rule till as late as 1856, when his territory was annexed by Lord Dalhousie. William Wilberforce Bird, who served as a judge in Banaras for a dozen years, was impressed with the quality of Muslim judicial officers, who dominated the native judicial ranks in that province. 'I do not think', he remarked, 'the Hindoos make so good judges as the Mahomedans; the Hindoos are very excellent in the way of keeping accounts and collecting revenue; but for judicial administration I should say the Mahomedans are much better.'[5]

According to the *Report* of the Parliamentary Select Committee of 1852, no Hindu judicial officer held the post of Principal Sadr Amin in the North-Western Provinces; the Muslims had 25 such appointments against 4 held by others in 1850. It was only as munsifs in the subordinate ranks of the judicial branch that Hindus faired relatively better, but there, too, Muslims were far ahead with 86 as against 25 Hindus and 8 others.[6] Hindus constituted nearly 86 per cent of the total population of the North-Western Provinces, including Banaras, but occupied only 18 per cent of the total number of judicial situations as against the Muslim share of 72 per cent.

In Bombay, however, the judicial officers were almost entirely Hindu. The position was different in Madras, where, out of a total of 186 situations, Muslims held 40 against 128 held by Hindus and 18 others. Thacker's *New Calcutta Directory* for 1858 showed that, although Muslims had lost their predominance of pre-British times, their position was still reasonably strong in the Company's judicial service in Bengal, Bihar and Orissa as a whole. The following figures indicate the position:

	Hindus	Muslims	Others
Principal Sadr Amins	13	12	8
Sadr Amins	15	9	2
Munsifs	112	82	7
Total	140	103	17

The extension of British rule affected Muslims not only politically but also in their preferential claim to employment in the judicial, executive and military branches of government. *Qazi*s and *mufti*s no doubt managed to continue as law officers authorized to give *fatwa*s or decrees in criminal cases, but the introduction of trial by jury in 1832 and the substitution of Persian by English in 1837 greatly reduced their importance even before the abolition of the posts of law officers. Muslims stood to lose in other ways, too. In pre-British days they enjoyed, for instance, the benefit of land grants for a variety of purposes, especially for the pursuit of religious and learned objectives. These became subject to resumption proceedings under British rule. And since Muslims had not historically controlled the country's revenue, commerce and finance, the Hindus continued to gain most with the expansion of trade and industry under the British. Hindus also gained from the Permanent Settlement of land revenue, which gave them a proprietary right in the soil and an opportunity to rise into affluence, providing them means not only to adopt western education, but also for their emergence as power elites. The resumption laws, on the other hand, deprived, in Bengal and Bihar, many from the Muslim learned classes of their means to maintain a decent living. The annexation of Oudh and the 1857 upheaval produced a similar effect in so far as the Muslims of Delhi and Lucknow were concerned.[7]

The fall of the Mughals was an important factor in the decline of the Muslim community. But its own backward-looking attitudes were no less responsible for such a condition. Hindus, on the other hand, were adept in Persian even under Muslim rule. Akbar's revenue minister, Todarmal, for instance, introduced Persian into the revenue accounts that had previously been maintained in the local vernaculars. Hindus competed with Muslim scholars in literary attainments, and by the time Shah Jahan came to the throne, frequently excelled them in this area. The result was that Saadullah, his distinguished minister, was replaced by Rai Raghunath and Chandra Bhan, who were known for their skill in letter-writing and powers of composition in Persian. 'One-half of the Persian literature of the eighteenth century', writes Blochmann, 'is due to Hindus. Their *diwans* [poetry] are as numerous as their *inshas* [model letters]; their Persian grammars and commentaries are most excellent, and they have composed the most exhaustive dictionaries and the best critical works on the Persian language.'[8] The increasing use of Persian and Urdu by Hindus gave a fillip to Indian elements in the Mughal services and counteracted the infiltration of foreigners. This trend continued unabated until Hindu officers became so numerous under Aurangzeb that they could defy any attempt on his part to uproot them from his administration.

Subsequently, while Hindus took full advantage of western knowledge by learning English, the Muslims believed that learning or using the language of non-Muslims would be an act of sacrilege, except when answering letters or combating religious arguments. With the abolition of Persian as the official language of the Company's government, official work began to be done either in English or in the local vernacular. Persian was consequently superseded by such developed vernaculars as Tamil, Telugu, Marathi, Gujarati and Bengali. In Bihar, the North-Western Provinces and Punjab Urdu came to be used, but in other regions the replacement of Persian by the provincial vernaculars threw out a considerable body of Muslim officers from the subordinate services of the Company.

However, the position of Urdu had not stabilized even in the North-Western Provinces, the sphere of direct Muslim influence. From a Report of its Director of Public Instruction (M. Kempson) dated 12 February 1870 it appears that the circulation of Urdu papers was limited to 'the official class and the inhabitants of large towns', because Urdu had come to be recognized as a court language. It lacked the appeal of Hindi, which was written in the Nagari script, although, as the D.P.I. observed, 'A really scholarly Hindee paper is still a desideratum for the use of schools.' The support of the Department of Education was extended to the 'Nagree version of the *Muir Gazette*, published from Moozuffernuggur', but it had to be withdrawn because of the failure of its editor to make it useful to schools.

However, a persistent discussion continued on the 'claims of Hindee to supersede Oordoo as the language of the courts'. Kempson's Report refers to the language controversy which, even at that early period, tended to divide Hindus and Muslims on communal lines:

The popular feeling is undoubtedly strong in favour of Hindee, in opposition to that of the official class, which holds on to existing usages. It may also be safely averred that the Hindoos, as a rule, advocate the claims of Hindee, and Mahomedans of Oordoo. Many Hindoo gentlemen of position have the subject near at heart, and the Maharajah of Benares went so far as to ask to be allowed to effect the change of Hindee for Oordoo within his own domain. The Rajah of Kashipore has stated his opinion very clearly in favour of Hindee; and the general wish is that a trial may be given to the change in suitable localities, such as Muttra or Bundelkhund. The change as considered by the most enlightened among its Hindoo advocates, is not so much one of language as of character.[9]

What led to communalism was, in fact, not the question of speech or language, but the scripts to be used—the question of Persian versus Nagari characters. Advocates of Hindi sought to justify their case with the argument that 'the Hindee character is the best generally known among the masses, and the most easily diffused; that it lends itself with facility to the representation of foreign vocables, and cannot well be mis-read or misunderstood'.[10] The science of language apart, the Rajput chief of Kashipur, Raja Sheoraj, who had formerly been a Member of Council, bewailed in the columns of the *Allygurh Gazette* 'the displacement of the old Hindoo vernacular by a mongrel tongue in foreign characters, which is rarely acquired in any degree of excellence, except by Mahomedans, and those Hindoo castes who look to government employment for a livelihood'.[11]

In a Report of February 1869 the Director of Public Instruction recommended the use of the spoken language in the Nagari character, adding that he 'would not stipulate for a return to the [erstwhile] archaic forms of speech and Sanskrit vocables, but accept all foreign additions which usage and the growing wants of the country have rendered current'.[12] In a letter of 1 April 1870 the Lieutenant-Governor of the North-Western Provinces supported the suggestion, although Muslims, despite Syed Ahmed's progressive writings in the *Allygurh Gazette* to the contrary, continued to express their opposition to English.

The employment opportunities of Muslims were affected by yet another measure of the Government of India. Till 1864 Muslims were fed with a belief that a knowledge of their own classics would be an acceptable qualification for government employment or admission to the legal profession. For until then candidates for the posts of munsifs and pleaders were permitted to take examinations either in English or Urdu. But a change was soon introduced, when it was declared that examinations for the higher grades of munsifs and pleaders would be held in English alone. Muslims were consequently put to a serious handicap.

The only alternative left for them was to participate in the existing educational system or to suffer the consequences. A committee was thus appointed in 1872 to report on the problems of better diffusion and advancement of learning among the Muslims of India. Syed Ahmed Khan Bahadur, who was one of its members, expressed himself in favour of devising an educational system on western lines that might contribute to the material advancement of Muslims without their children losing contact with their religion. The committee proposed the establishment of a two-pronged system: a 'general' system, which Muslims themselves were to organize, and 'special' official measures designed to enable them to benefit from the educational system adopted by the government.

The Muslims, who were till then highly critical of the British system of education, began to be drawn closer to it thanks to some measures taken by the Government of India after 1871 to promote, for example, the study of Persian in the University of Bombay. About this time, in 1870, Sir William Muir, the Lieutenant-Governor of the North-Western Provinces, took steps to have a branch of the Calcutta University Syndicate established at Allahabad, not only to spread higher western education, but also to secure the institution of chairs in the Oriental faculties, including Persian, a new development for the University of Calcutta which had so far remained mainly English-dominated.[13] Lord Mayo's government, in fact, marked an important phase in the development of British policy towards the education of Muslims and helped restore confidence in the community as a separate and distinct class. This led to the establishment of the Muhammadan Anglo-Oriental College in 1877 at Aligarh, which owed its establishment not to individual patronage, but to the 'combined wishes and the united efforts of a whole community'. Lord Lytton, the then Governor-General, laid the foundation-stone of the College, and the address presented to him expressed deep gratitude for encouraging Muslim education which was admitted as being 'different' from that of 'the rest of the population of India'. The address pointed out: 'We, the Mussulman subjects of Her Imperial Majesty, consider ourselves more particularly bound in gratitude to the Government of India for its having of late years shown so strong a disposition to advance the cause of education amongst our community, and for issuing directions to the provincial Governments to adopt special measures to supply our intellectual needs.'[14]

However, the 'special' measures taken to encourage Muslim education could not but encourage a parallel advance in the field of Sanskrit, and Hindi in the Nagari script. The revival and development of Indian classical literature, both Hindu and Muslim, inadvertently led to a widening of the communal gap. In the post-1871 period Sanskrit, Arabic and Persian were all admitted and allowed to take rank among other subjects of secular study, with languages forming part of the examinations for university degrees. Besides this, a Resolution of Mayo's government in 1871 had emphasized the need to develop a vernacular literature for Muslims, a task which was left to the discretion of Local Governments. The result of the Resolution of 1871, however, was a parallel growth of Urdu written in Persian characters and of Hindi written in the Nagari script, both being separately inspired by their respective sources in terms of cultural and political loyalties. Hindustani, which had been an amalgam of both Urdu and Hindi, was, in these circumstances, left to decay.

The only common bond of unity that remained with the passage of time was English, through which European knowledge filtered down to the Indian soil. It imparted a secular, humanist approach to both culture and politics. The Liberals in the Congress represented that trend. But the 'Extremists', who were fed on the classics or their vernacular derivatives, were socially exclusive but politically radical, caring more for immediate accession to power than a long-term national view of unity and integration. It was this trend in the Congress which started gaining popularity after the Age of Consent Act (1891).

Since the Muslims had failed to catch up with Hindus educationally, the influence of religion on politics was still greater in their case. During 1897–1902, only 60 *per mille* of Muslim males were literate, against 102 *per mille* of the general population.[15] Their revivalist approach to politics was in addition reinforced by a numerical inferiority in the public services, an inferiority which was, under a Resolution of 1885, sought to be played down on the score of their percentage of the total population and their inability to take advantage of the educational facilities made especially available to them.

In the North-Western Provinces and Oudh, it is true, the position of Muslims was reasonably strong. Their percentage in the total population was 13.4 in 1886, but they held 45.1 per cent of the total number of posts in the judicial and executive services of that province. The Hindus held only 50.2 per cent of these posts, although they constituted 86.2 per cent of the total population.[16] This proportion continued into subsequent years.[17] But in Bengal, which included Bihar and Orissa, Muslims held only 8.5 per cent of the posts in the judicial and executive services, although they formed 31.3 per cent of the total population. Their position was even worse in Assam where they held only 0.9 per cent of these posts, although they constituted 26.9 per cent of the total population.[18]

In 1903 the Government of India had a detailed statement prepared of all civil appointments held by Europeans, Eurasians, Hindus and Muslims since 1867, covering a period of 36 years in ten-yearly slices and comprising the establishments not only of the central government, but of all provincial governments also. The return, which included details of appointments with salaries of Rs 75 and over, was arranged separately for each department. It gives a clear picture of the numerical position of each of the four categories of incumbents.[19] The position of Muslims showed no marked improvement. There were several departments where they even remained unrepresented throughout the period of 36 years.

From a Note of W.S. Marris dated 20 March 1904, it appears that during the previous 36 years for which figures were available the total number of government posts in India had increased from 13,431 to 28,278. But the percentage of posts held by each of the four groups was:

	1867	1904
Europeans and Eurasians	55	42
Hindus	38	50
Muslims	7	8

As regards grades, more than half the appointments in India were and had always been posts on less than Rs 200 per month, where Indians, Hindu or Muslim, held most of the posts. The salary ceiling of the provincial services, where Indians usually figured, did not generally exceed Rs 800. The position of Muslims was negligible in the higher reaches of the judicial and executive services, both in the provinces and at the centre. Between Rs 800 and Rs 1,000, in 1867 there were four Indians in government employ, a figure which rose to 93 in 1903. In 1867 there were as many as 648 superior appointments carrying Rs 1,000 and above; out of these, 12 were held by Indians, all Hindus. In 1903, out of 1,370 such appointments, 71 were filled by Hindus and 21 by Muslims. The establishment of the M.A.O. College at Aligarh, which later grew into a university, as well as the encouragement given to Muslim education by the Government of India in the years after 1870, had helped restore them to a significant position in the public services, although it did not correspond to the status they had earlier enjoyed under the Mughals. The relative position of Hindus and Muslims in the provinces is indicated in the appendices to the government's statement of 1903. They make it clear that Muslims were more widely employed in the Punjab and United Provinces than elsewhere. The large Muslim population of Lower Bengal contributed very little to the supply of government servants: There, Brahmins and Kayasths retained their former hold. The proportional distribution of posts on monthly salaries of Rs 200 and above did not show any marked change in 1913 according to the Report of the Islington Commission: Muslims held 7 per cent of such posts in India as a whole and Hindus, Sikhs and Parsis held 31 per cent together.[20] In Bengal, Muslims had 8 per cent against the Hindu figure of 53 per cent.[21]

The Partition of Bengal in 1905 marked a tilt in British political policy, even though temporarily. It reduced Hindu Bengalis to minorities in both halves of that province. The swadeshi movement and the politics of terrorism, which followed in the wake of Partition, were both directed against the British Government but did not have the support of Muslims, who wanted the Partition to continue.

The well-known Muslim address[22] to Lord Minto was presented in these circumstances by the Aga Khan on 1 October 1906. Though mentioning the need for greater representation in government service, the address was mainly political and intended to have Muslim interests safeguarded by a due representation of Muslims in the projected scheme for the expansion of the legislative councils, both Imperial and Provincial. Its terms of reference did not signify any negation of the total national viewpoint. The Muslim community was to be treated as a part of India. The address clearly said:

It is true that we have many and important interests in common with our Hindu fellow-countrymen, and it will always be a matter of the utmost satisfaction to us to see these interests safeguarded, by the presence in our Legislative Chambers of able supporters of these interests, irrespective of their nationality. Still, it cannot be denied that we Muhammadans are a distinct community, with additional interests of our own which are not shared by other communities, and these have hitherto suffered from the fact that they have not been adequately represented even in the provinces in which the Muhammadans constitute a distinct majority of the population.

Under the existing constitution of legislative councils such representation as the Muslims enjoyed was almost without exception a result of nomination by the government. 'As for the results of election', the address pointed out, 'it is most unlikely that the name of any Muhammadan candidate will ever be submitted for the approval of Government by the electoral bodies as now constituted, unless he is in sympathy with the majority in all matters of importance, nor can we in fairness find fault with the desire of our non-Muslim fellow subjects to take full advantage of their strength and vote only for members of their own community or for persons who, if not Hindus, are expected to vote with the Hindu majority on whose goodwill they would have to depend for their future re-election.'[23]

Not being satisfied with the principle of nomination, the Muslim delegation sought a separate electorate of its own, and an electorate, too, where the proportion of Muslim representatives was not to be 'determined on the basis of the numerical strength of the community'. The demand was that Muslim representation be done on the basis of their 'political importance' in terms of their contribution to India's defence as well as 'the position which they occupied in India a little more than a hundred years ago, and of which the traditions have naturally not faded from their minds'.

A meeting of Muslims from all parts of India assembled at Dacca on 30 December 1906, and proceeded to form a political association, styled 'The All-India Muslim League'. Its declared object was to secure, with the help of the British Government, the protection and advancement of the rights of Indian Muslims without exciting any 'hostility towards other communities'. Coming so soon after the delegation of 1 October, the League became the latter's legitimate successor in organizational terms, and the Morley-Minto Councils were the immediate consummation of the goal it had sought to pursue with the Viceroy. The Muslims of India got the separate electorate they had pressed for.

The concession of separate electorates constituted a break with the past, a departure from the Indian policy hitherto pursued by the British which used nomination to provide political balance, and sought to strengthen Muslim education to ensure fitness for appointment to the public services. On the question of Muslim representation, a possible solution had earlier been suggested by the Secretary of State in a letter of 27 November 1908, in which he recommended a mixed and composite electoral college where both Hindus and Muslims were to vote together to return their own representatives in their due proportion. Its object was to keep the two communities together without any prejudice to the Muslim quota in the Councils. But since the Muslim League argued that the Hindus might in that case succeed in returning only pro-Hindu Muslim candidates, Minto's government did not act on the suggestion of the London authorities Minto, in fact, remained obsessed with the idea of providing a 'counterpoise' to what he regarded as the excessive influence of 'the educated and professional classes', a counterpoise which was additionally reinforced by the induction of the landed aristocracy.

The direct and separate Muslim electorates introduced under the

1909 Act could not immediately accomplish the object for which Minto had provided the counterpoise. The restricted franchise and limited constituencies recognized for the direct representation of some special class, interest or community could not by themselves redeem the lack of any real connexion between the primary voter and the member who sat in the Councils, a lack which, under the existing indirect electoral system for general seats, left little or no scope for political mobilization among the people. Out of twenty-seven elected members of the Indian Legislative Council, eighteen were elected to speak for sectional interests on direct voting, and nine who might be said to represent, however remotely, the views of the people as a whole. In the first category, the largest constituency returning a member directly to the Indian Legislative Council did not exceed 650 persons, and most constituencies were much smaller.[24] The constituencies, which returned the second category of nine representatives, were composed of non-official members of the various provincial councils, where the average number of voters did not exceed twenty-two. In the provincial councils, too, there existed the same division of members between those who were directly elected to represent special interests and those who were elected indirectly as representatives of the general population. For the latter, members of municipal and local boards either acted as electors or else chose electoral delegates to make the election; but in neither case did the constituency exceed a few hundred persons. The existing electoral system and the franchise were thus both unhelpful for mass mobilization on communal or caste lines. Despite the limitations imposed by the indirect electoral system, the results of the elections of 1909, 1912 and 1916 showed that lawyers had gained between 40 per cent and 45 per cent of the seats in the Indian Legislative Council in the first two elections and 54 per cent in 1916. They were returned by the non-official members of the provincial councils after excluding those returned by constituencies of the landholding and commercial classes as well as special Muslim representatives. Much the same state of affairs applied to the provincial legislative councils, where seats were also specially reserved for landholders and commercial men. This tended to reduce the strength of the legal profession in the councils as a whole. Even so, the percentage of lawyers among the elected members of all the councils together (excluding Burma) was 38 in 1909, 46 in 1912, and 48 in 1916. The number of lawyers was, however, especially marked in the constituencies formed by members of local bodies; out of 70 such constituencies 49, or 70 per cent, returned lawyers.²⁵ Landholders were losing to both lawyers and businessmen, a trend which was speeded up by the impact of World War I.

In fact, the compulsions of World War I established earlier than expected the weakness of the assumption on which the Morley-Minto system was founded, an assumption which had suggested the expediency of a 'counterpoise' and 'constitutional autocracy' to clip the wings of middle class nationalists. During the War the functioning of a rigid official bloc of Europeans in the legislature tended to give debates a racial twist, which, instead of creating divisions among Indian elected members, stifled their differences, if any, and drove 'them to a league against the Government, into which the nominated Indian members also tend to enter'.²⁶ The Government of India thus endeavoured 'to avoid contentious legislation during the War', and, before a Bill was introduced, made every effort 'to ascertain as far as possible non-official opinion'. The members of the Indian Legislative Council were, in the words of the Montford Report, 'animated' by a 'sense of responsibility' in the passage of even such 'controversial measures' as the Press Act and the Defence of India Act. The exigencies of the War induced 'a habit among the non-official members, nominated and elected, of acting together', more especially on racial questions, where it was 'natural' that all Indian members, irrespective of their religious or other differences, 'should not divorce themselves from the general Indian view'.²⁷

Despite the separate Muslim electorates and special electorates for the landholding and commercial classes, the elected and nominated members of the Central Legislative Council acted on the principle of joint action. A Memorandum submitted to the Viceroy by nineteen²⁸ elected members of the Central Legislature in 1916 on the subject of post-War reforms, affirmed that 'the Indian people have sunk domestic differences between themselves and the Government, and have faithfully and loyally stood by the Empire'. The result was that Indian soldiers freely joined the battlefields in Europe and internal peace reigned supreme throughout the country even in the absence of the British and Indian troops which had been sent abroad.

It was in the context of the War and the Bolshevik Revolution that, while the British Government made its political policy announcement of 20 August 1917, the Montford Report brought out the weakness of the Morley-Minto Councils which, within the unitary framework of financial, administrative and legislative powers being still centralized with the London authorities, could not meet the desire for political power which had been rapidly growing in the minds of educated Indians during the War. For the existing bonds of authority subjected Local Governments to the Government of India and the latter to the Secretary of State and Parliament.

The Montford Report thus recommended, on the basis of the announcement of 20 August, a definite, though limited, advance towards the principle of responsible government within the existing unitary framework itself, a framework which was considered essential for unity in diversity. The August declaration was an expression of a constitutional principle, purely territorial in concept, not tribal or communal, religious or racial. The authors of the joint Montford Report, too, did not recognize the inevitability of communal representation, even as a necessary stage in 'the development of a non-political people'. For even 'in its earliest beginnings', responsible government in Europe rested 'on an effective sense of the common interests', a bond 'compounded of community of race, religion and language', a bond which 'appeared only when the territorial principle had vanquished the tribal principle, and blood and religion had ceased to assert a rival claim with the State to a citizen's allegiance'. In the light of 'the history of self-government among the nations who developed it and spread it through the world', the authors of the Report were in principle against divided allegiance on communal lines, or 'the State's arranging its members in any way which encourages them to think of themselves primarily as citizens of any smaller unit than itself'.

The Montford Report recognized, not unjustly, that India, generally speaking, had not yet acquired 'the citizen spirit' in its territorial sense. A recognition of division by creeds would therefore mean the creation of separate political camps, prejudicial to the development of such a spirit. It was realized that any change from partisan politics to national representation would in that case be extremely difficult. Communal representation was even otherwise considered indefensible. 'A minority which is given special representation owing to its weak and backward state', it was felt, 'is positively encouraged to settle down into a feeling of satisfied security; it is under no inducement to educate and qualify itself to make good the ground which it has lost compared with the stronger majority.' The system of communal electorates was, therefore, regarded 'as a very serious hindrance to the development of the self-governing principle'.[29]

Even so, the authors of the Report recommended the continuance of the existing system. For they were obliged to face the hard facts: the concession of special representation with separate electorates in 1909 for Muslims and the Hindus' acquiescence embodied in the Lucknow Pact, and concluded between the Congress and the Muslim League in 1916. As the Muslims had come to regard communal representation with separate electorates as 'settled facts', 'any attempt to go back on them would rouse a storm of bitter protest and put a severe strain on the loyalty of a community which has behaved with conspicuous loyalty during a period of great difficulty'. From considerations of political expediency the British government was induced to go in for 'the creation of political camps organized against each other'. Its justification, too, was not far to seek. The Congress-League Agreement provided it, and made the task easier for the British to wriggle out of the anomaly. 'Much as we regret the necessity', the Report thus said, 'we are convinced that so far as the Muhammadans at all events are concerned the present system must be maintained until conditions alter, even at the price of slower progress towards the realization of a common citizenship.'[30]

The Congress-League Pact, which reinforced the decision of the Government to maintain separate communal electorates, provided for the representation of Muslims in the provincial councils in proportions that were much greater than their share of the total population. Thus, in Punjab the League was to have 50 per cent of the elected members, in the United Provinces 30 per cent, in Bengal 40 per cent, in Bihar 25 per cent, in the Central Provinces and Madras 15 per cent, and in Bombay 33 per cent.

As regards the Indian Legislative Council, the Pact laid down that one-third of the Indian elected members should be Muslims elected by separate Muslim electorates in the several provinces, in as near the proportion in which they were to be represented in the provincial legislative councils. The other provisions of the Pact related to an increase in the powers and functions of the councils at both provincial and central levels, as well as an emphasis on the transfer of authority from London to New Delhi.[31]

Taking its cue from the Congress-League scheme of 1916, the Morley-Minto system of representation was thus continued by the government. Its basic principle was applied under the Government of India Act (1919) not only to Muslims but also to the Sikhs in the Punjab and to certain other communities elsewhere. Apart from these special electorates, however, the elected members of a provincial legislature were returned through territorial constituencies by direct election. This principle was likewise extended to the Indian Legislature which, under the Act, consisted of two chambers, the Council of State and the Legislative Assembly. Both provided for communal and special representation for Muslims, Sikhs and Europeans. The interests represented through special constituencies were those of landholders and Indian commerce. The general constituencies, which returned most of the elected members, did so on a territorial basis and by direct election.

By 1919 the representation of Muslims through separate electorates had thus come to be recognized as a settled fact. But it is unjustifiable to attribute India's later division into two distinct sovereignties to separate Muhammadan electorates. For at no time did the British think of separate electorates in terms of a territorially divided India. Their integrated imperial interests, in fact, demanded a strongly united India as a bulwark of defence. They doggedly held to this view till the last moments of their rule. The case of the Muslims, though different, was not unrelated to the stability of that rule which alone could hold the balance between them and the Hindus who, by virtue of their traditional hold on the country's economy and their newly acquired lead in the government's employ, were emerging as a perceptible alternative to British power in India. All that the Muslim minority wanted was, in the context of history, a position of prestige to be determined by a due share in administrative and legislative bodies, the emerging source of political power, a position which was not to be a mere gift or a matter of discretion exercisable by the majority community, but one claimed as of right. The Morley-Minto arrangement, the Congress-League scheme and the constitutional provision made under the Government of India Act (1919) were all a recognition of that right.

The Indian Liberals, who firmly believed in the secular principles of representative government, appreciated the claims of the Muslim community and raised no objection to them. Gokhale lent them his full approval. The demand for self-government had, in fact, assumed a new significance with the Home Rule movement of Annie Besant, and even Tilak was emerging as an advocate of responsive co-operation with the British government. At the instance of Jinnah, a staunch supporter of Gokhale, both the Congress and the Muslim League held their annual sessions in 1915 at Bombay, where both organizations resolved to collaborate in formulating a common scheme of post-War reform, a political development which flowed perhaps from Turkey being at war against Britain, and brought the League closer to the Congress. The scheme formulated jointly by their Committees was approved by them at their annual sessions held at Lucknow in December 1916. It was in these circumstances that most of the Muslim members of the Indian Legislative Council had signed only a couple of months earlier a memorandum presented to the Viceroy by nineteen central legislators. They were all united in their demand for self-government.

Though claiming to represent a special interest as a distinct community, in the period before 1919 the Muslims had not thought in terms of a divided India. On the contrary, they were conscious of the existence of their many 'important interests in common with our Hindu fellow-countrymen'. They only emphasized that, apart from their important common interests with the Hindus, they had certain 'additional interests' of their own as a separate religious community whose importance should not be whittled down altogether in consequence of the developments which had taken place with the fall of the Mughals and the rise of British power in India. A reluctance on the part of Hindus to recognize these hard facts in the context of history and a persistence on the part of Muslims to keep harping on their past glory as a basis of special representation in government and administration constituted the main potential cause for rift between the two communities. An appeal to history was perhaps later employed by politicians to justify India's division as the only means to settle scores for power in a plural society.

Divisive Trends and Mass Politics 1919–36

As already noted, an important and immediate incentive for mass involvement in politics was the establishment of a bicameral legislature at the Centre and responsible governments in the provinces based on extended franchise and direct election, which created conditions for a relatively deep penetration of political influences. More significant in terms of mass politics, however, was the reciprocity of political and economic interests, whose significance was brought to the surface in the course of World War I, which quickened the pace of India's industrial development and led to the growth of a considerable working-class population that was given to organized agitation and strikes under political influence and pressure. According to Sir James Meston, the Lieutenant-Governor of U.P. during 1912–18 and Finance Member in 1918–19, a new India was in fact emerging during the War because of the increasing ambitions of the Indian merchant community to participate in the political life of the country and secure the promotion of its industrial development. Sir Dorab Tata, for example, advocated in 1915 at the Industrial Conference in Bombay that both politics and industry should be yoked together to force the Government of India to organize the local production of everything needed for the prosecution of the war as well as for the encouragement of Indian industries.[32] The establishment of the Industrial Commission in 1916 was a result of such political pressure. Later, such pressure obliged the government to regulate its import policy so as to encourage indigenous trade and manufacture.

The growth of a working-class population provided material for mass politics. It led to the establishment of an All-India Trade Union Congress in December 1920. The first Trade Union Congress which met in Bombay in 1920 was, surprisingly, inaugurated by Swami Shraddhanand and presided over by Lajpat Rai, two nationalist leaders of the Punjab Arya Samaj. Similarly, while aiming at striking a balance between capital and labour to promote national industries, V.D. Savarkar, the leader of the Hindu Mahasabha, thought in terms of Hindu interests which he would safeguard against non-Hindu aggression in both fields and factories.[33]

The main emphasis of development on the Muslim side, however, remained political and a preoccupation with the politics of pan-Islamism. With the defeat of Turkey in the War and the consequent threat to her territorial integrity, the pan-Islamists in India threatened agitation against the British and made common cause with other elements hostile to the government. From a secret memorandum[34] of the Political Office attached to the India Office in London it appears that the principal object of the Khilafat Movement in India was not limited to procuring the *status quo ante bellum* for Turkey. 'The primary object' of the movement was 'the independence of India, more especially Mohammedan India, with the object of uniting her to a revived Islam', the Khilafat Movement being 'only the visible cause by which Indian leaders are bringing the masses into a movement to overthrow British rule'. This became known from the activities of Mohammad Ali, the Khilafat leader, who visited Europe at the head of a delegation which met the British Prime Minister on 19 March 1920, apparently to plead the case of Turkey.

The delegation also visited various European capitals where it elaborated on the proposed phases of its projected non-cooperation scheme against the British in India. Mohammad Ali also wrote to Mustafa Kamal of Turkey in the name of Indian Muslims, promising to collect money and mobilize international public opinion in favour of the pan-Islamic political objective. Finding little or no support in France and elsewhere in Europe, Mohammad Ali turned to Russia which, according to the secret memorandum, 'promised freedom to Indian Muhammadans'. In fact, he was in touch with a number of Bolshevik agents who assured him of aid in 'the regeneration of all Muhammadan powers of the East'. The memorandum concluded that 'the Khilafat Central Committee... are out for the conversion of India into a Muslim State in order to link up with other Muslim countries. Their dream is a bloc of Muslim states leaning on Russia.' In other words, the Mullah-dominated Khilafat Committee was extra-territorial in its loyalty. However, it was not the sole representative of Muslim public opinion in India.

Though basically comprising a group of Ulemas, the Khilafat Movement came to include men of diverse political persuasions, nationalists, revolutionary nationalists and even communists. The movement was becoming increasingly organized and broad-based, swamping even the Muslim League through its pan-Islamic appeal,[35] although English-educated Muslims and the princely Muslim states generally remained outside the *concordat* that was developing between radical groups of Hindus and Muslims.

There was plenty of opportunity for a secular, economic approach to mass politics, based on a series of unprecedented strikes which occurred over a wide area, particularly in Bombay Presidency.[36] In a private telegram to the Secretary of State, the Viceroy admitted that their causes were primarily economic: the pressure of high prices; the profiteering of middleman and retail dealers; the knowledge that millowners were making large profits; the contrast between the wealth and comfort of the few and the comparative poverty of the mass of workers; the general shortage of industrial labour arising from increasing industrial establishments, and the reluctance of employers to increase wages until discontent with existing conditions manifested itself in the shape of strikes. There was, in addition, an 'epidemic strike fever' engendered by 'worldwide political unrest'. In India, as the telegram pointed out, these strikes were mostly spontaneous, the politicians appearing on the scene 'only after a strike [had] actually begun'.[37] Even so, radical nationalists adopted a religio-racist approach to organizing mass movements and boycotted the reformed constitution, the first phase of responsible government, introduced under the Government of India Act, 1919.

The political overtones which came to dominate the Indian situation towards the close of the War were mainly attributable to the rise of Bolshevism and pan-Islamism, both being a threat to the stability of the British in India, though from different premises. The Defence of India Act (1915), which took care of security and defence requirements during the War, had the full support of the Indian legislature. It invested the executive with powers to deport suspects or intern them as well as to create or authorize the creation of new offences. But given the anti-imperialist danger posed by the Russian Bolshevik Revolution, the Indian Muslim adherence to the cause of Turkey and the Khilafat, and the rising expectations of subject peoples and nationalities the world over, the British felt it necessary to enact further measures to curb possible danger to their position in India. The Government of India appointed in 1917 a Committee with Justice Rowlatt of the Supreme Court in England as president to investigate and report on seditious movements in India. Its recommendations were published soon after the publication of the Montford Report in 1918. The Committee favoured the enactment of special measures once the Defence of India Act ceased to operate. The Government of India thus introduced in Febuary 1918 the Rowlatt Bills, known as the Indian Criminal Law (Amendment) Bill No. 1 of 1919 and the Criminal Law (Emergency Powers) Bill No. 2 of 1919, providing against acts of conspiracy or sedition against the constituted authority of the State. The Rowlatt Bills became law on 21 March 1919.

Gandhi, who had begun gaining importance through official recognition[38] and the assistance of Mrs Besant's Home Rule League, chose to oppose both Bills and, at a meeting in Ahmedabad on 24 February 1919, launched a mass protest against their passage; this move led to the merger of the Home Rule League with his Satyagraha Sabha and established his claim to the leadership of the Congress which, after the withdrawal of the Moderates from it, had been languishing for some time. With Gandhi as its President, members of the Satyagraha Sabha took an oath to disobey the Rowlatt Act when passed and court jail. When the measure was enacted, Gandhi fixed 6 April 1919 as a day of general mourning, a development which led to a chain reaction, resulting finally in the Amritsar (Jallianwala Bagh) Massacre on 13 April.

Too much should not, however, be read into the Rowlatt Act controversy or even Gandhi's non-cooperation movement which he started with a communication addressed to the Viceroy on 2 August 1920.[39] For Gandhi had not during the satyagraha returned all the medals awarded to him earlier by the British; even while returning them on 2 August, he made it clear in his letter that he was doing so 'not without a pang'. Moreover, he communicated his notice of non-cooperation immediately after Tilak's death, which had occurred a day earlier on 1 August. Tilak's death strengthened Gandhi's position in the country's political leadership. He now cast himself wholeheartedly on the side of the revivalist Khilafat Movement and its pan-Islamic supporters. The Khilafatists not only provided a united Hindu-Muslim base for non-cooperation, but also a potential source of power for the reinforcement of Gandhi's political leadership. But how was Gandhi's politics oriented? What impelled him to adopt satyagraha or civil disobedience to oppose the Rowlatt Act, or even non-cooperation? In his letter of 2 August, Gandhi pointed out that 'the scheme of non-cooperation inaugurated today' was essentially 'in connexion with the Khilafat Movement' and that the Punjab question had merely given 'additional cause' for it. As for 'swaraj', the word found no mention at all in that letter and was not the reason for non-cooperation. What was it basically, then, that concerned Gandhi? A confidential government Note of May 1921 suggested that his motive was to save India from Bolshevism, a task which could not perhaps be accomplished except by assuming a radical political posture against the British government. The Note thus said:[40]

In June 1919 the Government of India received intimation that Mr Gandhi contemplated a renewal of civil disobedience, unless circumstances altered his plans. The experiences of the preceding two and a half months, he stated, had convinced him that nothing save Satyagraha, of which civil disobedience was an integral part, could save India from Bolshevism—and even a worse fate. He added that the exhibition of anti-British feeling during the second week of April was due, not to the Satyagraha movement, but to pre-existing causes, and that the Rowlatt legislation was the *causa causans* which he hoped would be withdrawn.

In fact, India's middle-class leadership and the British government were both united in their determination to see that international communism was not allowed to take root in Indian soil, for it was considered menacing to imperialism and Indian capitalism alike. Gandhi was no exception to this general thinking. But the task of post-War reconstruction, which called for resilience and understanding in striking a balance between imperial and national interests, became subject to political wrangles. Both official and nationalist groups were found wanting in restraint and the capacity to adhere to declared policies. The errors of Gandhi's thinly disguised political strategy of civil disobedience and General O'Dwyer's vindictive and indefensible conduct at Jallianwala Bagh were both examples of indiscretion which lent itself to a chain reaction that was prejudicial to constitutional development within the framework of mutual understanding.

Non-cooperation and Communal Politics

A more serious error of judgment on the part of Gandhi, which encouraged divisive trends, was his total and whole-hearted alliance with Mohammad Ali and Shaukat Ali, two of the foremost pan-Islamic Khilafat leaders who induced him to declare non-cooperation against the reformed Councils. The pan-Islamic contingent of that Movement attended the Calcutta session of the National Congress as delegates and formed part of the majority which adopted a resolution calling upon Indians to fight under the banner of the non-cooperation movement for the country's swaraj, with or without the British connexion. The erstwhile Home Rule League, which was opposed to a severance of that connexion, found itself transformed under pressure into a Swaraj Sabha of Gandhi's making. Jinnah and nineteen other members of the Home Rule League, who viewed these proceedings as both 'illegal' and 'unconstitutional', sent in their resignations[41] from membership of the Swaraj Sabha, of which Gandhi himself had become president. They raised serious objections to the methods he adopted to achieve political eminence. As part of his propaganda for non-cooperation, for instance, Gandhi went from one college to another in Gujarat and Bombay, asking students and others to vacate and shut down government schools and colleges, start national schools, withdraw lawyers from their courts, give up government service, surrender titles, withdraw from elections, and resolve to send no representatives to the Councils.[42] Little did he realize that in his exuberance for mass involvement in politics he might be setting examples which were not only disruptive and prone to division, but also, possibly, harmful for the generation to come. Gandhi claimed to have looked upon Gokhale as his guru, but acted in a manner that was very different from the political insight and deportment of that great Liberal leader, a person who was not only held in great respect by Jinnah, but also emulated by him.

In a Resolution[43] on the non-cooperation movement the Government of India, while explaining its policy for the guidance of provincial governments and administrations as well as for the information of the public, advised the use of restraint and instructed Local Governments not to take any action against those who advocated, simultaneously with non-cooperation, abstention from violence. The institution of criminal proceedings was therefore sought to be used only against those who indulged in open incitement to violence or tampered with the loyalty of the army or the police. The Resolution pointed out that the Government was reluctant to restrict the freedom of the press and speech at a time when the first phase of constitutional reforms was being heralded with a view to the ultimate realization of self-government. The chief consideration which was said to have influenced the Government in favouring moderation was the hope that 'the sanity of the classes and the masses would reject non-cooperation as a visionary and chimerical scheme, which, if successful, could only result in widespread disorder and political chaos'.

The official assessment of the situation was not unfounded. The Liberals and a significant body of educated Indian opinion, which later resulted in the formation of the Swaraj Party, had no confidence in the methods Gandhi adopted to win Swaraj. They were strongly opposed to the way in which he used the student population during a political campaign. The appeal made to the masses, on the other hand, produced a series of communal riots, of which the Moplah outbreak of 1921 was the most serious. With Gandhi's arrest in 1922 the non-cooperation movement was doubtless suspended. But the wave of communal riots did not cease. On the contrary, it increased, more especially during 1923–6. For while the settlement of the Turkish question made it unnecessary for Muslims to seek Hindu support, even educated and moderate Muslims now wanted to sustain and exploit for future political gains the populist methods of the Khilafat Movement. They saw the seriousness of the catastrophe that might otherwise overtake their community if the Hindu-dominated Congress alone seized power from the British by unconstitutional means. This realization began to haunt the Muslims, who noticed that most of the delegates at the 1924 Belgaum Congress, instead of attending their own meetings, were actively participating in the annual session of the Hindu Mahasabha being held simultaneously in Belgaum. The rising tide of Hindu-Muslim animosities led to the revival in 1924 of the All-India Muslim League, which, in the years following the Congress-League Pact, had, in fact, faded into the background.

The revival of the Muslim League proceeded from an apprehension of educated Muslims about the dangers of India's 'unregulated' or unconstitutional political advance, in which they could not compete with the majority community of Hindus. Their only safeguard lay in constitutional progress where they could, with the support of the government, secure a 'substantial guarantee' for the protection of their interests in both political and administrative matters. This attitude in turn appeared to Hindu nationalists to obstruct their quest for political advance. Thus, while Hindu nationalists, whether in the Congress or the Hindu Mahasabha, sought an acceleration in the pace of constitutional advance, the Muslim League insisted on regulated progress and reservation as a means for the protection of Muslims in India. The politics of economic interests was thus fraught with the danger of communal outbreaks in an unevenly balanced society.

Doubtless, the aftermath of the non-cooperation movement saw the emergence of an economic force sustained internally by the Congress and reinforced externally by the Communist International established in Russia. The Bolsheviks and Communists used the Khilafat Movement as an agency for the expansion of their activities in India. Their ideology and organizational operation came to light on two important occasions, first during the Communist Case at Kanpur in 1924, and next during the Meerut Conspiracy Case in 1929.[44] Wedded to a secular political philosophy, the communists sought to infiltrate Congress ranks through Jawaharlal Nehru, who was supposed to have international communist links in Europe. But other nationalists, who formed a powerful body of opinion on the side of Indian capitalism, refused 'to link up the Congress with the avowed enemies of the British Empire'. They could lend support to Nehru's lobby in the Congress only when that support was limited to passing resolutions calculated to secure more power. But they never showed any inclination to move against the British government to promote the cause of her 'avowed enemies'.[45] The dominance of the bourgeoisie as a class militated against the penetration of communist or socialist elements in the Congress. Socially and politically, the bulk of nationalist India remained oriented towards Hindu ascendancy. The revival of the revolutionary terrorist movement, mostly Hindu in complexion, reinforced the divisive communal trend all the more.[46]

The effect of the Khilafat and the allied non-cooperation movement did not remain limited to the outbreak of communal riots which, in turn, led to the revival of the Muslim League and the spread of the *Shudhi* and *Sangathan* movements under the Arya Samaj and Hindu Mahasabha as a counterforce. It was also reflected in the decision of the bureaucracy in India to slow down the country's constitutional advance, a decision which perhaps suggested the expediency of reservations to placate the minority communities as a counterweight to the Hindu demand for dominion status and full provincial autonomy.

Constitutional Advance Delayed

The whole trend of post-War constitutional developments was towards a shift in the balance of power from London to New Delhi, which in turn meant a corresponding devolution of authority from the Government of India to provincial governments. Consistent with this trend, the Joint Select Committee on the Government of India Bill (1919) recommended that the Secretary of State should only in exceptional circumstances be called upon to intervene in matters of purely Indian interest where the Government and Legislature in India were in agreement. Similarly, in matters where a provincial government and legislature were in agreement their view should ordinarily be allowed to prevail.

It was doubtless realized that as long as the Governor-General or the Secretary of State remained in law responsible to Parliament, as was agreed that he should, the suggested transfer of authority would involve a constitutional anomaly. The Joint Select Committee recommended a resort to convention as a way out. Montagu, the Secretary of State, suggested a mere affirmative 'Resolution' of both Houses of Parliament as being good enough to limit the directive and supervisory authority of the India Office. It was in fact on Montagu's own suggestion that Section 19A of the Government of India Act (1919) actually provided for the relaxation of control. It said: 'The Secretary of State in Council may, notwithstanding anything in this Act, regulate and restrict the exercise of the powers of superintendence, direction and control, vested in the Secretary of State and the Secretary of State in Council by this Act, or otherwise, in such manner as may appear necessary or expedient in order to give effect to the purpose of the Government of India Act, 1919,' which, as stated by the preamble, included increasing Indianization of the public services, promoting self-governing institutions, and introducing responsible government. Provision was likewise made under Section 45 A(d) for the transfer of reserved subjects to the administration of the Governor acting with ministers appointed under the Act, and for the allocation of revenues or monies to carry out such administration.

All that was necessary for this delegation in both cases was to make draft rules under the Act and have them approved by both Houses of Parliament without the formalities of any fresh enactment. Under this Act the civil services, too, were intended to be made subject to legislative control in India. Section 96B(2), for example, provided that the Secretary of State in Council, who was authorized to make rules governing classification, recruitment, conditions of service, discipline, conduct, pay and allowances, might 'delegate' his rule-making power 'to the Governor-General in Council or to Local Governments, or authorize the Indian legislature or local legislatures to make rules regarding the public services'. Every civil servant appointed after the passing of this Act could be made subject to such laws. Only those appointed before the commencement of the Act could retain their existing and accruing rights and claim compensation for the loss of any such rights.

It was within the framework of the constitution so provided that the bureaucracy in India was expected to function, a constitutional framework which reflected the pre-emptive superiority of the legislature over executive administration. In other words, it was a recognition of political influence on government—the basic principle of democracy or responsible government. In a letter of 13 May 1920 the Government of India actually proposed to the Secretary of State that he should divest himself of his controlling authority in respect of the transferred subjects in the provinces. It agreed that the draft rules suggested for the purpose be approved. But the Government of India had serious objections to such divestment of control in respect of executive authority at the Centre under Section 19A. For it feared that, in the absence of that final control of Parliament exercisable by the Secretary of State, the Government of India might become either a bureaucratic despotism or a mere tool in the hands of the legislature, thus endangering India's unity and stablility. Montagu none the less recognized the spirit of Section 19A, which provided for the Secretary of State being divested of control over subjects even other than the transferred provincial subjects. The Khilafat Movement and Gandhi's full support of its leaders, as well as his non-cooperation movement, were new developments which dictated the need for caution and a go-slow policy.

The rigidity of Gandhi's approach to politics became, in fact, a handle for the bureaucracy to justify delay in the relaxation of control. On 23 September 1921, Rai Jadunath Mazumdar Bahadur moved in the Legislative Assembly a resolution which, consistent with the spirit of the 1919 Act, demanded complete provincial autonomy in addition to responsibility at the Centre. The resolution was moved at a time when the non-cooperation movement was in full swing, with most prominent leaders in jail, apart from Gandhi. In the course of a conciliatory speech which led to an amendment to the resolution, the Home Member recommended to the Governor-General that he convey to the Secretary of State the view of the Assembly that the progress made in India in respect of responsible government called for a re-examination and revision of the constitution at an earlier date than 1929, the date provided by the 1919 Act.

The amended resolution was followed by interesting developments in connexion with the Prince of Wales' visit to India, a visit which was perhaps intended to assuage Indian feeling and make the new constitutional experiment a success. To avoid the unpleasantness of a planned strike on 24 December 1921, the day of the Prince's projected visit to Calcutta, the Viceroy (Lord Reading) was anxious to settle differences with Gandhi through a compromise. Pandit Madan Mohan Malaviya took this opportunity to use his influence with Gandhi on the basis of C. R. Das's telegram from jail saying that he would agree to call off the projected strike if proclamations against volunteers were withdrawn, prisoners released and a conference called to reconsider the reformed constitution enacted only two years earlier. The talks between the Viceroy and Malaviya promised hope, for the former agreed to meet a deputation of leaders on 21 December, and to 'invite members of the different political sections, including, of course, non-cooperators and moderates to attend a conference at Delhi probably in January [1922]', the object being 'to understand their practical propositions' relating to what they called Swaraj. He informed the Secretary of State about this in a 'very secret and very urgent' telegram of 18 December 1921.[47] Lord Curzon described the telegram as 'most formidable' and as indicating the Viceroy's readiness 'to compromise the whole India policy of the Government and to endanger British rule in India in order to purchase the ephemeral advantage of a good reception for the Prince of Wales in Calcutta'.[48] Gandhi's obduracy, however, helped resolve the dilemma of the London authorities, for he inserted a further condition to what had been stipulated, insisting 'that all *fatwa* prisoners, including those in Karachi, that is, the Ali brothers, should also be released'. Malaviya 'sent telegrams begging Gandhi to announce that he would cease defiance on broad and generous lines', especially when the Viceroy 'had told Malaviya that [he] should be prepared to deal generously with the matter if Gandhi made this forward move'. But Gandhi refused. He insisted on the release of the Ali brothers, which the Viceroy would not permit. The negotiations failed and the projected conference was abandoned.[49]

A new situation arose with Montagu's resignation which was followed by Gandhi's arrest in March 1922. The Swarajists soon emerged as a separate legislative wing of the Congress, but they did not reduce the sense of confrontation. For, instead of non-cooperation, which was suspended, they chose to contest elections and enter the Councils with the declared intention of wrecking the constitution from within. The advice of the bureaucracy to delay the execution of Montagu's plan of decentralization thus seemed prudent to the government, and a re-examination or revision of the reformed constitution before the statutory period of ten years was thus shelved.

Although Sir Malcolm Hailey, the Home Member, agreed in 1924 to institute an inquiry into the working of the reformed constitution in order to ascertain the feasibility of an advance within the 1919 Act, the majority report of the Reforms Inquiry (Muddiman) Committee recommended that there was no need for a further advance, for all the possibilities of the Act had not yet been exhausted. The Home authorities held the same view. Constitutional advance within the framework of the Act which Montagu had envisaged was thus scuttled. The Khilafat Movement and non-cooperation together released forces which destroyed the basis of understanding, an essential condition for further constitutional advance. This had to wait for the recommendations of the Statutory Commission.

The Reforms and Communalism

When the All-India Muslim League was revived in 1924 it had to contend with the Swarajists, the elitist and affluent upper-caste Hindus dominating the Congress and becoming increasingly strong in the legislatures after 1923. During 1923–8 communal conflicts were coming to a head, more especially in the Muslim majority provinces of Bengal and the Punjab, where the Congress acted more as a communal Hindu body than a national organization ready to recognize the claims of other groups. The reformed constitution, which led to the rule of a majority, increased the importance of Muslims in the Muslim majority provinces, and their demands, in turn, were dubbed communal by the Congress.

In the Punjab, for instance, the Congress represented mainly urban

Hindu interests with its influence largely limited to lawyers, the trading classes and moneylenders who, before the Reforms, had exercised by virtue of their educational and economic superiority a greater political influence than was warranted by their share of the population. Democracy appeared to them the antithesis of their privileged position. They were not prepared to recognize that democratic politics had to reckon with the principle of rule by numbers according to the strength of the communities in the population. It is true that the Unionist Party, which represented the interests of peasant proprietors with a Muslim majority, was guided more by the actual problems of administration and agricultural development than matters of political theory. As a major agricultural community, however, the Muslims had since the Punjab Land Alienation Act (1900) entertained apprehensions against the urban monied classes who frequently sought to acquire their lands in satisfaction of interest on loans. So long as the line of division in the Council was dominated by rural versus urban issues the two major religious groups co-existed in reasonable amity. But communalism soon raised its head when, at the instance of Fazl-i-Husain who was in charge of local self-government, the franchise qualifications were lowered and seats in the local bodies allocated on the voting strength of communities. This was justifiable in terms of democratic principle, but it aroused communal feelings; for it marked a departure from the earlier position where Muslim representatives on local bodies were far fewer than warranted by their strength in the population.[50]

The introduction of democracy, howsoever partial, under the constitution of 1919 produced jolts in the power structure not only between Hindus and Muslims but also between one caste and another. As the *Reports* of the Local Governments on the working of that constitution showed, castes vied for power to become equal with a dominant caste in a province: this resulted in tussles between Brahmins and non-Brahmins in Madras, between Brahmins and Marathas in Maharashtra, between Bhumihars and Kayasths in Bihar, and so on. The conflict of communal and caste interests was clearly reflected even in the distribution of jobs pledged in 1925 by Sir Alexander Muddiman, then Home Member. The Muddiman Pledge reserved through nomination one-third of recruitment to the public services for the minority communities which, apart from Muslims, included the 'depressed classes' who had begun to organize themselves into separate groups and wanted their separate entities recognized as a title to privilege in politics and administration. Questions raised in the Central Legislature on the subject of employment in the public services indicated the preoccupation of legislators with securing jobs for members of their own community, caste or religion.

The decade following the 1919-Reforms was thus marked by two main developments. While the working even of a limited democracy under the reformed constitution brought into foucs the divisive elements of India's segmented society which competed with each other for positions of prestige, the Swarajists, the most efficiently organized of the political parties functioning under the banner of the Congress, were heading towards a claim to speak for the whole country. The Report of the Nehru (Motilal) Committee (1928) was an expression of that claim. Even the Government acknowledged the Swarajists' organizational efficiency and the influence they had come to wield in the legislatures and outside. But the attempt they made to place party interest above everything else in a society which called for coalescence as a means to win total confidence, was bound to fail, as it did. It caused suspicion and put other parties and groups on the alert. The Muslim League, in particular, realized the dangers of the Congress emerging triumphant to the exclusion of others, and started building its organizational strength to provide a counterpoise. Even so, it was the Congress which gained most in the short term from a combination of a mass movement and legislative activity.

The Muslim and Constitutional Remedies

Experience of the operation of the reformed constitution suggested the expediency of so reshaping the provinces as to provide them a territorial compactness and cultural homogeneity that would help resolve the continuing communal problem, more especially when the whole weight of political opinion indicated that full provincial autonomy would form part of the next phase of constitutional advance. A resolution adopted by the All-India Muslim Conference held at Delhi on 1 January 1929 revealed that the Muslims were guided by a desire to create what they called a 'Muslim India' within the body-politic of India, with full local autonomy exercisable under a central government with limited functions. This idea of a Muslim India was elaborated by Muhammad Iqbal in his presidential address to the All-India Muslim League at Allahabad on 29 December 1930: he proposed amalgamating Punjab, the North-West Frontier Province, Sind and Baluchistan into a single Muslim state, with an emphasis on the continuance of a Muslim majority in the Punjab. He felt that a redistribution of this kind alone could 'finally solve the communal problem'. The Nehru Committee too had recommended a redistribution of provinces. But the basis of that redistribution was to be language, and there, too, only when demanded by a majority of the population in the area concerned; such demands were to be further subject to financial and administrative considerations.

The Muslim demand for territorial redistribution took its cue from the recommendations of the Statutory Commission which expressed itself in favour of an Indian federation and full autonomy for the provinces. Iqbal wanted the British government to carry out his proposed redistribution before the introduction of the new constitution. While the Nehru Committee demanded the grant of dominion status with provincial autonomy under a strong centre, which in principle meant continuance of the existing unitary system under Indian control, the Muslim League viewed this as an attempt to perpetuate the communal dominance of caste Hindus. Even Dr Ambedkar, who represented the 'Depressed Classes' was suspicious of this dominance and demanded separate electorates for the 'Depressed Classes' or joint electorates with reserved seats during a transitional period. It is true that the Poona Pact (1932) obviated the necessity of applying the terms of the Communal Award to the 'Depressed Classes'. The Pact, however, was poor recompense for the social indignities to which they remained subject despite the 'reservations' provided by the law and constitution. Hindu upper-caste dominance in politics and administration through the Congress was thus a cause of concern not only to Muslims but the weaker sections of Indian society as a whole.

The centrifugal tendencies inherent in the demands for full provincial autonomy on a linguistic or cultural basis posed a serious threat to India's unity. The Joint Select Committee of Parliament (1934) sounded caution on this subject: 'We have', it said, 'spoken of unity as perhaps the greatest gift which British rule has conferred on India; but, in transferring so many of the powers of government to the Provinces, and in encouraging them to develop a vigorous and independent political life of their own, we have been running the inevitable risk of weakening or even destroying that unity'.[51] The concern for unity was thus a crucial consideration for the government in framing a policy. It felt that provincial autonomy would be inconceivable unless there was an authority, 'armed with adequate powers, able to hold the scales evenly between conflicting interests and protect those who have neither the influence nor the ability to protect themselves. Such an authority will be as necessary in the future as experience has proved it to be in the past. Under the new system of Provincial Autonomy, it will be an authority held, as it were, in reserve; but those upon whom it is conferred must at all times be able to intervene promptly and effectively, if the responsible Ministers and the Legislatures should fail in their duty'.[52]

The Joint Select Committee was clearly thinking in terms of composite, though fully autonomous, provincial legislatures that promoted local or regional interests but also had to remain subject to the discretion of an authority that maintained a balance between conflicting interests and protected the weaker communities in the region. The principle of this constitutional plan, as the Joint Committee reiterated, was not to underrate the legislative function of a provincial government under the projected new scheme of constitutional advance, 'but [because] in India the executive function is, in our judgment, of overriding importance. In the absence of disciplined political parties, the sense of responsibility may well be of slower growth in the Legislatures, and the threat of a dissolution can scarcely be the same potent instrument in a country where, by the operation of a system of communal representation, a newly elected Legislature will often have the same complexion as the old'. Consistently with the opinion of the Statutory Commission, the Joint Committee also emphasized that each province must have an executive power which could step in immediately and save the situation before it was too late, a power which must be vested in the Governor.

Related closely with considerations of unity was the question of the Indian States and their constitutional relationship with British India, a question which started engaging the attention of the British government when the Princes expressed their willingness to join a projected All-India Federation at the first Round Table Conference in 1930. A search for political balance between the erstwhile unitary tradition and pressure for a devolution of authority from London to New Delhi gave rise to an arrangement based on reservations and safeguards, an arrangement which provided for unity even within the framework of provincial autonomy and a federal legislature composed of the representatives of autonomous units.

Under the principle of reservation, Defence, External Affairs and Ecclesiastical Affairs were to be the reserved departments of the Governor-General, remaining outside the ministerial sphere, which meant Dyarchy at the Centre. With respect to the reserved departments, the Governor-General was to be responsible to the Secretary of State and, thus, ultimately to Parliament. The object of 'safeguards', on the other hand, was to invest the Governor-General with the exercise of 'Special Responsibilities' in such matters as (a) the prevention of a grave menace to the peace or tranquillity of India, or any part thereof; (b) the safeguarding of the financial stability and credit of the Federation; (c) the safeguarding of the legitimate interests of minorities; (d) the securing for members of the public services the rights provided for them by the Constitution Act and safeguarding their legitimate interests; (e) the prevention of commercial discrimination; (f) the protection of the rights of the Indian States; (g) any matter which might affect the administration of a department under the direction and control of the Governor-General.[53] The Governor was to exercise the special responsibilities in his province on behalf of the Governor-General. These provisions were incorporated into the Constitution Act of 1935.

Widening of the Communal Gap

On 31 December 1929 at its Lahore session the Congress decided that Independence was to be its goal and not dominion status. It indicated that it was determined to forge ahead on its own and to use civil disobedience to seize power. For hopes had not only receded for the attainment of dominion status, which the Calcutta Congress had demanded in 1928, but Irwin's statement of 31 October 1929 made it still clearer that, although dominion status was implicit in the declaration of 1917, it was contemplated as a gradual development in India's constitutional advance, and to be attained only 'in the fullness of time'. The reason given for delay was that the Indian States should have the opportunity of finding their place in the new scheme, for it was 'desirable that whatever can be done should be done to ensure that action taken now is not inconsistent with the attainment of the ultimate purpose which those, whether in British India or the States, who look forward to some unity of All-India, have in view'. The entire issue was, therefore, proposed to be referred to a conference where representatives of different parties and interests in British India and the representatives of the Indian States could meet, either separately or together, to determine the nature and extent of advance in keeping with the country's unity.

In a private letter to the Viceroy on 20 June 1930, Wedgwood Benn, the Labour Government's Secretary of State for India, went a step forward to explain that the purpose of the Round Table Conference was to have 'the national view' of India duly 'presented', a view which 'an all-White Statutory Commission' could not reflect. 'I have read both its Reports', Benn added, 'and my criticism of them would be that, while they bear the stamp of Simon's high intellectual calibre, they lack colour and imagination' on account of their inadequate appreciation of 'the national feeling which has grown at a speed and has gained a momentum which have taken even qualified observers by surprise. An Indian would say, therefore, that we have, in the Reports, a too British view of the situation, and I think, in the main, he would be right'.

However, as the Congress claimed to speak for the whole country, it decided not to wait for the conference. Its declared goal of Independence was sought to be achieved by a boycott of the legislatures, a refusal to attend the projected Round Table Conference, and, above all, by a civil disobedience movement which Gandhi had initiated in February-March 1930 when he marched to Dandi to disobey the law governing the manufacture of salt.

The Lahore Resolution was partly a result of pressure put on the Congress by the left[54] at a time when thgere was a general economic depression and educated unemployment was on the increase. Acording to a report in the Times, the men of the Workers' and Peasants' Party, for instance, had 'mustered in force at Lahore, where they walked up and down with red flags attacking the Congress leaders with great bitterness for not going fast enough; and on two occasions they have made a mass attack on the Congress tent, being repulsed with difficulty by a strong force of "National Volunteers"—the Congress militia, who are policing the camp'.[55] Gandhi himself was anxious to avoid doing anything definite, for 'men of substance of all kinds [were] greatly concerned with the reddening colour of the Congress creed, and [were] appreciating the importance of organizing themselves and taking part in public life with a view to preventing such development'.[56] But, as he found it difficult to resist the pressure of young enthusiasts who dominated the left and centre of the Congress, he yielded to their pressure and decided on civil disobedience. Wedgwood Benn was as considerate to Gandhi as Montagu had been some years earlier. In a private letter of 22 April 1930, the Secretary of State thus cautioned the Viceroy against Gandhi's arrest:

It has seemed to me that we were faced with two quite different movements; one, the Gandhi movement—an attempt by the Mahatma to satisfy his conscience by taking a lead in civil disobedience; and the other, a revival of the terrorist activities by persons who have a contempt for Gandhi and who regard their movement as being hampered rather than assisted by Gandhi's more leisurely campaigns and non-violent professions. It appears to me therefore that we may come to the situation when Gandhi will pass out of the picture and the more serious and active campaign will have to be faced. The great advantage of leaving Gandhi to fade out instead of arresting him is that the public opinion behind him, or at any rate the feeling of honour in which he is held, will be thereby prevented from putting itself behind the activists, whereas if Gandhi were arrested and disorder followed, it would become merged in the terrorist organization and thereby strengthen it. It is perhaps too much to hope—and yet I do hope—that the Congress civil disobedience campaign will disappear and that it will be a straight fight with the revolver people, which is a much simpler and much more satisfactory job to undertake.[57]

On the question of Gandhi's arrest, official thinking in India held that the government should preferably not arrest him; but, equally, it could not let the impression gain ground that it was afraid of Gandhi. Gandhi was ultimately arrested and put in Yeravda jail in Poona. In a private letter of 3 July 1930 the Viceroy informed the Secretary of State that he had been in receipt of several indirect overtures from the Congress Party, the last being through Jayakar, who had asked the Viceroy to give a private assurance that he would support the Indian cause for dominion status; this might enable Jayakar to persuade Motilal Nehru to call the movement off. The Viceroy declined to accept such suggestions and concluded that the movement was flagging, and that Congress sympathisers themselves disagreed with Gandhi being invested with sole responsibility and the party itself being relegated to the background. In a state of political uncertainty, Indian businessmen, including Birla, viewed the movement with disapproval. The correspondence between the Viceroy and the Secretary of State during the civil disobedience movement, and the Prime Minister's policy declaration of 19 January, 1931 which prepared the ground for Gandhi's release early in February, were clear indications of a common concern and desire for peace. The Gandhi-Irwin Pact was in these circumstances concluded on 5 March 1931 and Gandhi agreed to participate in the second Round Table Conference as the representative of the Congress. The civil disobedience movement had in the meantime been 'discontinued' and 'reciprocal action' was taken by the government to facilitate Congress participation in 'further discussions' on the scheme of constitutional reform.

The scheme of constitutional reform which Gandhi was to discuss at the second Round Table Conference was not about 'Independence' but responsibility with reservations and safeguards under an Indian federation; a consensus had been emerging for such a federation to meet the demands of the minorities, the wishes of the Indian

Divisive Tendencies in a Plural Society 303

States and to preserve an all-India unity. The issue of Independence raised in the Lahore Resolution had not only intensified the communal problem, but also alarmed the Princes. Their proposed federation was intended to let them 'take a much more definite hand in the proceedings' of the Government of India than they could have expected earlier under the Chamber of Princes.[58] But it was the communal problem that persisted and refused to lend itself to a commonly accepted solution.

The Muslims in India were not a united force at this juncture. By 1928–9 the All-India Muslim League was divided into two rival factions, one led by Jinnah and the other by Mohammad Shafi; they worked separately till February 1930 and their groups were amalgamated only in 1934. Another important Muslim organization, which emerged on the occasion of the Aga Khan's visit to Delhi in 1929, was the All-India Muslim Conference, an organization of the upper-class English-educated Muslims that challenged the ulema who had earlier identified with the nationalist cause on the score of Congress support for the Khilafat. The Conference was established with Nawab Ahmad Said Khan of Chhatari as President and Sir Mohammad Iqbal, Dr Shafaat Ahmad Khan and Shafi Daudi as Vice-Presidents. It was designed to safeguard and promote the rights and interests of Indian Muslims at all stages of India's constitutional advance towards the goal of full responsible government. The Conference unanimously adopted the famous fourteen points which Jinnah formulated in regard to the projected scheme of constitutional reform. Briefly, these points envisaged the creation of a Muslim majority area in the north-west, with emphasis on the maintenance of a Muslim majority in Punjab and Bengal, as well as adequate and effective representation for Muslims in the legislatures, government and the public services.[59] The Conference stood firmly for separate electorates, and kept aloof from the civil disobedience movement of the Congress.

The nationalist Muslims, who generally supported the ulema, did not represent the English-educated professionals or the moneyed classes. In July 1929 men like S.A. Brelvi and Yusuf Meherally from Bombay formed a Congress Muslim Party with the blessings of Motilal Nehru, while an All-India Muslim Nationalist Party was formed soon after at a conference held in Allahabad on 27 July 1929 with Maulana Abul Kalam Azad as President. The Khudai Khidmatgars or the 'Red Shirts' of Abdul Ghaffar Khan of the North-West Frontier Province were likewise connected with the Congress.

Clearly, among Indian Muslims at this period there were only two main organizational categories, one consisting of an educated elite, stable and affluent, with high stakes in government and politics, and the other mainly comprising a group of traditionalists or even a few left revolutionaries and socialists whose sympathies lay with the Congress for anti-imperialist reasons. Like the Hindu elites who were contesting for power and a share in the services on the basis of a broad nationalism, their Muslim counterparts competed with Hindus on communal grounds. This development later filtered down to lower levels with the spread of modern education.

The Muslim case was presented in a Note by Fazl-i-Husain, Executive Councillor to the Viceroy who, in a private letter of 28 August 1930, passed it on as an enclosure to Wedgwood Benn, the Secretary of State, with a comment that the Note merited 'considerable sympathy'.[60] Fazl-i-Husain referred to the good relations that had developed in the past between the Muslim League and the Congress, relations which resulted in a compromise in 1916, when Muslims were still not well-organized. Under the Montford Reforms certain proportions of seats were allowed to Muslims, who remained in a minority even in the Muslim-majority provinces of Punjab and Bengal. But since under the Reforms there was an official bloc, the Muslims were protected from communal bias in most of the legislatures. When, in the course of the non-cooperation movement, the Congress boycotted the legislatures, the Muslims remained prepared to work the Reforms. A part of the administration thus passed into Muslim hands, particularly in the Punjab and Bengal where their legitimate aspirations could, as a rule, be satisfied with the help of the official bloc.

However, the Muslims feared that under the Simon Report their position in India would become much weaker than under the Montford Reforms. For, while under the Congress-League Agreement of 1916 they would be a minority even in the Punjab and Bengal legislatures, the absence of an official bloc in the provincial field as a whole under full provincial autonomy would everywhere subject them to the rule of a Hindu majority; in the proposed Federal Legislature, too, they risked being overwhelmed by a Hindu majority with the projected entry of the Indian States which were also mostly Hindu in complexion. The Note emphasized that:

> Muslims, therefore, have come to hold the view that further advance of [democratic] reforms on the lines of the Simon Report means definite loss in political power and prestige; and, therefore, if the reforms are to be on those lines, they will have nothing to do with them. The Indian Muslims do not feel that they are

called upon to commit suicide with the object of promoting the good of India. They are anxious to promote the good of India, provided it does not lead to their own loss of power, position and prestige, and to their ultimate elimination from the Indian body-politic. They feel that this is not for the ultimate good of India either.

Sir Fazl-i-Husain's conditions for political advance included the grant of full provincial status to the North-West Frontier Province and Sind, which was sought to be separated from Bombay to form a compact Muslim majority area that included the Punjab. In addition, Bengal was also to be recognized as part of the Muslim majority provinces vested with powers to send representatives to the Central Legislature on a federal basis and enter into statutory pacts with various communities for the protection of Muslim political and cultural rights.[61]

British Policy

As already noted, the policy of the Labour Government with Benn as Secretary of State was to ascertain a national Indian consensus through a Round Table Conference and to decide about reforms suggested in the Simon Report thereafter. The search for a national consensus, as Benn made it clear in a private letter of 20 June 1930, did not mean disregarding the majority:

We recognize that we have a duty to minorities, but that trusteeship must not be exploited in any way against the interest of the majority. That is to say that, while we cannot go away and tell the minorities to make the best terms they can with a powerful majority, we do not intend, on the other hand, to remain and deprive the majority of its rights, relying ourselves upon the support of one or more minorities. To put it another way, although we recognize the value of the support of the minorities, we will resist the temptation to coax or bribe them to give that support at the expense of what the majority ought to have.[62]

Benn's argument was, in fact, directed towards finding a solution by agreement and securing the participation of the Congress in the projected Round Table Conference, even if 'the Youth Leagues and terrorist organizations [had] really got control of the situation on the Nationalist side'. He believed that 'even Gandhi's support, with his doctrine of, and possible belief in, non-violence, might be worth having, and even necessary, to rally those who are yet untouched by what is becoming an ordinary violent nationalist insurrection'. In case no solution was possible along the lines so suggested and the Conference boycotted by the Congress, Benn added, 'then we shall have to stand up for their rights'. But in no case should safeguards for the minorities 'be made the excuse, directly or indirectly, for running India except in the interest of those to whom India is home'.[63]

The Secretary of State recognized that 'the greatest difficulty' of the Indian political situation was 'the question of the minorities and our position in relation to them. The suspicion in the Indian mind is that we favour disunion and attempt to bribe the minorities to oppose the national demand'. The admission by Sir Fazl-i-Husain of the existence of continued co-operation between the official bloc and Muslim representatives under the Montford Reforms could not but give some weight to that suspicion. And, while Benn himself was not convinced that 'such a plan' for 'disunion' existed, he felt that there was a trend towards unity, for the 'minorities' during his tenure of office had come to exhibit 'remarkable sympathy towards the nationalist cause'.[64] 'We have, however, to prove', Benn emphasized, 'that our position is *bona fide* and that the difficulties we find in conceding internal freedom to India are not resident in some desire to exploit them but are really inherent local problems.' And the *bona fides* of that position, it was realized, could not be proved except by the presence of the Congress at the Conference.

It is true that the presence of the Congress at the Conference could suggest that the government was placating the forces of organized disorder. It was also felt by some that concessions to the Muslims would ensure their support for British domination. But the Government of India believed that, however much the Muslims might be opposed to the Congress campaign of civil disobedience, they were, as a community, as eager for political advance as the Hindus. Benn, in fact, believed that, given a suitable opportunity and a proper function for the Conference, the Congress would abandon its campaign and participate in the deliberations. The Gandhi-Irwin Pact of 5 March 1931 was a result of this policy: Gandhi himself asked for an interview with the Viceroy under a letter of 14 February sent from Anand Bhawan in Allahabad.

Extended Democracy and the Unresolved Communal Problem

Earlier, at the first Round Table Conference in 1930, the Princes had led the way by indicating their choice for a federal set-up, which revised the Simon proposal for provincial responsibility and government control at the Centre. But the communal issue stood in the way:

the Muslims feared that provincial autonomy and responsibility at the Centre would be tantamount to Hindu Raj. Neither the Congress nor the Hindu Mahasabha was in existing circumstances prepared to recognize the communal issue as inherent in India's plural society, a reality which called for constitutional safeguards to ensure equity and justice in the distribution of governmental power and positions in the public services. It was unrealistic of them, if not hypocritical, to dismiss the issue as a product of the third party, namely, the British government.

As reported by Whitehall's proceedings of the Round Table Conference held on 19 November 1930, the Hindus, who were led by Moonje, were willing to concede Sind and a bare Muslim majority in the Punjab and Bengal; but in return they demanded joint electorates elsewhere. Jinnah was at the Conference as a Muslim representative, albeit 'under some suspicion from his colleagues'.[65] From an earlier private letter of the Viceroy dated 6 February 1930, it appears that Jinnah had not yet become a force in Muslim politics, which in itself was not united. 'Jinnah's Independent Party', the letter said, 'is broken up by personal divisions', its leaders doing 'Jinnah's bidding only when it suits them, while the four or five Muhammadan seceders from the Congress Party were seeking berths elsewhere' and not joining 'either the Muslim Centre Party or Jinnah's Party'. It seemed likely at one stage that most of the Muslim Centre Party would accept him as leader of a combined party, but Jinnah would not agree to the condition that it must be a purely Muslim party, as he was not prepared 'to throw over Sir Purshotamdas Thakurdas [a Hindu Liberal].' No organized and all-comprehensive Muslim party thus existed, though on communal questions there was a tendency among Muslims in the Central Assembly to 'work in unanimity'.[66]

According to the Simon Report, the Muslims would support a constitutional advance that transferred the 'reserved' subjects under Section 45A of the Government of India Act (1919). Before the Congress campaign of civil disobedience, Muslims were in general as eager as Hindus to reduce British domination. Even after the campaign, when the Round Table Conference opened on 14 November 1930, Jinnah put forward in the opening speeches a 'demand that the Conference should register the addition of another Dominion in the Federation of the Empire', a demand which fell flat, but none the less indicated the national perspective and secular frame of Jinnah's mind. The report of the opening session of the Conference clearly indicates that Jinnah made the demand despite his party's decision to the contrary. Commenting adversely on Jinnah's opening remarks, the official reporter thus said: 'He declined to give the Conference Secretariat a copy of his speech in advance as all the others had done. But then Jinnah of course was always the perfect little bounder and as slippery as the eels which his forefathers purveyed in the Bombay market. His was the only controversial note in the opening speeches, but it fell flat.'[67]

The obduracy of the Hindu delegation on the communal issue, however, led to a gradual hardening of the Muslim attitude, a development which tended to widen communal gaps, especially with the emergence later of a no-rent campaign in U.P. to reinforce Congress efforts to get into power by civil disobedience.

The proceedings of the Conference held on 27 November 1930 show that in and out of the talks about Federation there ran a constant current of communal dissension. The efforts of such Liberal leaders as Tej Bahadur Sapru and Srinivas Shastri remained fruitless. Despite Jinnah's 'usual yielding to the solicitations of his Hindu clients',[68] the Muslims were not prepared to compromise on the principle of communal electorates which had the support of the Government of India on the basis of the Simon proposals. In the words of the official reporter of the Conference proceedings of 27 November 1930, 'Not all the beef and ham pies of London lunches seem able to bring the two communities closer together.' According to the report of 10 December, it appears that the Hindus finally offered to let the Muslims have all their 'fourteen points' if they agreed to give up separate electorates. But the latter insisted on statutory guarantees to secure the fourteen points before they could commend to the Muslims in India the abandonment of separate electorates. This unusual Muslim rigidity proceeded perhaps from 'publication in certain papers such as the *Daily Telegraph* of statements that the Labour Government and particularly the Secretary of State are prepared to press the Muslims to abandon communal representation, and the latter take great alarm at this'.[69] When some Hindu delegates met the Prime Minister and asked him to arbitrate, the Muslims were embarrassed, for they already suspected that an alarmingly close relationship existed between the Labour Government and Hindus.

As the Conference report of 15 December showed, there followed a complete deadlock. While, on renewed pressure from India, the Muslims demanded both separate electorates and their fourteen points, the Hindus, led by Moonje, went back on their agreement to concede the fourteen points. Both sides tended to ignore the important problem of responsibility and power at the Centre, discussion of which had progressed around the projected Indian Federation, which, in turn, involved questions relating not only to communal legislation and the rights of the Indian States, but also the financial powers of the British government, especially in respect of exchange, tariffs and credit. In the absence of any communal settlement, the whole question of responsibility and power at the Centre was hanging in the balance, more particularty as the Prime Minister indicated that it would be impossible to place before Parliament any scheme of responsible government if the communal question remained unsettled.[70]

In these circumstances, Winston Churchill, the leader of the Conservative Party, said that his first desire was to smash the Conference itself: and Lord Reading, the Liberal leader, expressed in a letter to the Prime Minister his stipulations as to reservations and control at the Centre, which included complete reservation of defence and external affairs and control over finance, law and order and the maintenance of the services. In the meantime, it was 'heard that twenty-one Liberal Hindus had signed a document in which they agreed to give the Muslims 51 per cent in the Punjab and about 48 per cent in Bengal and, at the same time, to allow separate electorates'.[71] Moonje and Jayakar refused to be party to the document which, in fact, did not naterialize; but in the meantime the Prime Minister had begun to feel that 'he must, after all, proceed with the framing of the constitution'[72] on the basis of reservations and safeguards.

By the end of December 1930 opinion had crystallized on the question of provincial autonomy; Liberals and Conservatives both agreed that the Governor should, for all practical purposes, be made 'constitutional'. But differences arose on the question of responsibility at the Centre in a federation where the operation of parliamentary procedure would place the executive in its day-to-day functioning at the mercy of the legislature and lead to instability. The Conservatives therefore stressed at a meeting with the Prime Minister on 30 December 1930 the need for 'a separation of the Executive and the Legislature somewhat on the United States model'. The expedient of an 'irremovable executive' was, of course, not recognized, but provision was made, instead, to invest the Governor-General with overriding powers to enforce financial or public safety Acts, for example, which might be rejected by the legislature.

A policy statement of the Prime Minister made on 19 January

1931 reveals that he was anxious to achieve unanimity in Parliament concerning the principle governing India's constitutional advance. Although himself belonging to the Labour Party, he took care, as the Conference report of 31 December shows, to avoid a division in Parliament. 'If India felt', the report said about him, 'that there was a serious split in Parliament on the subject, then we as a Government would be greatly embarrassed. And there would even be an active danger that less responsible members of the Labour Party might ally themselves with Congress elements in India and produce some of the mischievous results which an alliance of this kind produced in the Irish situation.'[73]

The deliberations of the Minorities Committee produced the results necessary for a compromise on 2 January 1931. The main contrast was, as usual, between Moonje and Mohammad Shafi, representing the Hindu and Muslim points of view respectively, the former emphasizing joint electorates, and the latter enumerating their disadvantages. Despite Moonje's recalcitrance, the Indian Liberals, however, professed themselves willing to give way as far as possible to the Muslims, including their demand for separate electorates.[74] Satisfactory results were similarly achieved by the Federation Committee on 6 January on the question of responsibility and power at the Centre, which meant the grant of responsible government, subject to safeguards which the Indian delegation had admitted. While making the opening speech at the instance of the Lord Chancellor, Tej Bahadur Sapru, for instance, viewed with strong disapproval the existing form of government functioning at the Centre under the 1919 Act, and suggested the outline of a scheme of responsibility which conformed roughly to what the Prime Minister himself had earlier drawn out. Besides accepting reservations for defence and external affairs, he contemplated control over such other matters as law and order and finance.

Contrary to the proposals of Sir John Simon, who had nothing to do with responsibility at the Centre immediately, the Indian Liberals, too, now came down definitely on the side of responsibility with safeguards, the details of which were to be incorporated in the projected White Paper, and the principles of which were to be announced by the Prime Minister after the termination of the first Conference. But a hitch arose once again on the communal issue, for though the Liberals were quite generous to the Muslims in their offer, Moonje remained recalcitrant. This brought Shafi and Jinnah together; they felt that, if the policy declaration by the Prime Minister merely dealt

Divisive Tendencies in a Plural Society   311

with responsibility under the Federation, it would alienate Muslims unless they were explicitly given at least some guarantees about the allocation of seats in the legislatures, more especially in the Central Legislature where Jinnah's 'fourteen points' demanded not less than one-third of the total. They had no faith in responsibility even with safeguards; for, as the report of the Conference for 12 January 1931 pointed out, the Hindus 'know quite well that no safeguard stands very long in practice', and that, once they obtained responsibility at the Centre, they would be 'under no kind of pressure to give way to Muslims', even in Punjab and Bengal, particularly as the Secretary of State had earlier indicated his preference for joint electorates. Thus, Shafi and Jinnah insisted on specific guarantees and 'the advisability of stating in the Minorities Report that if agreement is not reached then Government must issue its own decisions which will in effect recognize the existing system of separate electorates and make those adjustments which are required in view of the absence of the official bloc, etc.' The Muslim delegates who saw the Lord Chancellor on the question of responsibility with safeguards on 13 January endorsed the views expressed by Shafi and Jinnah at the Conference the day before. They held out to him the possibility of serious trouble between the communities 'if the Muslims did not bring back from London the kind of guarantees they wanted'.

In these circumstances a fresh bid was made for a communal settlement on 15 January 1931. The Muslims came forward with their last attempt to attain practically 50 per cent of the seats in the Punjab and Bengal legislatures. The Hindus consented to negotiate, but, aided by Moonje, the Punjab Hindus and the Sikhs united to ensure that the Muslim vote was neither equal to, nor greater than, the combined Hindu and Sikh vote. The Prime Minister thus had no alternative but to include in his policy statement a clause that the Government would decide the communal issue on 'equitable terms' if no communal agreement was arrived at. This had to be done not only to reassure the Muslims but also to meet objections from Liberals and the Conservatives who wanted a communal settlement to precede any grant of responsibility at the Centre.

The first Round Table Conference thus evolved the lines on which the constitutional advance of India was to proceed. Its main principle was to meet through provincial autonomy the exigencies of regional development within the framework of all-India political unity, which was sought to be attained not by a unitary arrangement, but a federation of the 'two Indias' in a manner that brooked no weakening of the Centre. It was a definite advance over Simon's recommendation which concentrated on full provincial autonomy as an experimental measure. Its promise of responsibility with safeguards at the Centre not only alleviated the fears of the minorities and the Indian States, but also satisfied the demands of the British delegation and government, who were not unnaturally keen to preserve their commercial interests with India's security and financial credibility.

However, the Conference did not remove the Muslim fear of a possible Hindu Raj. The immediate cause was the growing feeling in the country 'that after the Prime Minister's declaration wisdom suggests that many people should make terms with their future masters',[75] the Congress. This feeling was reinforced by the tenor of speeches delivered by the Secretary of State and the Prime Minister in the House of Commons, where they pointed out that repression in India, though directed against 'a subversive agitation' was 'a regrettable business', a comment which indicated a tilt in favour of the Congress. Hindu big business in Bombay and Calcutta consequently sought peace and pressurized Gandhi to seek an interview with the Viceroy; the latter was willing to promote Indian business, which was anxious about the increasing influence of the left on the Congress.[76] As we have seen, on 14 February 1931 Gandhi sought an interview with the Viceroy, soon after the death of Motilal Nehru at Allahabad. 'As a rule', he said, 'I neither wait for onward prompting nor stand on ceremony, but straightaway seek personal contact with officials whenever I feel such contact is needed in the interest of a cause ... I have missed the guidance of the Inner Voice. But I have received suggestions from friends whose advice I value that I should seek an interview with you before coming to any decision.'[77]

Apart from the advice of Indian business, a more serious consideration underlying a search for understanding with the Government was, however, the success of the first Conference and the room left for further discussion to ascertain a national Indian consensus, a development which made continuing civil disobedience seem unnecessary. Contrary to the earlier feeling that the deliberations in London would be critically received in India, papers like the *Tribune* in the Punjab, the *Hindu* in Madras and even the *Leader* in U.P. were practically unanimous in observing that the Prime Minister's declaration should be taken seriously, for it offered 'a chance of constructive work by all parties'. The *Tribune* went further and 'advocated the abandonment of civil disobedience'.[78] Even a section of the Congress led by Malaviya shared these views of the Indian press.

The only question was whether civil disobedience should be called off before amnesty measures were announced or whether both should occur simultaneously. Gandhi's request for an interview with the Viceroy and the Pact concluded between them on 5 March 1931 were intended to resolve this problem. Among other terms[79], the Pact provided for the 'effective discontinuance' of civil disobedience as a precondition for reciprocal amnesty measures by the Government. By negotiting directly with Gandhi the Viceroy not only gave formal recognition to the former's importance, but also to the Congress. The Congress had gained in strength both during and after the civil disobedience movement, and the Government's undertaking of implementing 'reciprocal measures' was seen as a victory for Gandhi.

As a result of that Pact the Congress secured for itself a privileged position arising both from its organizational strength and official recognition, which encouraged it to assume the position of an intermediary between the Government and the people and treat itself as a body entitled to speak for them.[80] The position which the Congress aspired to hold was that of an alternative sovereign entitled to wield power by itself. It would not share authority with any other party or group which declined to toe its line. Even the communists, revolutionary socialists and terrorists, who had greatly added to Congress strength by providing a mass base, were thrown overboard in 1933. Civil disobedience was now reduced to individual satyagraha to be undertaken by individuals on the basis of a certificate of 'good conduct'. This novel Gandhian device was used to exclude non-conformists and clear the road for the assumption of power by the Congress. A total withdrawal of civil disobedience occurred in 1934 so that the 'reservists' could contest elections under the projected reforms, albeit on the make-believe grounds of wrecking the new constitution from within as the first step towards office-entry. Even moderate socialists like Jayaprakash Narayan and his associates were obliged to form a separate party outside the Congress fold. It was not unnatural, therefore, that the Muslims as a community looked upon political developments in the early 1930s as a Congress manoeuvre to establish its own party rule on the assets built by genuine anti-imperialist left forces after the Kanpur Bolshevik Conspiracy case of 1924. The stand which Gandhi took at the second Round Table Conference reinforced Muslim fears about a Congress move towards the monopoly of power. Addressing the Federal Structure Committee on 30 November 1931, for instance, Gandhi reiterated the Congress claim to speak for the whole of India.

His argument was that the Congress represented '85 per cent of the population of India', not excluding 'even the Princes', and that 'all the other parties' present at the meeting of the Committee 'represent only sectional interests. Congress alone claims to represent the whole of India, all interests', without distinction of 'race, colour or creed'. It was, however, a vague assertion. For, apart from the fact that most Muslims, and even the Depressed Classes denied that claim, Gandhi himself in his address to the Committee could not assert with confidence that the Congress had 'lived up' to its declared claim.[81] The minorities' proposals for the settlement of the communal problem thus naturally emphasized the need for statutory safeguards and the special claims not only of Muslims, but also the Depressed Classes, Anglo-Indians and the European community.

With the extension of democracy, the unity of India depended on a willing and spontaneous recognition of the realities of her plural society. But, obsessed by a strong desire to impose nothing less than its own party rule, the Congress refused to acknowledge this plurality. It attempted to side-track the communal problem and dismiss it as a creation of the British government. The Seond Round Table Conference, however, recognized those realities with a view to maintaining unity and a social consensus. The real achievement of the second Conference was to start the whole series of detailed inquiries that led to the Government's Communal Award of 16 August 1932. It was based on the invaluable reports of committees that went to India at the start of the new year—Lord Lothian's Committee, Davidson's Committee and Lord Eustace Percy's Committee.[82]

The Award recognized the principle that Muslim representatives in the Lower Chamber would be chosen by direct election through separate Muslim constituencies on the basis of their claim for one-third of the seats in the Federal Legislature. Sind was to be separated from Bombay and formed into a separate province, and the North-West Frontier Province was to be raised to the status of a full-fledged Governor's Province. Baluchistan, which comprised a large area of great strategic importance, was to enjoy special administrative autonomy. In view of the Poona Agreement, the Depressed Classes continued to be treated as part of the traditional Hindu social order. But other minorities, like the Muslims, were to be governed by the terms of the Communal Award itself.

A third session of the Conference was called in November 1932 at the insistence of the Indian Liberals to discuss such other outstanding problems as federal finance, financial relations between the Centre and the States and the proportions of princely and Muslim seats at the Centre. The Muslims insisted that of their 33 seats out of 100 in either House, 25 should be guaranteed to them from the British Indian quota, the remaining 8 seats being secured from the Indian States' quota. Moreover, the Muslim community was opposed to the grant of any weightage to the States; the Muslim population in the States was only 13.5 per cent of the total, and any weightage for the States would be to the advantage of the Hindu majority. At the end of the third Round Table Conference in December 1932 the British Government formulated its proposals for constitutional reforms in the form of a White Paper.

The concession of separate Muslim electorates and Muslim provinces in the east and north-west was no doubt considered necesary for accord between the two major communities, but it acted as an incentive for the spread of centrifugal tendencies, a condition for partition.

Chapter VI

THE POLITICS OF POWER AND PARTY RULE 1937–47: THE BASIC CAUSE OF PARTITION

The Partition of India, which took place in 1947, was the cumulative result of a variety of circumstances which had been developing over the years with the introduction of representative principles in the legislative organization of government, the basic source of an alternative sovereignty to that of British rule. The spread of democracy, which is prone to cause divisions in a plural society like India's, inexorably led to Partition. Broadly, however, it was the persistent Congress claim to speak for the whole country as the only alternative to British rule, that precipitated the crisis and made Partition inevitable. The Muslim League, aspiring to be the sole representative of Muslims in India, in reaction repudiated that claim. Even so, Indian nationalism was sought to be identified by the Congress in terms of its own party interest.

The Congress claim to represent all parties and interests in the country was, of course, not without some justification. It had taken the lead in the freedom movement in alliance with other groups and classes that stood for a united and independent India. But its determination to speak for all and sundry arose from its massive success at the polls in the General Election of 1937 that was held for the provincial legislatures under the Government of India Act, 1935. This occurred despite the repressive measures of Hoare and Willingdon, who had driven the party into the political wilderness for several years before the election. The Congress victory at the polls brought to it an absolute majority in six provinces, and with the help of independents it ultimately formed ministries in eight of the eleven British Indian provinces.

This accession to the strength of the Congress not only raised its stock, but also reinforced its claim to exclusive party rule. In terms of a parliamentary democracy it was difficult to fault its refusal to form coalition ministries, more especially with the Muslim League, which had not only failed to win a majority of the Muslim votes, but also become a source of humiliation to the Muslim community as a whole. In the early 1930s, the Muslims were hopelessly divided and the League in the 1937 elections gained no more than 4.8 per cent of the total Muslim vote. But a mere reckoning of votes betrayed a sad lack of statesmanship, a poor understanding of the communal problem in an historical setting, as well as a failure to recognize the problems of the party system in India's highly segmented society. The poor electoral showing of the Muslim League, which added weight to Congress adherence to party rule, was no real index of the Muslim mind, nor a correct assessment of the political situation: for the Congress primarily won on the basis of its anti-imperialist and radical socialistic posture. It made lavish promises to the peasants, which could not be fulfilled within the framework of the law and existing constitution. The election results of 1945–6, in fact, demonstrated the unsoundness of the 1937 Congress assumption, which had led to its refusal to form a coalition ministry with the League in the United Provinces, where not even half the Muslim votes had been cast in its favour. Indeed, this refusal made the League a strong and powerful force by the time of the next General Elections, held in 1945–6. In 1945–6 the League won 428 out of a possible 492 provincial seats reserved for Muslims, as against 109 in the elections of 1937. At the Centre, it bagged all 30 of the seats reserved for Muslims.

While the Congress emerged triumphant in the general constituencies which were predominantly Hindu, the Muslim League virtually established an exclusive claim to represent the Muslims in India, except in the North-West Frontier Province where Abdul Ghaffar Khan held the forces of communalism in check for a while. The League's progress in the Punjab was particularly remarkable: in 1937 it won only two seats, and of these one joined the Unionist Party which had as many as 90 seats. In 1946, on the other hand, the League secured 75 seats against 20 of the Unionist Party, of whom only 13 were Muslims. Of the 51 Congress seats, 34 were Caste Hindus, 6 Scheduled Castes, 10 Sikhs and only 1 Muslim. In Bengal, too, the League registered a major improvement in its position and popularity. The following figures give a comparative view of the League's province-wise position in 1937 and 1946:

	1937	1946		1937	1946
Assam	4	31	N.W.F.P.	nil	17
Bengal	40	113	Orissa	nil	4
Bihar	nil	34	Punjab	1	75
Bombay	18	30	Sind	nil	27
C.P.	5	13	U.P.	26	54
Madras	9	29			

The emergence of the League as the major representative body of Muslims in India and the electoral sanction it acquired in the Muslim-majority provinces, in Bengal and the north-west, created the necessary conditions for Partition, conditions which mainly flowed from the refusal of the Congress in 1937 to form coalition Ministries.

Provincial Politics and the Rise of the Muslim League, up to 1940

The terms on which the Congress was prepared to include in the U.P. ministry two League representatives were not acceptable to Chaudhury Khaliq-uz-Zaman, the League leader. According to the terms offered by the Congress in 1937, the Muslim League group in the U.P. legislature 'shall cease to function as a separate group'. Its members in the Assembly were, in other words, to become 'part of the Congress Party' and remain 'subject to the control and discipline of the Congress'. The Muslim League Parliamentary Board in U.P. was under these circumstances to be dissolved and no candidate was thereafter to be set up by it for by-elections.[1]

In *India Wins Freedom*, Maulana Abul Kalam Azad details the negotiations of the Muslim League with the Congress on the subject, and argues that the genesis of Pakistan lay in the failure of the latter to form a coalition with two Leaguers in the ministry. It is true that, in view of the poor performance of the League and the massive victory of the Congress there was, according to the principles of parliamentary democracy, no case for a coalition. But the U.P. problem called for special understanding in its historical setting and needed to be viewed in perspective. The Congress refusal to form a coalition was a serious error of judgement; it shattered the faith and confidence of the Muslims in the *bona fides* of the Congress. Muslims argued that, if the Congress could seem so self-centred in the limited sphere of provincial autonomy, Muslim interests were unlikely to be protected in the all-powerful central government under Congress rule.

Under the Montford Reforms, the Muslim aristocracy had enjoyed considerable political power. The Raja of Mahmudabad, one of the premier nobles of Oudh, was appointed in 1921 to the Executive Council of the Governor as Member of Law and Order. Five years later, he was succeeded by the Nawab of Chhatari, who held the portfolio for seven years and vacated the office at the end of 1933 after acting as Governor for six months. One minister of the U.P. Government during 1923–33 was always a Muslim landlord of eminence.

The exclusion from office of the Muslim aristocracy was thus deeply resented in 1937. The most influential Muslim landlords made common cause with Jinnah and joined the League to strengthen his leadership.

Another reason for the League's suspicion was the seemingly devious manner in which Rafi Ahmad Kidwai, a nationalist Muslim and friend of Nehru, was brought into the U.P. ministry through the University Graduates' constituency. Kidwai had not been elected in the general election from any of the Muslim constituencies; his inclusion in the ministry shook the confidence of Leaguers in the efficacy of the Communal Award and created a situation that was partly responsible for India's division.

The ministry-making in U.P. was by itself fraught with hazards. But Nehru's programme of direct mass contact with the Muslims alienated the League further. Far from being won over to the Congress side, the Muslims raised the cry of 'Islam in danger', interpreting Nehru's campaign as an attempt to wipe out the identity of their community, its religion and culture. Seized with such fears, they heeded the League's warnings and gave their support to it. In view of the Congress preference for a government solely from its own ranks, Nehru's campaign for direct contact with the Muslim masses seemed to them not only an expression of hypocrisy, but also a conspiracy against Islam itself.

Muslim apprehensions were fueled by the new class of Hindu political workers emerging with the extension of franchise on a direct electoral system. This class was thrown up in the rural areas by mass politics directed unconstitutionally against British imperialism. The process, which began with Gandhi's Non-cooperation Movement during 1920–1, gained momentum during the Civil Disobedience Movement in the 1930s. The formation of Congress ministries in the Hindu majority provinces provided political incentives to the growth of this group of workers—they were politically radical, but socially reactionary and backward-looking.

Unlike the English-educated middle-class person from the higher castes, the new emergent type mostly belonged to the non-English speaking middle and lower castes, who were wedded to the Brahmanic tradition and 'Hindu' culture, as opposed to 'Westernization'. Hindi in the Nagari script was being promoted under official patronage as a medium of vernacular education in the United Provinces—a late nineteenth-century development, which was additionally fostered by the Arya Samaj and Hindu Mahasabha. The Congress Government conceded equal status to both Hindi and Urdu. But the protagonists of Hindi threw out hundreds of words derived from Persian and Arabic, a tendency which found support even with Hindu officers of the government. The politically-oriented but ill-educated Hindi zealots were opposed to both English and Urdu.

These trends reinforced the fears of Muslim intellectuals and the intelligentsia. They were reminded of the Banaras movement in the last quarter of the nineteenth century, which had aimed to replace Urdu and the Persian script with Hindi and the Devanagari script. Such attempts had already alarmed the Muslim community and led Sir Syed Ahmad Khan to deplore the increasing communal rift on the score of language. The Muslim fear that Hindi would be imposed on their community later became an important factor in the creation of Pakistan. Jinnah, a one-time ambassador of Hindu-Muslim unity, soon hastened to make political capital of the linguistic outcry of U.P. Muslims. In his Lucknow speech of 15 October 1937 he referred to the possibility of a 'communal war' and the intentions of 'the majority community', which had 'already showed their hand that Hindustan is for the Hindus'. According to the Raja of Mahmudabad, the Chairman of the Reception Committee who lent his full support to Jinnah, the Congress refusal to form a coalition was viewed as a refusal 'to recognize even the existence of the Muslim community as such'.[2]

In his Lucknow speech Jinnah also launched a vitriolic campaign against the Congress, alleging attempts on the part of its leadership to liquidate the Muslim League, to impose Hindi as the national language throughout India and *Bande Matram* as the national song, to enforce obedience and respect for the Congress flag on all and sundry, and to 'masquerade' as the sole representative of Indian nationalism, able and entitled to speak for the whole country.[3] It is true that the omission of League representatives from the ministry embittered the subsequent history of Hindu-Muslim relations throughout India. But there is nothing to prove the allegation that the Congress Government deliberately followed a policy that was against the religious and cultural rights of Muslims. According to Coupland, many of the League's 'charges were exaggerated or of little serious moment'. He observes that 'many of the incidents complained of were due to the irresponsible members of the Congress Party, and that the case against the Government, as deliberately pursuing an anti-Muslim policy, was certainly not proved'.[4] No cases occurred for Sir Harry Haig, the U.P. Governor, to justify use of his special powers to safeguard the rights of minorities. The several 'atrocity' reports produced by Muslim intellectuals like Sir Shafaat Ahmad Khan of Allahabad University were thus viewed by officials as a recurrent feature of Hindu-Muslim antagonism.

Even so, Muslim fears were not without foundation, more especially on the Hindi-Hindustani controversy. Their suspicions were reinforced by Gandhi's Wardha Scheme of education, a Hindi-based scheme of national education, worked out in detail without any reference to the Muslims who viewed it as an instrument for the revival of Hinduism.[5] Altogether, the rise of Hindi extremism under the Congress seriously endangered the prospects of India's unity.

Apart from questions of religion and culture, which Jinnah exploited to his advantage by bringing the Muslims under the banner of the League, there was the personal factor of Nehru's seeming arrogance in his correspondence with Jinnah. Hindu-Muslim relations might have taken a slightly different course had the former shown greater politeness and courtesy in the conduct of negotiations. In a letter of 12 April 1938, Jinnah was constrained to remind Nehru:

> Your tone and language again display the same arrogance and militant spirit, as if the Congress is the sovereign power; and, as an indication, you extend your patronage by saying that 'obviously the Muslim League is an important communal organization and we deal with it as such as we have to deal with all organizations and individuals that come within our ken. We do not determine the measure of importance and distinction they possess', and then you mention various other organizations.

Jinnah did not want to negotiate from a position of subordination, but of equality. He thus reiterated that 'unless the Congress recognizes the Muslim League on a footing of complete equality and is prepared as such to negotiate for a Hindu-Muslim settlement, we shall have to wait and depend upon our inherent strength which will determine the measure of importance and distinction it possesses. Having regard to your [Nehru's] mentality, it is really difficult for me to make you understand the position any further'.[6] Indeed, Nehru was *made* to understand the position when the League revealed its strength in the elections of 1946, a development which divested the Congress High Command of its title to speak on behalf of the whole of India.

The Congress High Command had itself paved the way for the rise of the League with the resignation of Congress ministries between 27 October and 8 November 1939. The war in Europe broke out on 3 September 1939, and, instead of retaining the power and influence of office as an effective bargaining counter, the Congress High Command ordered its ministries' resignations, an act which enabled

Jinnah to seize the initiative and force the Congress to play into his hands. While the Congress started a futile and rancorous political dialogue with the Viceroy (Lord Linlithgow) on the subject of war aims, Jinnah, as President of the Muslim League, proceeded on 6 December 1939 to call upon Muslims all over India to observe Friday, 22 December, as a day of deliverance and thanksgiving, to celebrate the fact that the Congress governments had at last ceased to function.

Three days later, Gandhi appealed to Jinnah and Muslims in general to desist from observing the proposed celebration in the interests of communal harmony. But Jinnah refused to respond, and the 'day of deliverance' was observed all over India on 22 December. The rebirth of the League had begun. On the other hand, the resignations left the Congress during 1939–42 to sway aimlessly between anti-imperialist and theory-bound Leftist ideas and the Rightist desire to re-enter the portals of power through compromise. While Jawaharlal Nehru embodied the former, which proved a divisive course, Rajagopalachari represented the latter, which was more conducive to unity.

Perhaps Nehru acted on the assumption that the British could not carry on without the co-operation of the Congress. Little did he realize that the War had opened up wider opportunities for the stimulation of India's industrial development and employment in civil administration. It was proving to be a boon to businessmen and resulted in the increasing Indianization of the public services. The Congress thus found itself in an anomalous position. It had earlier been pressing for rapid Indianization of the defence services, but now began urging the people not to enlist in a war to which it had refused co-operation. However, this appeal for non-cooperation did not work. For the number of Indians with King's Commissions was fast increasing and moving towards about 20 per cent of the total strength of officers in the Indian army.

The idea of Pakistan and a divided India was not so long ago viewed as chimerical and impracticable; it remained consigned to obscurity till Jinnah and his Muslim League put the seal of approval on it at the Lahore session of the League, which passed its famous Pakistan Resolution on 23 March 1940. The Resolution asserted that Muslim India would not be satisfied unless the whole constitutional plan was considered *de novo*, and that no revised plan would be acceptable to the Muslims unless it was framed with their approval and consent. It laid down the basic principles of the constitutional plan of its choice by recommending that geographically contiguous units be demarcated into regions, with such territorial readjustments as might be considered necessary. It proposed that the areas where Muslims formed a majority, as in the north-western and eastern zones of India, be then grouped to constitute an independent state.

The Lahore Resolution thus envisaged the division of India into two parts, namely, a Muslim India called Pakistan and a Hindu India called Hindustan, the former consisting of two zones, the north-western zone including the Punjab, the North-West Frontier Province, Sind and Baluchistan, and the north-eastern zone comprising Bengal and Assam. The question of demarcating and defining the territories was to be taken up after the principle of division was accepted, but the existing boundaries of the proposed zones were not to be maimed or mutilated.

The Muslim League was thus forging ahead from a position of strength, while the Congress had led itself into the wilderness. In view of the Congress attitude of non-cooperation, the British turned for support increasingly to the League ministries functioning in the Muslim majority provinces. In these circumstances, the Viceroy reiterated in August 1940 that full weight would be given to the views of the minorities in any future plan of constitutional revision. 'It goes without saying', he assured them, 'that they could not contemplate the transfer of their present responsibilities for the peace and welfare of India to any system of Government whose authority is directly denied by large and powerful elements in India's national life nor could they be parties to the coercion of such elements into submission to such a Government'.[7] In a resolution of 2 September 1940 the Muslim League applauded the Viceroy's statement and recorded its satisfaction that 'His Majesty's Government have, on the whole, practically met the demand of the Muslim League for a clear assurance that no future constitution, interim or final, will be adopted by the British Government without the Muslim League's approval and consent'.

The Viceroy's statement was perhaps understandable in the context of the War and Congress intransigence. But in view of the 'safeguards' already provided by the Constitution Act (1935) it could be considered both hasty and uncalled for. The government's seeming tilt towards the League emboldened the latter to dictate terms to the Congress and the majority community. The Viceroy's statement could even be interpreted to imply implicit British sanction for India's partition if the League so insisted. The basic issue raised by the VicEregal statement, however, related to India's future constitutional advance, a federal question, and its relevance to India's unity as founded in the Constitution Act of 1935.

The Federal Scheme of 1935 and Its Rejection

Before 1935 India had been kept politically integrated under British rule by a unitary form of government, which remained operative for over 100 years before the Constitution Act of 1935 provided for a new polity based on a federal union of the British India provinces and the Indian States. The unitary form of government and the growth of local legislative authority had promoted a sense of nationality and the development of an alternative sovereignty. The White Paper containing the proposals of the British government for constitutional advance on the lines indicated by the Round Table Conferences recommended the creation of this new polity, which involved two distinct operations—the one a necessary consequence of the grant of provincial autonomy to British India, the other the establishment of a new relationship between British India and the Indian States. Both represented the consummation of an historical process.

The grant of provincial autonomy for federal purposes naturally assumed not only that the provinces no longer derived their powers and authority through devolution from the Central Government, but that the Central Government itself could not continue to be an agent of the Secretary of State. Both had to derive their powers and authority from a direct grant by the Crown, the only legal source of authority for a reconstituted Government of India. And as the previous relations between the Crown as paramount power and the Indian States had been subject to treaties and engagements of various kinds, the rights, authority and jurisdiction which were to be conferred by the Crown on the new Central Government could not automatically extend over any Indian State. In other words, the accession of an Indian State to the federation would only take place through the voluntary offer of its Ruler. The Constitution Act could not itself make any State a member of the federation. All that it could do was prescribe a method of accession. The Constitution Act thus provided for Rulers' Instruments of Accession, under which the Ruler of a State was to signify to the Crown his willingness to accede to the federation by executing an Instrument of Accession defining and delimiting his powers and jurisdiction in respect of those matters recognized by him as Federal subjects.

The Government of India Act (1935) thus established two distinct federations: one of the provinces of British India, the other an all-India federation of British Indian provinces and the Indian States. The former included the Governors' and Chief Commissioners' provinces functioning under a Central Government for British India. The powers of the Centre and the units were defined and demarcated, so that neither should encroach on the domain of the other. The Federal Court alone could adjudge cases of dispute, if any. The relationship of the federation of the provinces of British India and the Central Government was thus based on a permanent and unified system, their respective spheres being well demarcated and their sources of authority derived from the Crown.

However, this was not so with the all-India federation of Indian States and the British Indian provinces. The most important condition for its inauguration was that at least half the total population of all the States should accede to the federation before the Crown could proclaim it as established. It was specifically provided by the Instruments of Accession that such a declaration could not be made until the requisite number of Indian States acceded to the federation; and no State could become its member unless the Crown accepted each Instrument of Accession as valid.

The receptivity of the Princes to federal ideas was understandable, given their earlier subordination to the Political Department of the Government of India under its doctrine of Paramountcy. The Princes were in search of an escape from this subordination and found it in federation. The federal authority could now be exercised only by a federal legislature and executive, subject, in the event of a dispute, to the adjudication of the Federal Court. This, they felt, would dispense with Paramountcy and secure internal autonomy for the States, for the federating units were to retain in their own hands all powers except those which they willingly delegated to a common Centre, and over the control of which they themselves had a share through their nominees.

It is true that, despite the willingness of the Princes to accept a federal polity of an all-India nature, the distinction between the two projected federations did not disappear. The two units remained separate and did not form a single whole with a common spring of action. In the case of the States, the Prince himself, not the Crown, remained the ultimate source of authority in so far as his State was part of the federation. It is also true that, whereas the powers of the new Central Government in relation to the provinces were to cover a wide field on a uniform principle applicable to each province, the

Princes 'were not prepared to agree to the exercise by the Federal Government for the purpose of the Federation of an identical range of powers in relation to themselves'.[8] But the fact remains that the entire series of discussions on responsibility at the Centre and its connected issues, which engaged the labour and attention of the Round Table Conferences, had proceeded from the acceptance of federation by the States. What the States needed further were some assurances that the powers of the Princes would not be trampled upon by the country's nationalist leaders.

Concerned with the extension of its influence and patronage, the Congress and its supporters in the States were pressing hard to enforce elective principles in the choice of the States' representatives to the Federal Legislature. The quest for power was cloaked in the language of democracy. By a resolution passed at its session held at Haripura in 1938, the Congress thus rejected the new constitution of 1935 altogether:

> The Congress is not opposed to the idea of Federation; but a real Federation must, even apart from the question of responsibility, consist of free units enjoying more or less the same measure of freedom and civil liberty and representation by the democratic process of election. The Indian States participating in the Federation should approximate to the Provinces in the establishment of representative institutions and responsible government, civil liberties and method of election to the Federal Houses. Otherwise the Federation, as it is now contemplated, will, instead of building up Indian unity, encourage separatist tendencies and involve the States in internal and external conflicts.

None of these conditions was sought to be observed by the Congress when the States agreed to participate in the Constituent Assembly in 1946 under the terms of the Cabinet Mission Plan. But the Congress Resolution of 1938 had none the less not only alienated the States and shaken their confidence in the Federation, but also hardened the attitudes of the Liberals and Conservatives in England, which was partly responsible for delay in the implementation of the Federal Plan. In view of the 1938 Congress rejection of the proposed Federation, Linlithgow, too, could do little to smoothen matters, for the Congress held the reins of power in the greater part of the British Indian provinces. Any attempt to enforce the federal part of the Constitution was likely to lead to trouble with the Congress ministries, which, under the Resolution of 1938, had been asked to prevent the inauguration of the Federal Scheme. The Resolution stated:

> The Congress therefore reiterates its condemnation of the proposed Federal Scheme and calls upon the Provincial and Local Congress Committees and the

people generally, as well as the Provincial Governments and Ministries, to prevent its inauguration. In the event of an attempt being made to impose it, despite the declared will of the people, such an attempt must be combated in every way and the Provincial Governments and Ministries must refuse to co-operate with it. In case such a contingency arises, the All-India Congress Committee is authorized and directed to determine the line of action to be pursued in this regard.

The Constitution Act of 1935 had provided safeguards against possible disruptive tendencies arising from the grant of provincial autonomy. It had made institutional arrangements to hold the provinces together and preserve the unity and uniformity built up over the years under British administration. Its rejection by the Congress was to sap the foundations of that unity. The Congress no doubt acted in the belief that it was the only party in India with which the British ought to deal, but the Muslim League emerged as a force to challenge that assumption and bypass the federation with the creation of a separate Muslim State.

It is important to bear in mind that, while giving shape to the Constitution of 1935, great care had been taken to reconcile provincial autonomy with the demands of unity. The first step in the direction of unity was to provide against possible centrifugal tendencies arising from the grant of provincial autonomy. Simon had recommended a federal framework without providing a clear scheme for the Central Government, which he wanted to leave untouched for a time. His argument was that the issues involved in the reconstitution of the Central Government were fundamental and that it would arouse controversy on communal lines which would be difficult to resolve. While Hindu political opinion would want the Centre to be stronger than the provinces, Muslims would generally seek maximum independence for the provinces as their centre of gravity. Simon therefore advised caution and did not recommend any drastic change in the constitution of the Central Government, particularly as British public opinion treated constitutional advance in the provinces as merely experimental.

Simon's actual proposals concerning the Centre remained vague and ambiguous. They were neither in support of a wholly official executive nor did they propose an advance in the direction of responsibility. However, public feeling in India and pressure for some elements of responsible government in the reconstitution of the Centre, plus acceptance by the Princes of the idea of federation with the provinces at the first Round Table Conference, made some advance unavoidable.

In the light of discussions at the Conference an outline of a scheme for the Centre was framed by the British Prime Minister and the Secretary of State in consultation with Hailey, McWatters, Haig and Stewart. It was based on the assumption that a sufficient number of States would join the federation to make it a reality. In a secret telegram dated 29 December 1930 the scheme was sent to the Viceroy for his personal consideration.[9]

The scheme contained proposals for a bicameral legislature, it indicated the proportion of nominees of the Princes to representatives from British India, with effective powers being vested in the lower house. The executive was to be responsible to the legislature, except in matters of defence, foreign affairs and relations with the States, which were to be administered by the Governor-General responsible to the British Parliament. The Governor-General was to possess overriding powers in matters of law and order as well as finance and taxation. Under the scheme, ministers were to be appointed by him after consulting the leaders of parties in the legislature. He would himself remain the President of his Cabinet. The legislature was to deal with all matters, including those of only British India concern. Responsibility was thus sought to be introduced at the Centre within the framework of what might be called a dyarchical constitution. It was felt that this scheme was the minimum likely to be accepted by Indian representatives at the Round Table Conference, although it was not known precisely how they would react to it. 'The Government of India, however, recognized in its dispatch the advantages, from the point of view of satisfying public opinion, of a scheme which placed the primary responsibility for all subjects except defence, and foreign affairs and political relations with a ministry based in the legislature',[10] a principle of reservation which meant dyarchy. The scheme contained in the Secretary of State's telegram of 29 December 1930 went a step further and embodied safeguards to counteract possible disruptive forces prejudicial to unity at an all-India level.

It was on the score of unity that the soundness of constitutional advance at the Centre in the form of responsible government came into question. British opinion seemed to veer in favour of Simon's support for an irremovable central executive on unitary principles, at least during a preparatory stage. Lord Lothian, an admirer of Jawaharlal Nehru, even stated that 'Parliamentary responsible government cannot possibly work in a Dominion India, and that if it is adopted, government itself is bound to break down ... not because of any lack of capacity among Indians, but because the system itself is one which nobody could work successfully in Indian conditions'. In a personal letter of 9 December 1930 to Malcolm Hailey during the first Round Table Conference, Lothian recognized that, in view of the 'reiterated pledges' given since 1917 about 'responsible government' and 'Dominion Status', it might be too late to raise that question. 'None the less', he wrote to Hailey, 'if it is true that ordinary responsible government is a disastrous line of advance and if, as I think, there is an alternative road of advance—one which will give a better chance of the development of India on constitutional lines and make it easier for India to reach Dominion Status on an orderly and stable basis, we ought at least to consider that alternative before entering into commitments at this Conference'.[11]

Lothian's suggested alternative was based on the belief that 'the executive must have an existence independent of the legislature, so that it ceases to be one of the functions of the legislature to maintain an executive in existence. Then and then only will it be able to have that strong executive power which is the core of all stable government, in India, as elsewhere'.[12] His idea was based on the separation of legislative and executive powers, which Lothian considered as important as the separation of the judiciary from the executive. He took his cue from the classic examples of the United States of America and Switzerland. In his letter to Hailey, Lothian also suggested that the best way of securing the independence of the executive was to elect a President or Prime Minister either by the electorate or by some representative electoral college other than the legislature. Lothian's alternative suggestion for the provinces was that the Prime Minister should be elected by both Houses sitting in joint session; this Minister would be irremovable thereafter for the term of the session, except by the Governor, who might remove him on being satisfied that he had lost the 'confidence of the community'.[13]

As the federal scheme communicated to the Viceroy under the Secretary of State's telegram of 29 December 1930 showed, Lothian's thesis did not receive the approval of the British Prime Minister or his select advisers, including Hailey himself. However, it appears that Lothian's warning did not go altogether in vain, for the safeguards which the secret scheme included perhaps resulted from the arguments Lothian brought into focus in his letter to Hailey some three weeks earlier. Hailey's memoranda, which went into the shaping of different facets of the new federal Constitution, showed that he was not unaware of the weaknesses of responsibility at the Centre in a federal constitution with defined and demarcated powers under provincial autonomy. The elements of reservation and safeguards were the answer designed to counteract the proposed advance in responsibility at the Centre, an answer which the Governors' Conference at Simla also endorsed by way of a revision of Simon's recommendation of a virtual standstill or 'stopgap' arrangement in so far as the Centre was concerned.[14]

Hailey kept pressing for responsibility at the Centre with reservation and safeguards as the best answer to the grant of 'autonomous' powers to the provinces and the withdrawal from the Centre of powers of executive control over provincial subjects. Under his scheme ultimate responsibility in the event of a grave menace to tranquillity was to fall on the Governor-General personally in his discretion; whereas if Simon's recommendation was accepted and the existing irremovable executive retained, that responsibility would fall on the Governor-General acting in concurrence with his Executive Council whose Members, according to the Simon Commission itself, were to be increasingly taken from among the elected members of the legislature. The Executive Council so retained would have been subject to strong pressure from a legislature in which the Government would always be in a minority given the absence of an official bloc. The case for keeping the action of the Governor-General in his discretion outside the purview of the legislature was therefore considered a constitutional necessity.

In his Note[15] to the Joint Select Committee, Hailey made it clear that the disadvantage of retaining a Centre in its existing form was much greater with autonomous provinces functioning under it. He felt that it would suffer from weakness even as an agency for the administration of those subjects which would remain under it once provincial autonomy had been introduced. It might seem strong in form in so far as it was not responsible to the legislature; but the fact that it was both irresponsible and in a permanent minority would provoke persistent attacks and ensure frequent defeat in the legislature. This, he felt, would produce a general impression of weakness and impair the Centre's authority. He observed:

The atmosphere of authority which we have created, on the double basis of an impression of strength and of a general recognition of the value of our administration, has been our substitute for the reliance on armed forces, on which some systems have based themselves, or on a popular vote, in which others have sought their support. The destruction of that atmosphere puts the governments throughout India in a weak position, because they do not seek to rely on force and cannot as yet have a sure source of support in a popular vote. Again, the circumstances in

The Politics of Power and Party Rule 1937–47

which the Executive finds itself reacts on its capacity for pursuing its own policy. Only those who have experienced these conditions can realize the corrosive effect of persistent attack from a permanent and overwhelming majority.[16]

Moreover, it was felt that the retention of a fixed central executive would impair the efficient working of the autonomous provinces which dealt with all matters affecting most closely the great mass of the people. A strong supporter of a federal system which contained a substantial element of the Princes' nominees, Hailey supported the transfer of law and order to the autonomous provinces. In a Note to the Secretary of State he showed how the administration of law and order was interlocked with issues like students' agitations, local self-government, communal tension, caste riots, cow-sacrifice, land revenue, landlord-tenant relations, and irrigation, subjects which were proposed to be transferred to the provincial list. It was logical that, if these subjects were to be under provincial ministries, they must also control the agencies of law and order to check disorder. The reservation of law and order for the Centre would leave provincial ministers with subjects which were always liable to break into disorder, thus placing on Governors the burden of dealing with disorders only through the use of special powers which would involve the suspension of ministerial responsibility in a wide sphere of work.[17] Administratively, the retention of law and order with the Centre was thus not considered to be a practical proposition.

However, the transfer of law and order to the provinces did not signify complete lack of control by the Centre. Apart from the Governor-General being responsible in his discretion, provisions were made at the Governor's level in each province for control over the administration of police as well as the supply of information through an intelligence mechanism. In addition, the retention of some control over law and order by the Central Government was recommended by Hailey to strengthen the forces of unity in the proposed federal set-up.[18] He suggested the expediency of giving the Federal Ministers a Home Department to deal with terrorism, civil disobedience, communism or ordinary crime extending over several provinces, such as armed dacoity. A Home Department, with its close connexion with other departments and direct dealings with the provinces under the existing arrangement, would also be an excellent instrument for the collection and co-ordination of information, besides enabling the Governor-General to discharge his special responsibility.

A Home Department on the existing model was thus considered

necessary under the federal plan too. It was designed not only to give advice on policy, but also to take an all-India view detached from the pressure of local provincial considerations. However, unlike in the past, the Home Department in the federal set-up was not to live in the atmosphere of a reserved department. Its advice would have to be tendered to ministers, who alone were to judge its merit or the extent to which it was to be carried out. Under the federal scheme the Government of India did not itself suggest that 'it should do more than advise and protest', a view based on the assumption that in matters of law and order there would be 'identity of policy between federal and provincial ministers, and that the latter will require only information, guidance and advice'. Substantial differences could, however, occur as to the exact action to be taken in dealing with certain types of disorder, such as communal disturbances. Action might not in such circumstances be limited to mere advice and protest.

The ultimate responsibility for law and order doubtless vested in the Federal Government, for the army was a federal subject. But this was so only in a formal sense as otherwise the Centre would have had powers of control over law and order in the provinces, more especially over all those subjects giving rise to discord and demanding armed intervention in such fields as agrarian disturbance or student agitations. Hailey considered this undesirable, not because it would tend to cause a formal breach of federal principles, but because it would create friction between the Centre and provinces and blur the area of provincial responsibility with that of the Centre.

In future the Government of India would admittedly lose some powers with the break-up of the centralized unitary system that had made India into a unified political structure under overall British paramountcy. The constitutional scheme of 1935 also provided for the entry of the Indian States without compromising the unity already established. It represented a constitutional advance which, if worked in its totality, would have forestalled all possible efforts at division by any political party in British India. But the arrangement required to be worked with patience and understanding, which were found wanting in the country's national leadership. Despite the care with which the federal part of the Constitution had been shaped to preserve integrity and stability, the Congress rejected it without attempting to judge its constitutional merit. As the entry of the States was a precondition for the grant of responsibility even to British India at the Centre, the Federation did not come into being and the Government of India continued to function practically under the 1919 Act.

The Politics of Power and Party Rule 1937–47

Background to the Cripps Mission

The proposals made by the Cripps Mission in 1942 represent another attempt at constitutional advance in the direction of freedom with unity. Here again, the Congress-League conflit over power mainly determined the fate of the Mission's proposals. The Mission was dispatched to India in the background of certain developments, an awareness of which is essential for an objective analysis of the politics of the time.

Federation Delayed and Suspended

The delay caused in the inauguration of the federal part of the Constitution was a basic factor in the confused and unsettled state of politics that followed the introduction of provincial autonomy in 1937. It allowed the Congress and Muslim League time to seek to outmanoeuvre each other in their strategies of seizing power at the Centre. Under pressure from the left the former aimed at mass action and majority rule as its most effective weapon, while the latter ruled out the proposed federation altogether in preference for a separate Muslim state.

Linlithgow is criticized for what is generally considered to have been a leisurely approach to the Princes, which contributed to the postponement of the federation. But he had to confront obstacles both in India and in England. In the early period of his Viceroyalty Linlithgow was rather more enterprising in devising political initiatives than Zetland, his Secretary of State, who, in a letter of 8 February 1937 advised the Viceroy that in taking the Princes 'along the road towards federation ... we shall have to walk with circumspection'.[19] Later, when the Viceroy suggested the propriety of a royal visit to India for the proclamation of the federal scheme, Zetland stood in the way. 'I still think', he said, 'that the strongest argument against a visit next [1938] winter is the probability that we shall be approaching the inauguration of the Federation, and that the Congress, if they wish to make themselves objectionable, will make use of the cry which they used so effectively when the Duke of Windsor visited India in 1921, namely, that the Government were bringing out the King to bolster up their policy.' Besides, it was feared that Muslim opposition to the Federation would be strong as Moonje of the Hindu Mahasabha had publicly advocated 'the acceptance of the Federal proposals of the Act of 1935' on the ground that 'it would give the Hindus the whip hand at the Centre'.[20] Zetland sounded caution for yet another reason. He felt that 'since the acceptance of office by the Congress, the Princes were less inclined than ever to take what they regarded as a leap in the dark and come into a federation'.[21] The release of some thousands of detenus under Congress ministries enabled its left wing to come into the open and spread anti-imperialist propaganda,[22] which continued until the Congress finally decided at Haripura to reject the federal scheme of the 1935 Act.

Linlithgow was not uninfluenced by the strength of Congress unity, although he could not overlook the Muslims, or the Princes who were in special treaty relations with the British Government. In a private letter of 21 February 1939 he appreciated the action of the Secretary of State who proceeded now to take the case of the Princes to the Cabinet; but he also made it clear that the States should recognize the logic of changing times and take steps to improve their internal administrations. Speaking, for instance, of Jaipur, he said: 'My own feeling is rather definitely that, while we must avoid anything which would blacken the faces of the Durbar, or which could be represented as a climb-down in deference to Congress pressure, it is out of the question, given the turn that things now show signs of taking, to close down "progressive" activities in the State altogether, and that the Maharaja and his Durbar would be well advised to consider to what extent and in what manner they could best, consistently with the maintenance of their dignity and prestige, and of their control of the situation inside the State, slightly modify their attitude'.[23]

Linlithgow went a little further in his attempt to secure improvement in the internal administration of the States and, consistent with 'declarations in Parliament' and the views of the Secretary of State he even proceeded to 'urge the Ruler [for instance of Nandgaon] to establish a representative Assembly'. In his private letter to the Secretary of State, he thus added: 'I have myself been anxious that, so long as some other means of which we could approve and on which we could keep our eye was available, Residents individually should not be too closely associated with sanctions against Congress workers'.[24]

The Rajkot crisis, where Gandhi threatened to fast unto death, was significant. In a speech at Liverpool the Secretary of State made a reference to the Princes' position. Taking his cue from the carefully worded remarks made at Liverpool, the Viceroy spoke at Jaipur, Jodhpur and Udaipur, where he emphasized the importance which the British government 'now attach to administrative efficiency and to the provision of machinery to enable States' subjects to ventilate their grievances and desires'. In a personal letter of 7 March 1939 he informed Zetland that this advice had a marked effect on the Princes. 'Federation or no Federation', he observed, 'there is no question about it that at the stage things have reached Their Highnesses have no option but to buckle up in these matters, and that the greater the energy and the greater the promptness which they display in doing so, the less likely they are to be caught by pressure from outside and forced to make concessions of an order which they may have good reason for shrinking from contemplating'.[25]

The Indian States were thus being obliged to move steadily, even if slowly, in the direction of improving their administration and making institutional provision for the expression of popular grievances on the lines of British India. But the task of reducing the variety in their conditions to a principle of uniformity under the Instruments of Accession was not easy, especially in view of the strong and sustained Muslim opposition to the Federation and the pressure from the Congress left against the Act of 1935 itself. The result was that the federal issue dragged on until World War II broke out on 3 September 1939: a new situation then emerged which called for a review of the whole policy. The Secretary of State and the Viceroy had both decided 'to suspend the work in connection with Federation'.[26] And though the Viceroy was keen 'to secure the continuance of those forms of parliamentary government [which were then] in being, both at the Centre and in the Provinces, whether in Bengal and the Punjab or in the Congress Provinces', the resignation of Congress ministries left no room for reconsideration of the suspension already decided upon. Linlithgow now assumed an attitude of firmness towards the Congress and maintained the political *status quo*.

Interestingly, in the course of his meeting with the Viceroy at Simla on 26 September 1939, Gandhi is recorded to have remarked 'in parenthesis, that he was gradually picking up more and more information about the Government of India Act and that he had now reached the conclusion that whatever might be said against it, it was capable of expansion to the satisfaction of the full requirements of national life in this country'. 'Speaking as a private individual [the Viceroy added], he [Gandhi] wished to express his extreme regret that I should have failed to complete the constitution after having laboured so hard to build it and having done my best to implement the scheme of the Act. If Congress had done better [as Gandhi commented], and been wiser, in the last two years, he was prepared to admit it might have been possible before now to have implemented the scheme of the Act, and that in that event the difficulty of the Muslim and Princely position would have been less. But Congress like all others was fallible and it was no good crying over spilt milk.'[27]

These remarks of Gandhi represented a clear summing-up of Congress responsibility for the delay in the completion of the Federal Scheme of the Government of India Act (1935). The delay not only contributed to the suspension of that Scheme, but also led to political developments that were ultimately responsible for the division of India.

Political Developments

The political developments that followed the outbreak of the War resulted from a dichotomy which afflicted the contenders for power in the choice of strategy. Gandhi, as the Viceroy concluded from his interview with him, was 'very anxious to help' the Government by not insisting on 'stringent conditions'. But his difficulty was that he no longer felt 'strong enough to hold this [Congress] machine'. For 'the pressure of the Left [was] becoming more and more marked'. The Viceroy had serious doubt as to whether the right in the Congress was in a position 'to deliver the goods'.[28] Subhas Chandra Bose, the acclaimed leader of the left, had been re-elected as Congress President for a second term in 1939 despite Gandhi and Nehru. The Resolution which the Congress Working Committee adopted at Wardha on 22 and 23 October 1939 asking the British Government to declare its War aims and asserting India's right to frame her own constitution bore the clear imprint of the left, more especially of international communism, a political creed that was distasteful to Gandhi. In a letter to Zetland written after the Congress ministries had resigned, the Viceroy appreciated the extent to which 'the Mahatma' had contributed to the 'peacefulness' of the political atmosphere 'by bringing these [Congress] Ministries out of office', which 'contrived to silence the Left Wing',[29] without affecting the prospect of the right to return to power if and when summoned at an early opportunity.

However, Sir Harry Haig, the erstwhile U.P. Governor, advised a go-slow policy and against evolving an alternative solution to Section 93 of the Act, under which Governors had taken over in the Congress provinces. He emphasized that 'it would be a great mistake in any statement by Governors to underline the fact that we were merely anxious to keep the seat warm for Congress to return'. The Viceroy agreed with Haig who held that, as a matter of political expediency, the Congress 'will no doubt go out of office expectant of being called back at a very early stage as a result of some formula being devised, or a clear indication of our intentions in this matter'. The main object of a go-slow policy was to ensure that the supporters of the government were not discouraged, or that the police, magistracy and lower ranks of the public services fear Congress reprisals should the party soon return to office. The Governors felt that members of the Congress 'detest the thought of leaving office', and thus suspected that 'if they [the right in the Congress] are prepared to eat a certain number of words and to come at any time to the Governor in a Section 93 situation with an assurance that they have a majority and are prepared to carry on the government, the Section 93 situation automatically comes to an end'.[30] Linlithgow thus did nothing to encourage the Congress that it would return to power in the provinces and held on to a policy of maintaining *status quo*.

However, there was additionally little or no room for a change so long as the Congress attitude towards the Muslim League remained inflexible. The former insisted on constitutional advance at the Centre to precede the settlement of the Muslim question, while the latter emphasized that provincial difficulties must be cleared before reforms at the Centre could be considered. Gandhi was fighting to have Congress Muslims recognized as representatives of the Muslims. As Linlithgow pointed out, 'The real idea in the Mahatma's mind and those of his friends is to put us in the position of forcing Congress domination upon the Muslims, for they know very well that they could not do it by themselves in a century'.[31] In view of the Congress move to non-cooperate with the War effort, the Viceroy was not going to oblige it on this front, particularly as Jinnah had definite terms for dealing with the Congress. These terms, which were to be applicable to the period of the War only, included coalition ministries and an assurance that no measure would be forced on a provincial legislature if two-thirds of the Muslim members of the Assembly opposed it. In an interview with the Viceroy on 13 January 1940, Jinnah additionally emphasized that the Congress flag was not to be flown on public institutions, that *Bande Mataram* must be abandoned as its proposed national anthem and that the Congress was to give up its confrontational tactics against the Muslim League.

There had, in fact, been a move by the Viceroy to restore the Congress to power in the provinces, but, as indicated above, this was resisted by Sir Harry Haig and the bureaucracy in general. In attempting to seek the co-operation of leaders of public opinion, Linlithgow also had a series of talks with Sir Sikander Hayat Khan,

Fazlul Huq, Jinnah and Gandhi between 3 and 6 February 1940. In a personal letter of 6 February to the Secretary of State the Viceroy reported that both Sikander and Fazlul Huq expressed their readiness, though perhaps against their will, to accept coalition governments in their provinces into which the Congress would be admitted as part of a general scheme of coalition governments. Such a scheme, if worked in the right spirit, might have contributed to understanding and unity.

But his talks with Gandhi on 5 February 1940 turned out to be a disappointment for the Viceroy. After the meeting, Gandhi made a statement which, Linlithgow said, brought out 'still more clearly' that Gandhi was sticking to his old point of view, for he emphasized

that India (by which he means Congress and their friends) must be left at liberty to make up her own mind as to what she wants; that she must be warned of the risks involved in the adoption of any particular course but if she desires to adopt that course having been so warned, she must be allowed to do so and to pay the penalty if she has made a mistake. The minorities, Princes and the like are to be steam-rollered. If the whole business is handed over to the Congress, with the assistance of a Constituent Assembly, we may be certain that no irrelevant or improper decisions would be taken by that body if it is a really representative body.[32]

Gandhi seemed to be insisting on official sanction to reduce the minorities and the States to domination by the Congress and to claim for his party a potential title to paramountcy. No government could concede such a demand within the framework of the law and constitution and, besides, there was the threat of serious communal disorder. In these circumstances the Governors of provinces advised the Government against helping the Congress return to power in the provinces.

The threat of further civil disobedience could not, however, be ruled out. For Gandhi now had a Congress President of his own choice in Abul Kalam Azad, who had been elected by a majority on 16 February 1940, defeating M.N. Roy, the candidate of the left. The Government, too, was aware of the fact that Gandhi often assumed a radical posture to outmanoeuvre the left. The potential danger of civil disobedience was therefore recognized as real. But the League's Lahore Resolution of 23 March 1940 produced in the meantime a moderating effect on Gandhi. On 6 April, while suggesting that civil disobedience was a possibility, he made it clear that civil disobedience would not be started unless he declared conditions to be fully ripe for such a movement.

In an interview with the Viceroy towards the end of June, Gandhi suggested that, because of his continuing differences with the Congress on fundamental issues of ideology, he could not guide the organization in policy matters, although he himself would be satisfied with the grant of Dominion Status immediately after the War. The position, however, changed in July 1940 when the Congress Working Committee passed a resolution at New Delhi demanding the unequivocal declaration of India's independence and immediate formation of a 'National Government' at the Centre with the closest co-operation of responsible governments in the provinces. Jinnah at once denounced the resolution as an attempt to foist Congress Raj and a Hindu-majority government on India, which would never be accepted by the Muslims. In this political context the British government authorized the Viceroy to make a policy statement. It was issued in August 1940 and invited a certain number of Indian representatives to join his Executive Council and establish a War Advisory Council. The latter would meet at regular intervals and include representatives of the States and of other Indian interest groups.

In political terms the 'August Offer' of 1940 implied a mere expansion of the Viceroy's Executive Council, and no more. Its real significance, however, lay in a veto given to the Muslim League concerning the shape and nature of India's future constitutional advance, although the statement also held out the promise of a constitution-making body to be constituted 'without the least possible delay' after the conclusion of the War. As already noted, the statement made it quite clear by implication that the British government would not only refuse to transfer power to any government whose authority was denied by the Muslim League, but would also not be party to the coercion of Muslims into submission to such a government under Congress control.

The Congress, though not free from responsibility for the emergence of a situation of this kind, was not wrong in its criticism that the August Offer was not only undemocratic, but also prejudicial to the evolution of a free and united India. By its resolution of 22 August the Congress Working Committee not only rejected the offer, but recommended its condemnation in general by holding public meetings and using other methods. For the Viceroy's statement did not indicate political neutrality, but an obvious slant in favour of the Muslim League. By a resolution of 2 September 1940, the League recorded its satisfaction that the British Government had met its demand with a clear assurance that no future constitution for India would be adopted without its approval and consent. It is difficult to free Linlithgow and the London authorities from responsibility for the subsequent aggravation of communal tension which intensified the forces of separatism.

Developing War Position

In December 1941 Britain and the U.S.A. declared war on Japan, which was making speedy inroads in South-east Asia towards Burma and India. After the fall of Singapore on 15 February 1942, Bengal lay exposed to invasion. 'When Rangoon fell on March 7', says Coupland, 'it seemed as if the tide of Japanese conquest, which had flowed so swiftly and irresistibly over Malaya and then Burma—only yesterday a Province of the Indian Empire—would soon be sweeping into Bengal and Madras. Refugees poured out of Calcutta. If there was less immediate alarm in Delhi or Bombay, it seemed nevertheless quite possible that Japanese armies might be able to penetrate as deep and quickly into India as they had into Malaya and Burma.'[33]

The Japanese advance towards India gave rise to a new prospect for constitutional advance in the direction of what the Congress envisaged as independence. The Muslim League, however, envisaged the realization of its cherished goal of Pakistan. The first indication of a further move towards Indian freedom was given in the House of Commons by Sir Stafford Cripps, Leader of the House at the time. In a debate on the War he expressed the British Government's concern for Indian unity and looked forward to a full debate on how best the Government could contribute to maintaining it.

The Cripps Proposals and Indian Reactions

Consistent with Cripps' suggestion, the Prime Minister (Winston Churchill) made a statement in the House of Commons on 11 March 1942.[34] It represented the views of the British War Cabinet, which Churchill described as 'a just and final solution' of the Indian question, a solution which undoubtedly constituted an advance on Linlithgows' August 1940 offer. It was now made clear that the British Government would 'avoid the alternative dangers either that the resistance of a powerful minority might impose an indefinite veto upon the wishes of the majority or that a majority decision might be taken which would be resisted to a point destructive of internal harmony and fatal to the setting up of a new Constitution'. This principle of check and countercheck underlined the proposals contained in the Draft Declaration of the War Cabinet, which Sir Stafford brought to India for discussion with Indian leaders. The object was 'the creation of a new Indian Union which shall constitute a Dominion, associated with the United Kingdom and the other Dominions by a common allegiance to the Crown, but equal to them in every respect, in no way subordinate in any aspect of its domestic or external affairs'. With this end in view, the British Government made the following declaration:

(a) Immediately upon the cessation of hostilities, steps shall be taken to set up in India, in the manner described hereafter, an elected body charged with the task of framing a new Constitution for India.

(b) Provision shall be made, as set out below, for the participation of the Indian States in the Constitution-making body.

(c) His Majesty's Government undertake to accept and implement forthwith the Constitution so framed subject only to:

(i) the right of any Province of British India that is not prepared to accept the new Constitution to retain its present constitutional position, provision being made for its subsequent accession if it so decides.

With such non-acceding Provinces, should they so desire, His Majesty's Government will be prepared to agree upon a new Constitution, giving them the same full status as the Indian Union, and arrived at by a procedure analogous to that here laid down.

(ii) The signing of a Treaty which shall be negotiated between His Majesty's Government and the Constitution-making body. This Treaty will cover all necessary matters arising out of the complete transfer of responsibility from British to Indian hands; ... but it will not impose any restriction on the power of the Indian Union to decide in future its relationship to the other Member States of the British Commonwealth.

Whether or not an Indian State elects to adhere to the Constitution, it will be necessary to negotiate a revision of its Treaty arrangements, so far as this may be required in the new situation.

(d) The constitution-making body shall be composed as follows, unless the leaders of Indian opinion in the principal communities agree upon some other form before the end of hostilities:

Immediately upon the result being known of the provincial elections which will be necessary at the end of hostilities, the entire membership of the Lower Houses of the Provincial Legislatures shall, as a single electoral College, proceed to the election of the Constitution-making body by the system of proportional representation. This new body shall be in number about one-tenth of the number of the electoral college.

Indian States shall be invited to appoint representatives in the same proportion to their total population as in the case of the representatives of British India as a whole, and with the same powers as the British Indian members.

(e) During the critical period which now faces India and until the new Constitution can be framed His Majesty's Government must inevitably bear the responsibility for and retain control and direction of the defence of India as part of their World War effort, but the task of organizing to the full the military, moral and material resources of India must be the responsibility of the Government of India with the co-operation of the peoples of India. His Majesty's Government desire and invite the immediate and effective participation of the leaders of the principal sections of the Indian people in the counsels of their country, of the Commonwealth and of the United Nations. Thus they will be enabled to give their active and constructive help in the discharge of a task which is vital and essential for the future freedom of India.

On 29 March 1942 Cripps held a press conference where he explained that the term 'Dominion' in the Draft Declaration did not signify any legal or constitutional limitation and that the contemplated Constituent Assembly might start even with a declaration of independence. However, on the subject of the Constituent Assembly he made it clear that, while it was obligatory on the part of the British Indian provinces to send their representatives to the constitution-making body, none of the Indian States were to be forced either to join it or, in case they decided to participate, to choose their representatives by election. Their adherence to the new Constitution was to be subject to revision of their Treaty arrangements.

Even so, the Mission's proposals constituted a definite and positive advance which, if accepted as a whole, could have preserved India's unity. The only potentially divisive provision in the Draft Declaration was the right given to a province of British India to accept or not the new Constitution to be framed by the projected Constituent Assembly. In principle this recognized the League's demand for Partition. But unlike the 1940 August Offer, the Cripps proposals did away with the possibility of an indefinite veto by the Muslim League against the wishes of the majority for constitutional progress. Moreover, the Draft Declaration had provided for only one constitution-making body: in case a province declined at the end of its proceedings to accept the new Constitution or join the Indian Union, it could keep out, as Cripps pointed out at his press conference. But it could do so only if 60 per cent of the members in its Provincial Assembly decided by a simple majority against accession. In case the percentage was less, the minority could demand a plebiscite of the whole province in order to ascertain the wishes of the people. This provision led the Muslim League to reject the proposals on the ground that the principle of separate electorates was altogether absent, both in the Constituent Assembly and in the plebiscite issue, as a bare majority was to be considered enough for any decision. By a resolution of 11 April 1942 the League thus turned down the Mission's proposals: in the absence of separate electorates it stood little chance of winning either in any legislative assembly or on the plebiscite front, even in Muslim-majority provinces. As Cripps added, the non-acceding provinces could, if they so decided, combine to form a separate Union through a separate Constituent Assembly. But they had to be geographically contiguous. Secession was thus hemmed in by several restrictions, to become difficult or even impossible.

Another Congress Error

The leadership of the Congress, however, erred again. By its resolution of 11 April 1942 the Cripps Mission's proposals were finally rejected. Congress objections doubtless related to the non-representative nature of spokesmen from the Indian States as well as to the Mission's acceptance of the principle of non-accession for a province.[35] It was argued that the proposals might lead to the creation of 'enclaves' or separate centres of power for foreign armed forces to stay on in India and operate as a perpetual menace to the whole country. But it was not these future possibilities which were responsible for the final rejection of the proposals by the Congress. In view of the imminent threat of Japanese invasion, the Congress felt 'it is the present that counts', which meant full political control over military operations as a condition for co-operation with the British War effort. In matters of defence, however, the British could naturally not transfer to the Congress any substance of sovereign political control during the War.

During negotiations with Abul Kalam Azad, the Congress President, Cripps suggested the need for a discussion with the Commander-in-Chief to explore the possibility of solving the controversial question of control over defence. It was proposed that the defence portfolio under the Commander-in-Chief be split into two Departments, namely, Defence and War. The object was to provide for the inclusion of an Indian minister of defence with the responsibility of discharging functions previously vested in the Defence Co-ordination Department under the Commander-in-Chief; under this plan, the latter, as War Member of the Executive Council, would have full control over the armed forces in India and act as the channel of communication between the Government of India and the British government on all operational and policy matters.[36] The Congress, however, was not interested in any adjustment of organizational or functional details, even those suggested later by Colonel Johnson, President Roosevelt's personal envoy in India. The Congress leadership wanted the entire defence responsibility to be transferred to Indian hands, and certain aspects to be then voluntarily surrendered to the Commander-in-Chief for the duration of the War.[37] In his letter of 4 April the Congress President had already made it clear to Cripps that the Congress Working Committee was interested in the main political issue, the transfer of sovereign power. 'I do not think', Abul Kalam Azad said, 'it is necessary for me to send a note about organizational details. We are interested as you know in the political aspect of the problem, the full popular control of defence as well as all other departments of administration. . . . Problems of higher strategy may well be controlled by inter-Allied Cabinets or Councils, but the effectual control of the defence of India should rest with the Indian National Government.'[38]

The breakdown of negotiations thus finally proceeded from the reluctance of the British government to immediately concede the establishment of a responsible national government with full powers to control defence, a government which, for all practical purposes, was bound to be the government of the Congress party in the existing situation. Agreement to such a transfer would imply coercing the Muslim League to submit unconditionally to the Congress, which the Draft Declaration had never contemplated; it would also involve facing the united and organized opposition of the Muslims and other minorities who had been supporting the British War effort. Cripps was naturally recalled: his effort failed for no fault of his.

There were other reasons for the refusal of the British government to transfer defence to full popular control, including a lack of confidence in the *bona fides* of the Congress promise to resist possible fascist aggression against India. Apart from past experience of Congress politics, the lack of confidence was also attributable to the political pressure which the Congress intensified at a time when a Japanese invasion was imminent. It was hardly the moment for handing over defence to a popular government consisting of men not only lacking in administrative experience but also divided in their counsels. Even an avowed opponent of the Axis powers like Nehru would find it difficult to keep the Congress on Britain's side, especially in the face of Subhas Chandra Bose whose pro-Axis broadcasts from Berlin carried considerable anti-British appeal among young Indian nationalists. There was, in addition, the fear that Gandhi himself, on the score of pacifism, might prevent his Congress supporters from helping the War effort.

It is indeed difficult to explain the Congress rejection of the Cripps Mission plan except on the ground that the party was obsessed by the belief that Britain was on the verge of defeat in the War. Instead of depending on an 'uncertain future', or what Gandhi described as a 'post-dated cheque on a failing bank', the British were being pressed to transfer power immediately. Had the Congress accepted the Mission's offer and entered the portals of power at the Centre and in the provinces, the influence of the Muslim League would have been weakened, if not eliminated, under the influence of the Congress's political patronage and administrative control. Alternatively, its rejection of the proposals and the 'Quit India' Movement which soon followed, contributed to intensifying communalism and reinforced the strength of the Muslim League. As the 1946 election result showed, the Muslim-majority provinces came under Jinnah's direct control, a development which the Congress sadly failed to foresee.

The Congress assessment of the Muslim League suffered from serious errors of judgement. The secret reports of intelligence officers for March 1942 show that news of the Cripps Mission had inclined Jinnah to reach an understanding with the Congress for the duration of the War, without any prejudice to his Pakistan scheme. His demand for parity with the Congress at the Centre still remained unaffected, but a report from Bombay unmistakably indicated that, in the event of the acceptance of the Cripps offer by the Congress, Jinnah would be obliged to become yoked to the Congress.[39] Jinnah was, in fact, worried at the time; for he knew that Cripps tilted in favour of Congress. But, more importantly, a reference in the Prime Minister's statement made it clear that no minority would be allowed to exercise a veto on the wishes of the majority. Jinnah was indeed in a quandary, especially as Japan seemed pro-Hindu. The Congress rejection of the Mission's plan for India's constitutional advance thus came as a great relief to Jinnah and the Muslim League. Once more, the Congress had led itself into the wilderness.

The failure of the Cripps Mission was followed by a political stalemate which continued for nearly three years, until 1945. The breakdown of negotiations was at the same time marked by differences between Gandhi and Nehru. While Nehru held that, regardless of the failure of the Cripps Mission, India should continue to maintain her antifascist attitude towards the War, Gandhi's mind had begun to work in a manner which was interpreted by the Japanese as an invitation to invade India. In a statement issued in the *Harijan* Gandhi had pleaded for the complete withdrawal of British and Allied troops, and added: 'If the Japanese really mean what they say and are willing to help to free India from the British yoke, why should we not willingly accept their help?' He realized that 'aggressors' could hardly be 'benefactors', but insisted that Britain should withdraw from India and leave India to face the consequences.

The Congress Working Committee which met at Allahabad on 29 April was faced with a double crisis, which included not only the ideological polarization between Nehru and Gandhi, but also a tug-of-war between Gandhi and Rajagopalachari, who was determined to press for coming to terms with Jinnah on the basis of conceding the principle of Pakistan. The whole tilt of the resolution adopted at Allahabad was, however, on the side of Gandhi. For it said: 'The present crisis as well as the experience of negotiations with Sir Stafford Cripps has made it impossible for the Congress to consider any schemes or proposals which retain even in a partial manner British control and authority in India. Not only the interests of India but also of Britain's safety and world peace demand that Britain must abandon her hold on India. It is on the basis of independence alone that India can deal with Britain or other nations.' The whole basis of the resolution was, in fact, an assumption that Japan was going to win and deliver the control and sovereignty of India to the Congress, an eventuality which would automatically put an end to the Muslim League.

The resolution which the Working Committee adopted at Allahabad recommended a policy of non-violent non-cooperation as resistance to the Japanese, a sop to Nehru's anti-fascist utterances. But Rajagopalachari's resolution which recognized the League's claim to self-determination was altogether rejected. It was superseded by Jagat Narain Lal's counter-resolution opposing a principle or plan that might lead to India's disintegration. The general consensus was, of course, in favour of a National Government to mobilize patriotic resources to resist aggression. But while Rajagopalachari pleaded for a united national front by accepting the principle of Pakistan or Muslim self-determination as a means to awaken a general consensus for India's defence, Nehru chose to toe the Gandhian line and allow things to drift despite the crisis which had resulted from the failure of the Cripps Mission. President Roosevelt's envoy in India, Colonel Johnson, who was later succeeded by William Phillips, endeavoured without success to see that discussions were duly maintained between the Congress and the British government. But while the Congress remained keen to exploit the developing war situation to its own advantage, both Cripps and the Secretary of State, L.S. Amery, reiterated in a debate in the House of Commons on 28 April their arguments about the impracticability of forming a national government of the kind the Congress wanted as well as the charge about that the party wanted to dominate such a government.

In these circumstances the whole situation was heading towards a climax when Cripps, reacting to Gandhi's shrewd demand for a transfer of power either to the League or to the Congress, firmly held out by saying: 'We are not going to walk out of India right in the middle of the War though we have no wish to remain there for any imperialistic reasons.' Placed in a dilemma through its faulty assessment of British war capabilities, the Congress was left with no option but to stand on prestige; it took a decision at Wardha on 14 July 1942 to launch a mass movement if the British did not withdraw from India. The decision was approved under Gandhi's influence by the All-India Congress Committee in its meeting held in Bombay on 8 August. This marked the beginning of the Quit India campaign which Gandhi was to lead with a call to the masses to 'do or die'. The Governor-General, who had decided on the action to be taken, had a communique ready in justification of his policy. It was issued in the early hours of 9 August, when Gandhi, Maulana Azad, Sardar Patel, Jawaharlal Nehru and other members of the Congress Working Committee were arrested and taken to Poona by a special train. Dr Rajendra Prasad was arrested in Patna. The movement in fact led to arrests over the whole country. In a statement Jinnah regretted what he considered a most dangerous movement which Congress had chosen to launch despite numerous warnings and advice from various parties, individuals and organizations.

Though controversy exists concerning the real motives[40] of the movement, the lawlessness caused by it is generally acknowledged. It is important to note that the arrest of Congress leaders marked a climb down for Gandhi: his correspondence with the Viceroy amounted in effect to an explanation of the violence that broke out in the wake of the India-wide arrests of Congress leaders. He pleaded not guilty, and expressed a desire for release, which the Government firmly rejected. He then proceeded to fast, but had to remain confined until he was released for health reasons on 6 May 1944. The effects of the Quit India campaign weakened the Congress which, in the absence of any clear line of policy and proper guidance, became disorganized and frustrated under stringent official measures to maintain law and order. During this period of political deadlock the Muslim League grew steadily in strength, a development which contributed in no small measure to India's division.

Even an avowed anti-imperialist writer like R. Palme Dutt characterized the Quit-India resolution of 8 August as absolutely 'ill-judged'. For it provided grounds for an immediate British reaction 'to launch its attack' and use the resolution as an opportunity to expose the Congress leadership as 'pro-Fascist', 'pro-Japanese', and bent on 'sabotaging the war effort of the peoples of the United Nations'.[41] The Government of India thus projected the national leaders' seemingly pro-Japanese move as a justification for its suppressive measures.

After the arrests of Gandhi and other Congress leaders in August 1942, the initiative on the Indian side for resolving the deadlock passed into the hands of a group led by Tej Bahadur Sapru who invited prominent public figures, including Rajagopalachari, to a meeting at Allahabad. The meeting occurred on 12 and 13 December: no formal resolution was adopted, but the gathering sought to resolve the political deadlock and provide a basis for an agreement that might prove generally acceptable. This called for the early summoning of a conference of all parties, including representatives of the Congress and the Muslim League.

It was recognized that the proposed All-Parties Conference had no chance of success unless the British announced that the provisional government to be set up following a possible agreement would have full powers and authority over administration, without any prejudice to the position and authority of the Commander-in-Chief in the efficient prosecution of the War; and that the release of Gandhi and all Congressmen should be ordered to enable them to participate in the proposed Conference. However, the British were not prepared to relent unless Gandhi and the Congress recanted. On his own, Sapru then cabled the Prime Minister, Churchill, pleading, only as a war measure, for the restoration of popular governments in the provinces and a provisional national government at the Centre. The latter was to consist entirely of non-officials drawn from all the recognized parties and communities, and it was to be in charge of all portfolios, subject only to responsibility to the Crown. Sapru's plan was rejected.

There was a serious constitutional difficulty involved in the establishment of a provisional national government with an executive responsible to the Indian Legislature, for the existing Legislature functioned virtually under the 1919 Act. Rajagopalachari thus suggested that the Legislature be based on the British Indian part of the proposed federal legislature, and elected in the same way as the latter. This meant an Assembly (Lower House) of 250 members elected by the Provincial Assemblies, and a Council of State (Upper House) of 156 by direct election. It was an ingenious device—a federal arrangement without the Princely States.[42] But as the participation of the Princes was a necessary condition for the establishment of a federal legislature, Rajagopalachari's suggestion, though worthy of consideration, did not receive the attention it deserved, more especially when the Muslim League and the Princes were both opposed to responsibility at the Centre without 'safeguards'.

It was thus acknowledged that the political deadlock could not be resolved except by settling the issues involved in the League's demand for a separate Muslim state. Gandhi's release from jail in May 1944 afforded an opportunity for a discussion with Jinnah to find a mutually acceptable solution. Knowing, as Rajagopalachari did, the divergence in the views of the two leaders on the subject of Pakistan, he devised a formula intended to preserve unity with consent and understanding and a recognition of the aspirations of the Muslim community for virtually a separate homeland. For some days in September 1944 it formed the basis of prolonged discussions between Gandhi and Jinnah. The formula[43] which Rajagopalachari evolved stipulated that both Gandhi and Jinnah were not only to agree to it, but also to endeavour to get the Congress and the League to approve it. Subject to the terms set out in the formula as regards the constitution for free India, the Muslim League was to endorse the Indian demand for independence and to co-operate with the Congress in the formation of a provisional interim government for the transitional period.

The proposed settlement laid down that, after the War

a commission shall be appointed for demarcating contiguous districts in the north-west and east of India, wherein the Muslim population is in absolute majority. In the areas thus demarcated, a plebiscite of all the inhabitants held on the basis of adult suffrage or other practicable franchise shall ultimately decide the issue of separation from Hindustan. If the majority decide in favour of forming a sovereign State separate from Hindustan, such decision shall be given effect to, without prejudice to the right of districts on the border to choose to join either State.

It was to be open to all parties to advocate their respective points of view before the plebiscite was actually held. And in the event of separation being decided upon by the plebiscite so held, mutual agreements were to be entered into to safeguard defence, commerce and communications and other essential purposes. The transfer of populations, if any, was to be on an absolutely voluntary basis. And, above all, these terms were to be 'binding only in case of transfer by Britain of full power and responsibility for the governance of India'.

The formula remained inoperative. While the nationalists viewed it as a formal sanction for India's vivisection, Jinnah characterized it as putting the Lahore Resolution out of shape and mutilating it; for that Resolution pertained to contiguous Muslim-majority provinces and not Muslim-majority districts, which Rajagopalachari had suggested. Jinnah held on to his own interpretation of the Pakistan Resolution throughout the unsuccessful negotiations.

Consistent with the terms of Rajagopalachari's formula, Gandhi was willing to concede self-determination to Muslim-majority areas, for he seemed to have pinned hope on projected 'common arrangements between the two areas in regard to defence, commerce and communications', etc., which were to be entrusted 'to a common central administration, however loosely knit'. But Gandhi's self-determination was vitally linked with the formation of a national government for the interim period. He wanted Jinnah to associate himself (a) with the demand for the immediate declaration of independence, to become operative immediately after the War; (b) the formation of a real national government, except for reservation in regard to defence, and (c) the release of Congress leaders. It was hard for Jinnah to agree to any of the three terms.

The Gandhi-Jinnah talks in Bombay naturally failed to produce any positive result. The stalemate continued, although in a letter of 24 September 1944 Gandhi sought to explain his position by offering terms which modified Rajagopalachari's formula. The appointment of the commission for the demarcation of Muslim-majority areas was to be approved by the Congress and the wishes of the inhabitants were to be ascertained through adult franchise. In case the voting went in favour of separation, these areas could form a separate state only after India was free of foreign domination. Even so, as the letter added, 'There shall be a treaty of separation which should also provide for the efficient and satisfactory administration of foreign affairs, defence, internal communication, customs, commerce and the like, which must necessarily continue to be matters of common interest between the contracting parties'. It was laid down that, on the acceptance of this agreement, the Congress and the League 'shall [both] decide upon a common course of action for the attainment of independence of India', even though the League had not agreed to participate in 'direct action'.[44] Though somewhat vague, Jinnah's response to matters of common interest did not seem to rule out some co-operation and understanding with the Congress for an interim period.

The Congress continued to be obsessed with the idea of a national government responsible to the Legislature during the War, even though the viability of the idea was being questioned on both constitutional and pragmatic considerations. This obsession not only resulted in Congress leaders being cast into prison without any resulting gain, but also led them to negotiate from a position of comparative weakness with Jinnah, who had already time and again dubbed responsible government as a device of 'Congress Raj' These negotiations added to his importance without any gain to the Congress. A practicable measure in the existing situation was the formation of a national government consisting of representatives of all the parties, which would be irremovable by the Legislature during the interim War period, though remaining technically responsible to the Crown. This would of necessity have left no alternative to the League but to get yoked with the Congress in the formation of such a government at the Centre, with the prospect of Section 93 being replaced by popular Congress ministries in the provinces. But Gandhi remained wedded to a political approach to constitutional advance, where no Congress leader was a match for Jinnah, whose strength additionally flowed from his proximity to the ruling clique in the government and bureaucracy.

The Wavell Plan and Simla Conference

The Wavell Plan and the Simla Conference constituted yet another attempt to resolve the deadlock in the summer of 1945. It represented no constitutional advance, but sought to prepare the ground for it within the framework of the proposals made by Cripps in 1942. Limited in its scope, the Plan related to the formation of a provisional interim government at the Centre, a subject which had become the hub of political controversy in India, particularly since Gandhi's release from jail in May 1944. The world situation had changed a good deal since 1942. Statements by Churchill, Roosevelt and Stalin at the Yalta Conference on 12 February 1945 and the assured victory of the Allied forces in the European war were soon followed by a thaw in the rigidity of British policy towards India. Wavell, who had been sworn in Viceroy on 20 October 1943, left for London on 22 March 1945 for consultations with the British authorities.

The Plan in its final form indicated the manner in which the interim arrangement was to be made. It was proposed 'that the Executive Council should be reconstituted and that the Viceroy should in future make his selection to the Crown for appointment to his executive from amongst leaders of Indian political life at the Centre and in the provinces, in proportions which would give a balanced representation of the main communities, including equal proportions of Moslems and Caste Hindus'. In Wavell's own words, his Plan 'is not a constitutional settlement, it is not a final solution of India's complex problems that is proposed. Nor does the Plan in any way prejudice or prejudge the final issue. But if it succeeds, I am sure it will pave the way towards a settlement and will bring it success'.[45]

In a broadcast on 14 June 1945 the Viceroy explained the policy announcement of the Secretary of State (L.S. Amery), who made no reference to the Congress demand for 'independence', and pointed out that, although the Plan had been formulated under the existing constitution, it represented a definite advance on the road to self-government; for the complexion of the Viceroy's Executive Council was to be entirely Indian except for the Viceroy and the Commander-in-Chief who would retain his portfolio as War Member and the Viceroy his responsibility in the choice of his Council.

Consistent with the policy indicated in the Wavell Plan, members of the Congress Working Committee were released on 15 June, so that they could attend the Conference scheduled to be held at Simla. Invitees to the Conference included, besides Gandhi, Jinnah and the Congress President, Abul Kalam Azad, the 'Premiers' of provincial governments, past and present; the leaders of the Congress and the Muslim League in the Central Legislative Assembly and the Council of State; the leaders of the Nationalist Party and the European group in the Assembly; and Rao Bahadur N. Siva Raj, representing the Scheduled Classes, and Master Tara Singh, representing the Sikhs. With Wavell in the chair, the Conference began at Simla on 25 June 1945.

Differences, however, soon arose between the Congress and the League. The understanding earlier reached between the parliamentary leader of the Congress in the Assembly (Bhulabhai Desai) and the parliamentary leader of the Muslim League (Liaqat Ali Khan) was on the basis of Congress-League parity in the formation of a provisional government at the Centre. But the principle of parity included in the Plan was a Caste Hindu-Muslim parity. It meant that the Congress would either have to get relegated to the status of a Hindu organization, or, by claiming one of the Muslim seats for a Congress Muslim, violate the basis of the parity with the League. Alternatively, the acceptance by the League of a Congress Muslim to one of the Muslim seats was viewed as surrendering parity and accepting an inferior position to the Congress.

The Conference was adjourned a number of times informally to bring about agreeement between the Congress and the League. But none of the efforts could produce any positive result. The representatives of the parties invited to the Conference were then asked to submit their respective lists of names for the Executive Council. The Viceroy was to make the final choice in consultation with their leaders. All the parties represented at the Conference submitted their lists, except for the European group and the Muslim League.

In a letter of 6 July 1945, Jinnah informed the Viceroy that his panel of names should be chosen in a personal discussion with him, but that all the Muslim members must without fail be chosen from the League itself, and not from any other organization, Hindu or Muslim. The Viceroy met Jinnah on 9 July and discussed the matter with him. But he could not guarantee that all the Muslim members would be from the League alone. Jinnah refused to submit any list. The Viceroy appeared to make his own selection, but, instead, met Jinnah again after a couple of days to tell him that he was prepared to have four members of the League in addition to one Muslim member from the Unionist Party of Punjab. But Jinnah would not agree to any Muslim except from the League. The Viceroy did not acquiesce, but his tame handling of the situation had already given Jinnah a long rope which he used to throttle the Plan. In fact, Jinnah used a veto to create an impression that no constitutional progress was possible in India without the League's consent.[46]

Jinnah was perhaps apprehensive that, if he accepted the interim arrangements provided by the Wavell Plan, the case of Pakistan might be weakened or even shelved should communal harmony develop in the course of working together and sharing the experience of administrative realities.[47] Jinnah, therefore, considered it tactically advantageous to keep the League away from the seats of power by an intransigence that encouraged a resurgence of Hindu communalism. Wavell's surrender to Jinnah's tactics thus fuelled the forces of communalism, and the sharpening of social divisions became the easy road to power for some politicians. Whether this surrender formed part of British policy at the time, or whether it was a mere index of the Viceroy's own weakness needs to be ascertained. But it seems indisputable that he allowed the League to use a veto against the majority's wish to form a provisional government according to his own Plan, a government which, in his own words, was intended to pave the way for India's unity and self-government. The failure of the Simla Conference, which made its own contribution to the aggravation of communal tension, helped create favourable conditions for the League, which won a landslide victory in the Muslim-majority areas in the 1945–6 elections. It provided an electoral sanction to its demand for separation. In the face of the League's electoral success, the Labour government now in power in Britain, could not put the clock back on the side of Indian unity.

The Congress Move Towards Power

While the Muslim League was seeking to achieve a separate sovereignty on communal lines, the Congress proceeded to regain its importance by taking advantage of the situation emerging from a new social challenge that demanded increased employment, education and peacetime economic development. This challenge had to be met in order to make sure that extremist left elements did not exploit the situation to their own benefit. A pamphlet addressed to Congress Committees pointed out that the worsening economic condition in India, more especially the desperate food problem, was a 'credit item', an asset to the Congress, in that 'these mounting miseries will drive the people deeper into the ranks of the rebellion and create those forces of chaos and disorder which when canalized and manifested in organized political action will bring the foreigner's rule to a standstill'. Congressmen were therefore advised not to participate in any of the food committees set up by the government or its supporters so long as 'a real Swaraj Government' was not established.[48]

The Congress pamphlet recognized at the same time the impracticability of the 'Quit India' movement, which had depleted party resources in terms of both men and money. Indian capitalists, who earlier contributed liberally to the Congress exchequer now 'thought it bad investment to apply funds to an enterprise that does not appear to them to be profitable any more'. Thus, although the Congress still stood by its August Resolution, it saw no advantage in holding to it 'in the absence of immediate mass response'.[49]

British policy, too, was not without its own compulsions which called for a change of attitude. It could not but take into account the likely desperate state of Britain's post-War economy, the serious danger of unemployment arising from demobilization and, above all, the weakening of the European arm of the Indian Civil Service which, besides being already depleted thanks to the stoppage of recruitment during the War, needed to be reinforced to a degree which Britain could not afford. Moreover, India which had in the past been a debtor country, was emerging in the course of the War as a creditor, her sterling balance having increased from about £55,500,000 on 31 March 1939 to about £745,000,000 at the end of May 1944.[50] These developments cried aloud for collaboration with India, more especially for peace with the Congress, as most British capital investment in India existed in the Hindu-majority areas where the Congress happened to be the ruling force.

Consistent with a policy of collaboration, the Viceroy appointed Sir Ardeshir Dalal in June 1944 to his Executive Council as Planning Minister, to secure the execution of a Five-Year Plan he had in mind. Sir Ardeshir belonged to the House of Tata, and along with seven other fellow industrialists and associates, had produced the famous Bombay Plan to modernize India in the course of fifteen years. Sir Ardeshir's appointment had a political significance; despite his loyalty to the Congress even in 1942, he was expected to establish a link with Tatas.[51] Through this appointment, Wavell wanted to gain the co-operation of nationalist leaders in such matters as food supply and famine relief, agricultural production and industrial development. In fact, the Simla Conference was an expression of his wish to reach a political settlement for national collaboration.

The collaboration between the government and the Congress was not to be confined only to the social and economic spheres. Their joint efforts were directed on a political plane against radical leftists, more especially the Communist Party of India (CPI), by far the most efficiently organized of the Indian political parties, which, since the lifting of the ban on it in July 1942, had been emerging as a counterforce to the Congress.

The legalization of the CPI proceeded from its 'People's War' slogan which the Government exploited in the mobilization of its War effort and the maintenance of internal security. This, however, led to considerable increase in its strength, especially among industrial workers, poor peasants and college students.[52] But the Party in its legal form later extended its membership in the armed forces, a development which seriously worried the government. A secret report of early 1945 showed that communist cells had already started functioning in the army, though known communists took special care to be good soldiers.

The victory of the Allied armies in Europe, which restored confidence in the British capacity to defend India against aggression, created a situation where communist help in maintaining internal security could easily be dispensed with. But more important than the Allied victory was Britain's anti-communist swing in policy. It was feared that, with the end of the War the CPI would 'look around for ways of fishing from the troubled post-War waters' advocating 'communization', with such 'twists' as might attract '(a) labour and (b) peasants'. The government and the Congress both shared this view and even suspected possible trouble from the demobilized Indian soldiery.[53]

It is significant that these shifts in policy were taking place noticeably between May 1944 and June 1945, the period that began with Gandhi's release and ended with the Simla Conference. This was precisely the period when Gandhi was attempting to prove the *bona fides* of his intention to co-operate with the government and the London *Times* was emphasizing the need to promote the principle of equal 'partnership' between Britain and India, and to 'broaden the whole content of the Indian national programme'. During this period Congress leaders started speaking against the communists and excluded them from the newly constituted Congress Committees, especially in the Congress-dominated provinces. While dealing with Congress-Communist relations a Home Department confidential survey of 30 June 1945 thus pointed out:[54]

> The Congress attitude towards the Communist Party, never very friendly, became progressively hostile during the period under review. By the end of January 1945, the communists were debarred from membership of the newly formed Congress organizations and from participation in the Congress 'constructive programme' in all the provinces except those where Congress was in a relatively weak position. In Bengal, Sind, Orissa, Assam and the North-West Frontier, no formal ban was imposed upon the communists, but there prejudice against them was no less prevalent. . . . The communists might have found themselves excluded altogether from the Congress but for a press statement issued by Sarojini Naidu on 18 December [1944] categorically affirming that nobody who signed the Congress pledge could be debarred from primary membership; she was not prepared, however, to oppose the exclusion of communists from *ad hoc* Congress Committees. Later, the utterances of Jawaharlal Nehru, Acharya Narendra Deo and others left little room for doubt that the communists had incurred the serious displeasure of the higher Congress leadership.

The Politics of Power and Party Rule 1937–47

This hostile Congress attitude towards the CPI proceeded principally from two premises, namely, the organizational strength which the Party had acquired by 1945, and the danger of its emerging as a potential competitor for power in the post-War period. During this period there had also developed an identity of interest between right-wing nationalists who controlled the Congress, and the government, which was concerned with the security of British capital investment in India. Both looked upon the communists as their common enemy. The victory of the Labour Party in July 1945 and Wavell's assurance to Gandhi that the situation arising from the failure of the Simla Conference would be reviewed, suggested a new move towards collaboration with the Congress. The Viceroy was later invited to London for a review of the situation in terms of both its economic and political implications. Lord Pethick-Lawrence, the Secretary of State in the new Labour Government, had reiterated in a statement of 7 August his conviction in an 'equal partnership' as the basis of collaboration between Britain and India; and at the opening of Parliament his Government expressed its determination to promote, in consultation with Indian leaders, the early realization of full self-government for India. An announcement was accordingly made to hold elections to the Central and provincial legislatures in the coming winter. On 21 August the Viceroy, too, announced that elections would be held. Wavell left for consultations in London on 24 August 1945.

The Labour Government discussed with Wavell the steps to be taken to achieve its declared objective. Apart from holding elections, the steps formulated in the discussions included (1) the formation immediately after the elections of a constitution-making body, either on the basis of the Cripps offer of 1942 or some other method; (2) preparatory discussions between the Viceroy and representatives of the Indian States about the mode of their participation in the constitution-making body; (3) consideration by the Government of the contents of the treaty required to be concluded between Britain and India; and (4) bringing into being, just after the elections, of an Executive Council with the support of the main Indian parties, so that Indians themselves might immediately proceed to deal with the country's social and economic problems. The basic policy objective and the steps proposed to accomplish it were announced in a broadcast by the Viceroy on 19 September 1945. It created hope for a better future and held out the promise of power for those who had earlier been incarcerated in connexion with the Quit India movement.

Limitations of the Measures Announced

The measures announced on 19 September 1945 suffered from certain limitations which, though well-intentioned, perhaps also led to an aggravation of the communal frenzy underlying India's later division. The formation of the constitution-making body was, for instance, left to be decided either on the basis of the Cripps offer which, as said before, had been turned down by the League, or by some other alternative or modified scheme, which, instead of being speltout, was left to be evolved by a process of discussion. Unlike the Cripps plan, the mode so suggested not only suffered from vagueness, but encouraged communal political propaganda in the coming elections, with little or no prospect of agreement between the two major parties competing for power.

Perhaps seeking an earlier seizure of power, the Congress declared that the announcement of 19 September was 'vague, inadequate and unsatisfactory'. For while the formation of the proposed Executive Council after the projected general elections was welcomed by the League, the Congress wanted popular governments to precede and not follow the general elections. By a resolution adopted four days after Wavell's broadcast of 19 September, the Congress Working Committee in Bombay expressed concern at the British government's intention of going ahead with its plan to hold general elections with a limited franchise. The Congress emphasized that the existing franchise should be fully revised and the electoral constituencies reformed by popular governments to be installed immediately at the Centre and the provinces. The Congress perhaps feared that, in the existing poor state of its organization, and in view of the political situation which was developing on communal lines, the results might tilt the balance in favour of the League and prejudice its own case. It therefore wanted to strengthen its position before a general election was ordered, and this could not be achieved except by immediate restoration of the popular ministries in the provinces and the formation of popular government at the Centre. The British government declined to concede this demand for fear of appearing partial to the Congress, although its rejection amounted to a concession to the forces of communalism which were building up steadily.

Despite the League's opposition, the Congress sought to forge ahead and build mass pressure to wrest the substance of independence. Thus, its defence of the Indian National Army's accused and the revolt of naval personnel caused considerable political heat and feeling. But its road to power was blocked by the refusal of the British government to effect any interim arrangement, more especially of the kind that the Congress wanted, before the elections were over and the results known. In a statement of 4 December 1945 the Secretary of State made it clear that the question of 'the future governance of British India' was to be decided 'by, and in consultation with, the elected representatives of the Indian people'. This clearly meant that independence was to follow, not precede, the framing of the Constitution by India's elected representatives. The Congress demand was the other way round. It wanted the substance of independence transferred to a provisional government acting as the precursor of a full independence that would emerge from the deliberations of the projected constitution-making body. Britain did not concede this demand from considerations of constitutional propriety; for it meant transferring *de facto* sovereignty to India with *de jure* sovereignty still remaining with the British Parliament, an arrangement which involved divesting responsibility from power.

The General Elections and Partition

The general election made its own contribution towards Partition. The results not only provided an electoral sanction for the League's claim to a separate sovereign Pakistan in the Muslim-majority areas, but also reinforced the Congress demand for a united India. The questions of unity and division were both approached from a communal angle, and the elections involved the common people in the power politics of political parties. The election result showed that only the Congress and the Muslim League had emerged as competing forces to reckon with. Their attempts to outmanoeuvre each other on the score of their electoral successes in their respective spheres of influence could not but lead to Partition.

The election result to the Central Legislative Assembly became available towards the close of December 1945.[55] The Congress came out triumphant in the General constituencies, securing 93.3 per cent of the votes cast in non-Muslim constituencies. The Muslim League won every reserved Muslim seat, securing 86.6 per cent of the total votes cast in such constituencies. The Nationalist Muslims forfeited their deposits in several cases. The final figures were: Congress 57; Muslim League 30; Independents 5; Akali Sikhs 2; and Europeans 8, making a total of 102 elected seats. In the previous Assembly the figures at the time of dissolution were: Congress 36; Muslim League

25; Independents 21; Nationalist Party 10, and Europeans 8. While the Nationalist Party of Madan Mohan Malaviya disappeared, the Independents lost heavily in the political polarization and Hindu Mahasabha candidates and other opposition nominees withdrew in most cases. The Congress and the Muslim League were thus the only parties that counted in the country.

In the provincial elections[56] the Congress emerged victorious in all the General constituencies, while the Muslim League held its ground in the reserved Muslim constituencies. Congress ministries were thus formed in seven of the eleven provinces—Assam, Bihar, Bombay, C.P., Madras, Orissa and the U.P. In the North-West Frontier Province the Congress won 30 seats (of which 19 were Muslims) as against 17 captured by the League, but the province was later overwhelmed by communal propaganda and the political influence of the League. As we have already seen at the start of this Chapter, the Muslim League swept the polls in all reserved seats in the provinces, winning 428 out of a possible 492 provincial seats reserved for Muslims, as against 109 in the elections of 1937. A secret report of the Punjab Government for the second half of February 1946 commented thus: 'The League successes, the failure of nationalist Muslims and Ahrar candidates and the virtual elimination of the Unionist Party have been an unpleasant surprise to Congress, and Vallabhbhai Patel has complained that "all our efforts and resources have been wasted and hopes given were false and calculations and expectations were wrong". However, he has found some comfort in the possibility of a Congress-Akali-Unionist Coalition.'[57] A Congress-Akali working alliance was doubtless formed, but it did not last long. Power politics gave way to shifing party loyalties which, in turn, enabled the League to gain an absolute majority and form its own government.

In the Bengal Assembly, of 250 members the Muslim League won 113 of the total 119 Muslim seats. The Congress obtained 87. H.S. Suhrawardy was invited to form a ministry. He held negotiations with local Congress leaders to form a coalition ministry, but the Muslim League eventually formed a ministry with the support of Independents, including Europeans, when Suhrawardy's negotiations with the Congress failed. The League had made a similar offer of a coalition with the Congress in Punjab, too. Here negotiations also failed, for the Congress claimed parity in the ministry for the Congress-Akali alliance and freedom to nominate as ministers those belonging to any community, and not only Hindus or Sikhs as the League insisted. A third point of contention was Congress opposition to any discussion of an extra provincial question like Pakistan, which was of vital interest to the League. The wisest course in the interests of unity and understanding would have been to accept the League's offer of a coalition arrangement on its own terms and then proceed to work that arrangement in a manner capable of creating confidence. But power politics, in the absence of a broader national perspective, did not permit such a course and the League went its own way to achieve Pakistan.

The Cabinet Mission

Consistent with his earlier announcement, the Viceroy was confronted with the task of drawing up a programme of action following the declaration of the election results. The Viceroy had prepared an analysis of India's political situation, in which he described the Congress as the only well-organized party and one that commanded the support of practically all Caste Hindus, a party with the financial backing of all the business magnates, who were anxious to ensure a peaceful solution of India's constitutional problem. As regards the Muslim League, Wavell felt that, if Jinnah again refused to participate in the interim government, he would tell him that the Government would be compelled to go ahead without the League.

The Secretary of State, Lord Pethick-Lawrence, and the Cabinet agreed with the Viceroy generally, but were sceptical about his proposal to negotiate in distinct stages. They argued that, once negotiations with the leaders began, they were bound to cover simultaneously all the stages as a whole. Instead of these negotiations being left to be conducted by the Viceroy alone, as declared by the policy statement of 19 September 1945, the British government decided to send a mission of cabinet ministers, of whom the Secretary of State was to be one, to conduct the negotiations in association with the Viceroy. The decision to send a Cabinet Mission to India was announced on 19 February 1946 simultaneously by Lord Pethick-Lawrence in the House of Lords and Prime Minister Attlee in the House of Commons. The members of the Mission were to be Pethick-Lawrence, Sir Stafford Cripps, President of the Board of Trade, and A. V. Alexander, First Lord of the Admiralty.

During a debate in the House of Commons on 15 March the Prime Minister made it clear that his cabinet colleagues were going to India to help the country attain freedom as speedily and as fully as possible. It was for India to decide on the form of government that was to replace the existing regime. But he concluded with a significant remark, saying: 'We are mindful of the rights of the minorities and the minorities should be able to live free from fear. On the other hand, we cannot allow a minority to place their veto on the advance of the majority.'

The Cabinet Mission arrived in New Delhi on 24 March 1946. After conferring with the Viceroy on their future programme and acquainting themselves with the Indian situation generally, its members held a series of interviews with Indian political leaders and other important people. The interviews did not produce promising results and personal contacts with individual leaders proved equally fruitless. Jinnah remained firm in his demand for Pakistan, and opposed to any proposal for a common legislature or executive. A three-tier scheme was later evolved, with a federal Indian Union to be composed of two parts legislating for optional subjects, each unit with two legislatures under a Union Executive and Legislature above them for compulsory subjects, such as defence, foreign affairs and communications.

The scheme so formulated by the Mission was based on certain fundamental principles. The future constitution of British India was, for instance, to comprise: (a) a Union Government dealing with defence, external affairs and communications, and (b) two groups of provinces, one predominantly Hindu, and the other predominantly Muslim, dealing with all other subjects which the provinces in the respective groups desired should be dealt with in common. The provincial governments would deal with all other subjects and would have all the residuary sovereign rights. Under this arrangement, the Indian States were to take their appropriate place in the structure on terms to be negotiated with them.

The Cabinet Mission met at Simla and discussed the scheme with representatives[58] of both the Congress and the League between 5 and 12 May 1946. But disagreement between these two parties continued. In these circumstances, the Mission was obliged to announce for acceptance by the British government its own ideas on the basic form which the new constitution should take.[59] These were part of a statement issued by the Cabinet Mission and the Viceroy on 16 May 1946. They recommended that:

(1) There should be a Union of India, embracing both British India and the States, which should deal with foreign affairs, defence, and communications, and which should have the powers necessary to raise the finances required for these subjects;

(2) The Union should have an Executive and a Legislature constituted of representatives from British India and the States. Any matter involving a major communal issue in the Legislature should be decided upon by a majority of representatives, present and voting, of each of the two major communities, as well as a majority of all the members present and voting;
(3) All subjects other than the Union subjects and all residuary powers should vest in the provinces;
(4) The States were to retain all subjects and powers other than those ceded to the Union;
(5) The provinces should be free to form Groups with executives and legislatures, and each Group could determine the provincial subjects to be taken in common;
(6) The constitution of the Union and of the Groups should contain a provision whereby any province could, by a majority vote of its Legislative Assembly, call for a reconsideration of the terms of the constitution after an initial period of ten years and at ten-yearly intervals thereafter.

As the Mission had not succeeded in making the two major communities join in setting up the constitution-making machinery which they proposed be brought into being immediately, they recommended that representatives of the provincial legislatures, elected roughly in the ratio of one to a million of population and allocated to each community in proportion to its population in the province, meet at New Delhi as one body, together with representatives of the Indian States. After electing a chairman and completing the necessary preliminaries, these representatives were to separate into three sections as follows: Section A, comprising Madras, Bombay, U.P., Bihar, C.P. and Orissa; Section B, comprising Punjab, the North-West Frontier Province, Sind and Baluchistan; Section C, comprising Bengal and Assam.

These three sections, which were to form the Constituent Assembly, would decide on the constitutions for the provinces in their group. They were also to decide on whether any Group constitution should be set up, and, if so, to determine the subjects with which the Group as a whole should deal. It was provided that a province would have the power to opt out of a Group by a decision of its new legislature when the new Union constitution came into force. The three sections of the Constituent Assembly would, after the settlement of the Group constitution, reassemble, together with representatives of the Indian States, to frame the Union Constitution. This Constituent

Assembly was to negotiate, in its capacity as a sovereign body, a treaty with the United Kingdom to provide for such matters as would arise from the transfer of power.

The Viceroy was advised forthwith to request the provincial legislatures to proceed with the election of their representatives according to the following table of representation:

Section A

	General	Muslim	Total
Madras	45	4	49
Bombay	19	2	21
U.P	47	8	55
Bihar	31	5	36
C.P.	16	1	17
Orissa	9	0	9
Total	167	20	187

Section B

	General	Muslim	Sikh	Total
Punjab	8	16	4	28
N.W.F. Province	0	3	0	3
Sind	1	3	0	4
Total	9	22	4	35

Section C

	General	Muslim	Total
Bengal	27	33	60
Assam	7	3	10
Total	34	36	70

Total for British India	292
Maximum for the Indian States	93
	385

Apart from the fact that most of the representatives of the Indian States were non-Muslim, the percentage of Muslim representatives from British India in the 292-member Constituent Assembly was only 26.7. However, the possibility of a province opting out of its Group after the Union constitution came into force could threaten the stability of Sections B and C; here the non-Muslim minorities in Jinnah's projected Pakistan of six provinces was bound to be very considerable, as the following figures based on the 1941 census show:

North-western Area		Muslim	Non-Muslim
Punjab	...	16,217,242	12,201,577
N.W. Frontier Province	...	2,788,797	249,270
Sind	...	3,208,325	1,326,683
British Baluchistan	...	438,930	62,701
Total		22,653,294	13,840,231
Percentage		62.07%	37.93%

North-eastern Area		Muslim	Non-Muslim
Bengal	...	33,005,434	27,301,091
Assam	...	3,442,497	6,762,254
Total		36,447,931	34,063,345
Percentage		51.69%	48.31%

These figures led the Cabinet Mission to conclude that the setting up of a separate, sovereign state of Pakistan in the north-west and north-east on the lines claimed by the Muslim League would not solve the communal problem: nor did it see any justification, on Jinnah's own logic for a two-nations theory, to include within a sovereign Pakistan those districts of Punjab, Bengal and Assam in which the population was predominantly non-Muslim. A smaller sovereign Pakistan confined to the Muslim-majority areas alone was, on the other hand, regarded by the Muslim League as impracticable; for it involved the exclusion of (a) the whole of the Ambala and Jullundur Divisions of Punjab; (b) the whole of Assam, except the district of Sylhet; and (c) a large part of West Bengal, including Calcutta where Muslims formed only 23.6 per cent of the population. Besides, in view of their common languages, cultures, shared history and traditions, the Mission felt that any radical partition of Punjab and Bengal would be contrary to the wishes and interests of a very large proportion of these provinces. Neither a larger nor a smaller sovereign state of Pakistan was thus recommended as an acceptable solution for the communal problem.

In fact, the Cabinet Mission ruled out the division of the country into two sovereign states and recommended only one Constituent Assembly as a sovereign body invested with the making of India's constitution. While the constitution-making process was in progress, administration was sought to be carried on by setting up an Interim Government with the support of both the Congress and the League, a Government in which all the portfolios, including that of War Member, were to be held by Indians. Despite the League's reluctance to join it, the Mission and the Viceroy went ahead with the formation of the new Government. On 16 June they issued a statement announcing their determination to set up an Executive Council of fourteen persons, all of whom were mentioned by name. Six belonged to the Congress, including a representative of the Scheduled Castes, five to the Muslim League, there was one Sikh, one Indian Christian and one Parsi. The list included the names of both Nehru and Jinnah. The statement made it clear that, if any of those invited were unable to join the Council, the Viceroy would invite another person to take his place. It was also emphasized that, if the two major parties, or either of them, proved unwilling to join, the Viceroy would proceed with the formation of an Interim Government which would be as representative as possible of those willing to accept the statement of 16 May.

Difficulties soon arose as the Congress insisted on including a nationalist Muslim as part of its own quota. The Muslim League would not accept this, although it later accepted the statement of 16 June. The Congress broadly accepted the statement of 16 May, but on 29 July the League resolved to reject it. Since both statements together formed an integral whole, constitution-making was delayed. In the mean time, the Mission left for London on 29 June.

The statement of 16 June was not based on a Hindu-League parity. Under it, the Congress could in principle include a nationalist Muslim. And since it had accepted the Mission's scheme of 16 May, the Viceroy could under the 16 June statement go ahead with the formation of the Interim Government. To block this possibility, the Muslim League decided to observe 16 August as 'Direct Action Day', with processions and meetings in almost all the major towns. Suhrawardy, the Bengal Premier, even stated that Bengal would declare independence if the Congress were put into power. The situation so created led to the unprecedented Calcutta killings, arson and loot.

Against this background, Nehru was invited to make proposals for the formation of an Interim Government, over which the Viceroy was to retain his veto. Jinnah's co-operation was sought, but without success. The Viceroy thus decided to go ahead without the League, although efforts to secure its participation were never given up. On 24 August a press *communique* was issued stating that the King had accepted the resignations of the existing members of the Governor-General's Executive Council and that, in their place, his Government had appointed the following: Pandit Jawaharlal Nehru, Sardar Vallabhhai Patel, Dr Rajendra Prasad, M. Asaf Ali, C. Rajagopalachari, Sarat Chandra Bose, Dr John Mathai, Sardar Baldev. Singh, Sir Shafaat Ahmad Khan, Jagjivan Ram, Syed Ali Zaheer and Cooverji Hormusji Bhabha. The *communique* added that two more Muslim Members would be appointed later and that the Interim Government would take office on 2 September.

The Calcutta holocaust and the formation of the Interim Government produced a sobering effect on Jinnah who then expressed his desire to co-operate to the Viceroy. The Viceroy made it clear that entry of the Muslim League in the Interim Government would be subject to Jinnah's acceptance of the long-term plan of 16 May. Jinnah agreed to this and on 14 October sent to the Viceroy the names of five nominees, namely, Liaqat Ali Khan, I.I. Chundrigar, Abdur Rab Nishtar, Ghazanfar Ali Khan and Jogendra Nath Mandal, a Scheduled Caste member who and been a minister in the Muslim League ministry of Bengal. Since two Muslim seats were vacant, the three additional League nominees were accommodated in place of Sarat Chandra Bose, Sir Shafaat Ahmad Khan and Ali Zaheer, who were allowed to retire. Both the Mission's plan for a Constituent Assembly and an Interim Government thus seemed to be materializing.

India Divided

Although the Cabinet Mission's plan was geared to an acceptance of an Indian Union, India was divided. The circumstances responsible for division had been developing over the years, more especially since 1937, when the Congress rejected proposals for coalition governments in the United Provinces and Bombay. Division, however, became immediately perceptible with the formation of the Interim Government. The elections of 1945-6 had reinforced by means of electoral sanctions the mutually exclusive claims of the two major parties; in their efforts to outmanoeuvre each other, they then proceeded to employ extra-constitutional methods through their respective volunteer organizations acting as private armies and providing physical force to back up the demands of their parties.[60] The League's 'Direct Action Day' of 16 August marked the beginning of the new trend, which produced its chain reaction in other parts of the country. A strong government was needed to curb the growing lawlessness, but the Interim Government was a house divided against itself and the authority of the British in India was already on its way out. Sane perspectives were the need of the hour, but the quest for power dominated the vision of India's political leaders. The Constituent Assembly resulting from the Cabinet Mission's plan opened on 9 December 1946 under the competitive stress of power politics. Dr Sachchidanand Sinha from Bihar was its temporary Chairman, and he made over to Dr Rajendra Prasad a couple of days later as permanent Chairman. On 13 December, Jawaharlal Nehru moved a resolution outlining the objectives of the Assembly in terms of a sovereign, independent republic of India, and indicating the reluctance of the Congress to wait for the League to join in. Realizing the danger to India's unity as a consequence of that resolution, M.R. Jayakar moved an amendment to postpone its consideration until representatives of the League and the States came in. The Assembly consequently adjourned on 23 December until 20 January 1947.

A controversy had in the meantime developed on the interpretation of Grouping, a basic feature of the Cabinet Mission's statement of 16 May. By grouping Sections B and C together, the plan provided for a second federation to be invested under their Group Constitution with all such powers as did not belong to the Indian Union—an ingenious compromise and a substitute for a separate, sovereign Pakistan. This basic cornerstone of unity, which was reiterated by the British government in a statement of 6 December, only three days before the opening session of the Constituent Assembly, was challenged by the Congress Working Committee in its meeting of 22 December. It declared Grouping to be a violation of the fundamental principle of provincial autonomy. In a meeting at Karachi on 29 December, the League's Working Committee reacted strongly and characterized the Constituent Assembly assembled in Delhi as 'a rump', an illegal body totally different from that contemplated by the Cabinet Mission. The League decided not to convene the proposed meeting of its Council to reconsider its earlier decision against participation in the Constituent Assembly. It objected to 'the advice reported as having been tendered to Assam by Mr Gandhi that that province should withdraw from the Eastern Section'.⁶¹ Jinnah viewed this as an attempt to recognize the right of veto for a province included within a Section. The two parties thus moved to positions of no retreat. The only alternative left was to decide issues by a trial of strength which, in the given situation, meant a degeneration into organized communal violence.

The League's refusal to enter the Constituent Assembly also perhaps proceeded from Congress manoeuvres to win over the States on communal lines. The Cabinet Mission had provided that in the preliminary stage the States would be represented in the Assembly by a States' Negotiating Committee which was not to be too large, but large enough to give every State the confidence that its interest would be duly protected. In consultation with the Princes the Political Department of the Government finalized a representative committee of sixteen with the Nawab of Bhopal as Chairman. In the organization of this States' Committee 'Mr Nehru even had to concede that the monarchical form of government in the States was not being questioned and that negotiations for the representation of the States in the Constituent Assembly would be on this basis'.⁶²

However, the Rulers did not stand by their Negotiating Committee, which led to the disintegration and loss of their bargaining position. The Maharaja of Baroda took the exclusion of his Dewan from the Committee as 'an insult to his State, so he decided that he would deal direct with the Constituent Assembly' to obtain 'a better bargain'. With growing dissension between members of the Congress and the League in the Interim Government 'it was being whispered into the ears of Hindu Rulers that they might be ill-advised to continue negotiations through the States' Committee of which the Chairman and spokesman, the Nawab of Bhopal, was a Muslim. Soon, one by one, helped at times by the advice of Dewans whose future might depend on the goodwill of Congress leaders, and led by the Maharajas of Bikaner and Patiala, the Rulers followed Baroda and decided to join the Constituent Assembly without conditions. The States' Negotiating Committee soon ceased to exist.'⁶³ This was a serious blow to the Muslim League, for if the States had entered the Assembly as a bloc, their representatives plus those of the League would have formed nearly half the total in the Assembly and the possibility of a compromise constitution being evolved would have induced the League to participate. The collapse of the Princes ruled out such a possibility; despite the League's non-cooperaton, the Congress had started forging ahead as a superior force.

The chain reaction of communal riots which emerged in the wake of the League's 'Direct Action Day' called for a new approach to political negotiations, especially in the context of the British decision to withdraw. In an interview with George VI on 17 December Prime Minister Attlee made it clear that, although Wavell had done good work, it was doubtful that he had 'the finesse to negotiate the next step when we must keep the two Indian parties friendly to us all the time'.[64] Attlee's doubts about Wavell's political insight dated back to the time of the Cabinet Mission. Wavell had then suggested that, should an outbreak of violence arise in the event of negotiations breaking down, the Hindu-majority provinces should be handed over to the Hindus, and British troops and officials withdrawn into the existing Muslim-majority provinces included in Sections B and C, with attempts made to hold on to Delhi. The object of his scheme was to continue British rule in the Pakistan part of the Indian provinces and to protect it with adequate land, sea and air force from Britain. In his recommendation he was guided in fact by military and strategic considerations, which applied specifically to areas in north-east and north-west India in terms of British imperial interest. Wavell felt that it would be possible for the British to remain in those areas indefinitely. Sir Claude Auchinleck, the Commander-in-Chief, did not agree with the Viceroy. His immediate objection was that Muslim troops of the Indian army would be treated as foreigners among a hostile non-Muslim population if they were withdrawn, for example, into Bengal and Assam; and if the Viceroy's contemplated plan to retain British authority in these areas was accepted, there would be no option left but to hold them with the help of British troops which, in effect, would mean supporting Pakistan against India. This was precisely what the Commander-in-Chief was strongly opposed to. While he was thinking in terms of a broad Commonwealth interest, Wavell's mind was working in the old imperialist groove.

The Commander-in-Chief emphasized that, from a broad Commonwealth point of view, Britain's strategic interests lay in the Indian Ocean area, with its western entrance controlled through the Red Sea and its eastern entrance through Singapore and the Malacca Straits. In addition to the ability of Commonwealth countries to use air routes across Arabia, Iraq, the Arabian Sea, India, Burma and Malaya, these interests included a regular supply of oil from Persia and Iraq, and the control of Ceylon for a port of call and a naval and air base. His fear was that if India became 'unfriendly or liable to be influenced by a power, such as Russia, China or Japan, hostile to the British Commonwealth, our strategic position in the Indian

Ocean would become untenable and our communications with New Zealand and Australia most insecure'.[65] He therefore emphasized the importance of India's unity, the need for an undivided country. He believed that, if a separate Pakistan was formed, even as part of the British Commonwealth, without India or against India's will, the latter might be a source of danger to both Pakistan and Britain. India would in that case throw in her lot with Russia to conquer and absorb Pakistan to restore her unity. It would then be difficult to ensure protection for Pakistan against India.

Even strategically, India's unity was of paramount importance. 'As atomic energy develops', Auchinleck argued, 'and weapons of all sorts, whether on the sea, on the land or in the air, improve, depth in the defence and adequate space for dispersion of base installations, including industrial plants, must become increasingly essential in war'. A united India possessed these qualifications as well as an independent 'Hindustan' under Congress rule. But this was not so with Pakistan even if it was to include north-east India. Apart from its inadequate raw materials and industrial production, its territorial depth even in the north-west would hardly exceed 300 miles. If the Indian Ocean area was to be kept free from possible foreign influences, it was an independent 'Hindustan' that could help, not Pakistan. The Commander-in-Chief therefore reiterated his point by saying: 'If we desire to maintain our power to move freely by the sea and air in the Indian Ocean area, which I consider essential to the continued existence of the British Commonwealth, we can do so only by keeping in being a united India which will be a willing member of that Commonwealth, ready to share in its defence to the limit of her resources'.[66]

The Cabinet Mission plan incorporated the principle of Auchinleck's suggestion and emphasized the importance of India's unity which, in face of the conflicting claims of the Congress and the Muslim League, was sought to be maintained by introducing a Grouping system acceptable to both. This new trend of British policy, which arose from its determination to withdraw, clashed with Jinnah's way of thinking and Wavell's continuing adherence to the League as a British ally against the Congress. The efforts of Cripps and, to an extent, of the Secretary of State, to win the confidence of Gandhi and the Congress in their last few days in Delhi was clear indication of the Mission's desire to preserve India's unity. However, the worsening political situation in India and the rapid decline of authority called for 'the next step' in political negotiations, a task which, as the Prime Minister suggested to the King on 17 December, was to be entrusted to Viscount Mountbatten, the King's cousin, and not Wavell.

On 20 February 1947 the Prime Minister announced in the House of Commons the definite intention of the British government 'to effect the transfer of power into responsible Indian hands' by June 1948. Mountbatten, who was appointed to effect this transfer, had a full discussion with the Prime Minister and his colleagues of the Cabinet Mission before his arrival in India on 22 March 1947. The general guidelines that emerged in these discussions were in keeping with what Auchinleck had already suggested, with an emphasis on a united India, though with the consent of both the major parties, without the use of compulsion.[67]

But confronted with a communal bloodbath in the Punjab and the virtual paralysis of the Interim Government, Mountbatten was soon obliged to think in terms of a partition. It is not within the scope of this work to detail the manner in which Partition came about, and there exists already a considerable body of published work on the subject.[68] It is necessary, however, to appreciate the Viceroy's limitations which underlined his recommendation for division as the only alternative to a 'major civil war',[69] a recommendation which was carried on 2 May for approval to London through Lord Ismay, his Chief of Staff. These limitations proceeded from a policy of desisting from the use of compulsion in the conduct of negotiations. Nehru, for instance, stood firmly opposed to compulsion of any kind being used to enforce 'Grouping', a concession given to the League to ensure an all-India unity. An A.I.C.C. resolution of 5 January 1947 had even threatened that if any attempt was made at such compulsion, a province or part of a province could 'take such action as may be deemed necessary in order to give effect to the wishes of the people concerned'.[70] The Muslim League, on the other hand, was not prepared to accept the Constitution framed by the existing Constituent Assembly, nor would the British government permit its enforcement in those parts of the country unwilling to accept it. The only alternative left under these conditions was to leave the provinces invested with the right to determine their own future. In other words, on the basis of Nehru's own logic and the resolution of the A.I.C.C., Madras, Bombay, U.P., C.P., Bihar, Orissa and Assam were to be formally required to reaffirm that they wished to remain within the Indian Constituent Assembly. Bengal and Punjab were similarly to determine their own course. A referendum was to be held in the North-West Frontier Province, while Sylhet was given the option of joining a partitioned Bengal.

Mountbatten's plan appeared to reject the Cabinet Mission's scheme for a united India as the successor to power; in its place he invited acceptance of the claims of a number of successor states who were to be permitted to unite into two or even more separate sovereignties. H.S. Suhrawardy had started campaigning for an altogether independent Bengal. It meant Balkanization, and when Mountbatten showed the draft to Nehru on 10 May, he vehemently opposed the whole scheme. Nehru had presumed that the proposals Lord Ismay carried to London were based on the principle of self-determination which, in the existing context, envisaged a partition of Bengal and Punjab. He did not realize that once the use of compulsion was dispensed with, it would mean freedom not only for Assam to opt out of Bengal, but also for other provinces to unite with a Group of their own choice based on regional interests. As revised by the Cabinet, the Viceroy's draft plan for partition could not but recognize the rights of the provinces to unite in their own way, and not according to the dictates of Nehru, who insisted that 'Hindustan' should constitutionally be regarded as the successor Government of India, with Pakistan to be recognized as having the legal status of a mere seceder nation.[71]

Nehru's reaction, however, led the Viceroy to postpone till 2 June his proposed conference of leaders, which was originally to be held on 17 May. A way out of the impasse was in the mean time suggested by V.P. Menon, then the Reforms Commissioner, who had formulated a plan of his own for the transfer of power to two separate central governments on the basis of Dominion Status, not on what Nehru idealized as independence. In a lengthy discussion with Vallabhbhai Patel earlier in January, Menon had convinced him that it was better to divide the country rather than allow it to gravitate towards civil war. And by consenting to accept division and Dominion Status, Jinnah's demand would not only be limited to those portions of Punjab, Bengal and Assam which were predominantly Muslim, but the Congress too would stand to gain. It would ensure a peaceful transfer of power and its acceptance would be warmly welcomed by Britain in the interests of the Commonwealth. Besides, British assistance and goodwill would be of great help in building an efficient army, where the higher ranks were still largely officered by the British, with the Indian navy and air force being virtually non-existent.

Menon first won over Patel to his point of view. He later secured the approval of Jawaharlal Nehru. Once the principle of his plan received the assent of the Viceroy after consultation with the parties concerned, the former thought that the deadline for handing over power could very well be advanced. Mountbatten proceeded to London with Menon and had the plan finalized towards the close of May. He also persuaded the British government to introduce the necessary legislation in Parliament in conformity with the final plan announced on 3 June, on the assumption that power would be handed over to two successor states on the basis of Dominion Status, with the right to secede from the Commonwealth if they so determined under the Statute of Westminster (1931).

The Viceroy's time-table for the transfer of power on 15 August 1947 under the announcement of 3 June constituted a crash programme. The Radcliffe Boundary Commission, which was appointed to demarcate the continuous partition line in both Bengal and Punjab, was required to work with the greatest speed and present its Award within the time already fixed: the work was completed in six weeks. The Boundary Commission consisted of two groups, one for Bengal, the other for Punjab. Each group contained four members, two in each representing India and Pakistan. Nehru and Jinnah both agreed that the members should be High Court judges, and that the Viceroy should arrange with the British government to make a suitable person available to function as Chairman. Sir Cyril Radcliffe was selected with the full consent of both parties.

While the Bengal part of the Commission met at Calcutta, the Punjab part met in Lahore. As it was not possible for Radcliffe to attend all their public hearings, he made his own headquarters in Delhi, where he received both sets of proceedings. In addition, he paid two visits of several days each to Calcutta, and a similar visit to Lahore. He made it clear during these visits that a boundary line agreed by Indians themselves would have much greater weight than a line drawn by himself. But in the absence of agreement between the contenders, Radcliffe had to make the Awards. To avoid political complications before the transfer of power on the due date, Mountbatten raised with the Chairman the possibility of holding back the submission of the report until after 15 August. Even though Radcliffe insisted on its submission according to plan on 13 August, Mountbatten did not relent. The Viceroy was convinced of the advantages of holding back the report until 16 August, and of giving it to the rival parties simultaneously and together. This was accordingly done, and both sides, despite initial outbursts of bitter complaints and counter-complaints, finally agreed that the Awards be announced and implemented forthwith.

India thus became independent, but it was divided. Though reduced and truncated, Pakistan came into existence as a sovereign state. The Indian Independence Act (10 & 11 Geo. VI, c. 30), passed on 18 July, formally set up with effect from 15 August 1947 two independent Dominions to be known as India and Pakistan, each of these successor authorities being invested with powers to enter into treaties between themselves and with the British government over matters arising from the transfer of power.

The constitutional arrangement announced on 3 June related to British India only. It did not extend to the Indian States, which were concerned about the lapse of Paramountcy. But, a great number of States had already decided to come into the Constituent Assembly unconditionally. Barring a few exceptions, the accession of all the others occurred as a matter of course with the establishment of a States Department (India) under Vallabhbhai Patel as Member and V.P. Menon as Secretary. Acting with the greatest speed and efficiency, the newly formed States Department drafted the necessary documents and by 31 July organized the necessary propaganda intended to persuade all the Princes to accede by 15 August. With the active co-operation of Mountbatten,[72] their accession became a comparatively easy task, except for Hyderabad, Jammu and Kashmir and a few other Muslim States who were dealt with separately. The Indian Political Service was wound up from the midnight of 14 August and its functions transferred to the States Department under Patel.

The accession of the States strengthened the importance of India, but it could not repair the loss to the subcontinent arising from its partition. Partition could not perhaps be avoided in the context of a deeply plural society, and the emphasis on power as the main guideline for public and political conduct. Acting under the influence of a segmented social system and motivated chiefly by the quest for power, the Congress took several erroneous decisions which paved the way for Partition. In its anxiety to curb political radicalism and ensure imperial stability by seemingly effecting a communal balance, the British government rendered itself from time to time subject to certain policy-deviations which appeared to act as a counterpoise to the rising Hindu elites. In turn it led to the increasing intransigence of the Muslim community competing for power to protect its own economic and cultural interests. The conflict for power created a situation where Partition appeared to be the only alternative to civil war. Above all, the contiguity of Muslim-majority areas in the north-west was perhaps the key to the Muslim urge for a separate homeland, an urge which, as shown in our study, was not without justification. Partition was, in fact, a consequence of cumulative developments which flowed from the spread of democracy.

NOTES

Chapter I. THE INSTRUMENTS OF TERRITORIAL INTEGRATION

1. Ilbert, *The Government of India*, p. 42.
2. Sir Charles Fawcett, *First Century of British Justice in India*, p. 66.
3. See Long, op. cit., pp. 23, 32, 64–5.
4. Forrest, *The Life of Lord Clive*, II, 120.
5. See Malcolm, *The Life of Robert Lord Clive*, II, 126–7.
6. See Misra, *Judicial Administration of the East India Company*, chap. II.
7. Keith, *Speeches and Documents of Indian Policy*, pp. 36–7.
8. *Hansard*, Third Series, 10 July 1853, V(19), p. 508.
9. Ilbert, *The Government of India*, p. 62.
10. 24 Geo. 3, c. 25, s. 33.
11. Ibid., s. 18.
12. Prior to Pitt's India Act which reduced the Council to subservience, I have, in conformity with the provisions of the Regulating Act, used the form Governor-General and Council, not Governor-General in Council, which signified the supremacy of the head of executive government.
13. *Parl. Hist. of England*, XXIV, p. 326.
14. 24 Geo. 3, c. 25, s. 34.
15. Ibn Hasan, *The Central Structure of the Mughal Empire*, p. 65.
16. This term is used in para 41 of the Court's dispatch of 10 December 1834 to the Govt of India with the 1833 Act as an enclosure.
17. Court's dispatch of 10 Dec. 1834, para 43.
18. Ibid., para 9.
19. Sharp, H. (ed.), *Selections from Educational Records*, Part I, p. 99.
20. See Anderson and Subedar (eds.), *The Development of an Indian Policy (1818–1858)*, p. 6.
21. Ibid., pp. 5–6.
22. Ibid., p. 43.
23. 3 & 4 Will. IV, c. 85, s. 59.
24. See Misra, *The Central Administration of the East India Company, 1773–1834*, pp. 34–55.
25. I.O. Records no. (19) 1085, *Papers Relating to the Constitution of Indian Governments*, p. 1. Minutes by Members of the Civil Finance Committee can be seen on pp. 3–36. Under the Act the overgrown Presidency of Bengal was to be two distinct Presidencies called the Presidency of Fort William and the Presidency of Agra, each with a Governor and Council of its own. But the provision never came into operation. It was suspended by an Act of 1835 (5 & 6 Will. IV, c. 52) and the suspension was continued indefinitely by the Charter Act of 1853 (16 & 17 Vict., c. 95, s. 15). The Supreme Government was, however, authorized in 1853 to appoint during the period

of suspension a servant of the Company to the office of Lieutenant-Governor for the North-Western Provinces.

The Supreme Government of India was doubtless divested under the Act of any local charge for a separate Presidency. But in keeping with Bentinck's views the home authorities agreed that the Governor-General of India in Council should also act as the Governor in Council of the Presidency of Bengal. The Court of Directors were in this regard authorized to reduce the strength of the Council or even suspend the Council altogether and vest the executive control in a Governor alone. This political arrangement applied to the administration of all subordinate Presidencies with a Governor and Council, its object being to raise the status and authority of the head of the executive government.

26. *The Cambridge History of the British Empire*, vol. V (1858–1918), p. 489.
27. Ibid., p. 392.
28. *Report on Indian Constitutional Reforms*, chap. X, paras 296–7.
29. See Ludlow, *Thoughts on the Policy of the Crown Towards India*, 1927 (Letter VII, pp. 72–82).
30. Ibid., p. 72.
31. Cited by Ludlow, op. cit., p. 71.
32. Philips, Introduction, *Correspondence of Lord William Cavendish Bentinck*, vol. 1, p. xxx.
33. Ibid., p. xxxix.
34. House of Commons' Return on Adoptions, p. 101 (Ludlow, op. cit., Letter VIII).
35. See Anderson and Subedar (eds.), *The Expansion of British India, 1818–1858*, pp. 193–6.
36. Jarrett, *Ain-i-Akbari*, ii, 37. As the *nazim* was a civilian magistrate, penalty of death was not enjoined by quranic law to be inflicted by him. This power of the qazi was exercised in the case of murder, unless the party of the victim agreed to accept what was usually called 'the price of blood'. Sarkar, *Mughal Administration*, pp. 139–40.
37. These were Calcutta, Murshidabad, Burdwan, Dinajpur, Dacca and Patna.
38. See Misra, *The Judicial Administration of the East India Company, 1765–1782*, pp. 158–74.
39. According to Pelsaert the laws administered by the qazis 'contain such provisions as hand for hand, eye for eye, tooth for tooth'. Pelsaert, *Jahangir's India* (Moreland's tr.), p. 67.
40. See chap. V.
41. See Misra, op. cit., pp. 139–43.
42. Ibid., pp. 266–7.
43. Ibid., note 3, pp. 354–5.
44. Cited in *The Cambridge History of the British Empire*, V, 379. Cornwallis had even earlier provided that British-born subjects in the interior were to execute bonds to subject themselves to diwani adalats in civil cases. Besides, European magistrates were authorized to arrest British-born subjects not in the Company's or King's service, and send them under custody to Calcutta

for trial. Other Europeans were to be amenable to the country courts of adalt. See Misra, *Central Administration*, p. 37.
45. Anderson and Subedar (eds.), *The Development of an Indian Policy (1818–1858)*, p. 45.
46. In 1834 Macaulay first arrived in Madras and proceeded in a palanquin from there to Calcutta. The three civil servants were Cameron, Sir John Macleod and Anderson. After a few months, Millett was appointed by the Governor-General in place of Anderson who left for reasons of health. See *The Life and Letters of Lord Macaulay* by George Otto Trevelyan (Popular Edn., Longmans Green, 1889), p. 299.
47. Ibid., p. 299.
48. *Collected Works of Macaulay*, xi, 555.
49. Ibid., p. 584.
50. 3 & 4 Will. IV, c. 85, s. 43.
51. See *Proceedings of the Legislative Council of India*, i, 2.
52. Ibid., p. 2.
53. Ibid., pp. 3–4.
54. See Canning to Wood, 24 March 1861, Halifax Papers, I.O. Mss. Eur. F. 78 (55), p. 2a.
55. Parl. Papers (HC), no. 479 of 1852–3, p. 51.
56. Leg. Council Proceedings, 1856, pp. 581–93.
57. See Misra, *The Administrative History of India, 1834–1947*, pp. 25–6.
58. See Anderson and Subedar (eds.), *The Expansion of British India (1818–1858)*, p. 191. See also *Christian Intelligencer*, June 1860.
59. See Home (Public) Letter to Se. of State, 26 Jan. 1861.
60. See Home (Public) Progs. 8 Sept. 1860, nos. 9–18.
61. Section 10.
62. Wood to Canning (9 April 1861), I.O. Mss. Eur. F.78 (Letter Book), vii, 130.
63. Wood to Canning (26 April 1861), ibid., p. 210.
64. Halifax Papers, I.O. Mss. Eur. F.78(55), Box no. 2, p. 2a.
65. Hastings to Barwell, 22 July 1772 (Gleig, *Memoirs*, i, 316).
66. Hastings to Court of Directors, 11 Nov. 1773 (Gleig, *Memoirs*, i, 368).
67. They were in the arrangement of 1769 designated as 'Supervisors' who were simple lookers-on, without trust or authority; they later became Collectors, and ceased to be lookers-on. Gleig, *Memoirs*, i, 268.
68. See Misra, *District Administration and Rural Development in India*, chap. I.
69. See Misra, *The Central Administration of the East India Company, 1773–1834*, chap. II, and *The Administrative History of India, 1834–1947*, chap. II.
70. Martin, Wellesley's Dispatches, ii, 339. See Misra, *Central Administration*, chap. V on the Civil Service.
71. See Misra *Central Administration*, pp. 391–2.
72. See Trevelyan, *Life and Letters of Lord Macaulay*, f.n. p. 585.
73. See Misra, *The Bureaucracy in India*, p. 79.
74. Trevelyan, op. cit., p. 586.

75. Ibid., p. 585.
76. Ibid., pp. 586–93.
77. Macaulay's colleagues were Lord Ashburton, the Rev. Henry Melvill, Principal of Haileybury College, Mr Jowett, and Sir John Shaw Lefebre.
78. See Misra, *The Bureaucracy in India*, pp. 74–80.
79. See Home (Public) Letter to Sec. of State, 26 Jan. 1861.
80. Cited in the *Cambridge History of India*, i, 64.
81. H. H. Wilson, Sec. to the Asiatic Society, edited *Mackenzie Collection of Oriental Manuscripts*, in two volumes (Calcutta, 1828).
82. Clements R. Markham, *Memoir on the Indian Surveys* (London, 1871), p. 59.
83. Ibid., p. 61.
84. See *Asiatic Researches*, vol. xii. For early attempts to measure the Himalayan peaks see also Murray's *Discoveries in Asia*, ii, 382; Baillie Fraser's *Journal*, p. 323; Buchan Hamilton's *Nepal*, and *The Quarterly Review*, no. 34.
85. In 1829, Herbert started a periodical called *Gleanings in Science* and became Astronomer at the Lucknow Observatory in 1831, when James Prinsep took over as the editor of his journal. In 1832, his *Gleanings* was converted into *The Journal of the Asiatic Society of Bengal*. Herbert died at Lucknow on 24 Sept. 1833. See Markham, op. cit., p. 66.
86. See Markham, op. cit., note on p. 65. Buchanan's survey cost Rs 30,000. His Mss were sent to London in 1816, and it was not before 1838 that Montgomery Martin was permitted by the Court of Directors to publish extracts in three volumes.
87. See Markham, op. cit., pp. 145–69.
88. See *The Evolution of the Army in India* (Govt. Publication, 1924), pp. 23–4. The changes which cleared the way for union are indicated on pp. 22–3.
89. Ibid., p. 5.
90. Ibid., p. 9.
91. Ibid., p. 11.
92. Ibid., p. 14. In 1796, as said above, the total strength of European troops was 13,000 against total Indian troops of 57,000. In a period of nine years the Presidency armies increased a little less than three-fold. This was mostly a period of territorial expansion under Wellesley.
93. Ibid., p. 15.
94. See H. Compton, *European Military Adventurers of Hindustan*.
95. Cited by R. G. Burton in *The Mahratta and Pindari Wars*, p. 4.
96. Ibid., p. 12. The British armies amounted to some 111,000 men (the Grand Army, over 40,000; and the Army of the Deccan, 70,000). This was in addition to the followers who numbered 300,000. See also Badenock, *The State of the Indian Army*, London, 1926.
97. Auber, *Rise and Progress of British Power in the East*, vol. I, p. 151.
98. See Misra, *The Indian Middle Classes*, chap. IV dealing with technological progress and social change.
99. For the earlier history of postal communication see Misra, *Central Administration*, pp. 415–49.

100. Alexander Mackenzie, *History of the Relations of the Government with the Hill Tribes of the North-East Frontier of Bengal*, Calcutta, 1884 (Preface, p. iii).
101. See Ruthnaswamy, *Some Influences that Made the British Administrative System* (London, 1939), pp. 215–16.

Chapter II. THE DELIMITATION OF INDIA'S FRONTIERS

1. See Fort William Secret Dept. Progs. nos. 3–7, 17 June 1808.
2. See Fort William Secret and Separate Progs. for 4 and 11 July 1838.
3. See Fort William Foreign and Secret Cons. Aug. to Dec. 1855.
4. The political and financial reasons for the use of caution as well as the manner of Russian advance in Central Asia from 1953, including a general description of the hill tribes, are all detailed in the Foreign-Political Department proceedings of the Govt. of India, Jan., 1869, nos. 50–678.
5. Mayo to Argyll, 1 July 1869 (Parl. Papers, 1878–9, LVI, 466).
6. The Indian section of the Afghan Delimitation Commission which was headed by Lieut. Col. J. W. Ridgeway, left Quetta on 19 Sept. 1884. A small survey party was attached to the Commission. It consisted of Capt. St. G. Gore and Lieut. M. G. Talbot, assisted by three Indian surveyors. See Charles E. D. Black, *Memoir on the Indian Surveys, 1875–1890* (London, 1981), p. 172. The Indian section met the English section of the Delimitation Commission in the valley of the Hari-Rud westward from Herat. Ibid., p. 176.
7. Ibid., p. 172.
8. Ibid., p. 175.
9. Black, op. cit., p. 128.
10. See Fort William (Foreign) Progs. no. 8 of 27 June 1851.
11. See Fort William (Foreign) Progs., nos. 10–14 of 27 June 1851.
12. Younghusband, *India and Tibet*, p. 23.
13. See Alexander Mackenzie, *History of the Relations of the Government with the Hill Tribes of the North-East Frontier of Bengal* (Calcutta, 1884), p. 11.
14. Ibid., p. 12.
15. *Records of the Survey of India*, op. cit., p. 161. The information given about Nain Singh's account of his route-survey from Leh to Lhasa and his return to India via Assam is based on the report drawn up by Trotter. Ibid., pp. 160–80.
16. 'According to the Indian survey maps the boundary between Ladakh and Tibet' as Trotter points out, 'is a good deal to the west of the line as given by the Pandit. The latter states that the stream of the Niagzu valley which flows southwards near the meridian of 79 degrees from Mandal to the Khurnak Fort is the true boundary. The one given on the survey map, viz., the watershed to the west of the above-mentioned stream, is derived from Major Godwin Austen's plan-table survey of the country to the north of the Pangong lake in 1863. This survey extends to within a few miles of Noh, and the details of it generally agree most satisfactorily with the Pandit's route survey from

Lukung to Noh, although there is this discrepancy in the position of the boundary line.

I find on a reference to Mr Walker's map of the Punjab and Western Himalaya which accompanies General Cunningham's well-known work on Ladakh that Niagzu is there also given as the boundary between the two countries, but that south of Niagzu the watershed to the east of the Niagzu or Chang Parma river is shown as the boundary. The Ruang or Rawang stream which enters the main valley north of Niagzu is there shown as belonging to Tibet, but it appears from the text of the Pandit's narrative that he ascended the Rawang stream and found there huts and a grazing ground belonging to the people of Tankse.' Ibid., p. 162 (note).

17. Ibid., p. 179.
18. Ibid., Part II (1879–1892), pp. 238–87. Kishen Singh started his work from Darjeeling and Kalimpong. After spending some time in Lhasa he started for Mongolia.
19. Younghusband, *India and Tibet*, p. 42.
20. Ibid., p. 43.
21. The annexation of Upper Burma towards the close of 1885 involved Chinese claims to local feudatory states; the Panjdeh crisis involved a clash with Russians during 1885–6; and, in addition, troubles in Khartoum as well as in Ireland were all series matters that resulted in a stalling of British plans.
22. See Govt. India to Sec. of State (Foreign Dept.) no. 1 of 21 Oct. 1889 with Enc. (Papers Relating to Tibet, Cd. 1920, no. II.)
23. See Enc. in no. 5 (Govt. of India to Sec. of State, Foreign Dept., 25 March 1890), ibid., pp. 6–7.
24. Govt. of India to Sec. of State, Foreign Dept. no. 9 of 4 July 1893; ibid., pp. 8–10.
25. Ibid., p. 11.
26. Ibid., pp. 22–3 (Annexure I).
27. Ibid. (Annexure nos. 31–6 and enclosures), pp. 113–17, containing press reports and correspondence regarding the Tibetan (Dorjieve) mission to Russia during 1900 and 1901, which obliged Curzon to establish direct communication with the Dalai Lama. See Annexure no. 37 (Govt. of India to Sec. of State, 25 July 1901), ibid., pp. 117–18.
28. See Annexures and enclosures to nos. 37–8 (Viceroy to the Dalai Lama and Sec. of State, Aug. 1901); ibid., pp. 120–3.
29. Lord George Hamilton, Sec. of State, to Curzon, Gov.-Gen., 27 Feb. 1903, ibid., no. 78, pp. 183–5.
30. See Viceroy to Sec. of State (nos. 123–9, 26 Oct. to 23 Nov. 1903); ibid., pp. 214–21.
31. See Lamb, *McMahon Line*, vol. i, pp. 258–64 (App. VI).
32. *India and Tibet*, p. 336.
33. See Lamb, *McMahon Line*, i, p. 15.
34. Lamb, *McMahon Line*, i, 230.
35. Ibid., ii, 599 (App. XI).
36. Ibid., App. XII (Sir James' Memorandum to the Wai-Chiaopu, dated Peking, 17 Aug. 1912).

37. Lamb, *McMahon Line*, ii, 478.
38. Ibid., App. XVI.
39. See Mehra (ed.), *The North-East Frontier*, i, 45–60 (Confidential Note by Chief of General Staff, 1 June 1912).
40. See Lamb, op. cit., ii, chap. xxvi.
41. Ibid., App. XVII.
42. See M. E. Willoughby, 'The Relation of Tibet to China', in *Journal of the Central Asian Society*, XI, 1924.
43. See Jean Fairley, *The Lion River: The Indus*, chap. VI on disputed frontiers.
44. Ibid., p. 56.
45. See Jean Fairley, op. cit., p. 56.
46. The hill tribes of the state of Manipur—Nagas, Kukies and others, were suggested as the best carriers for the transport of goods across the mountains.
47. The Chefu Convention of 1876 was a result of this murder.

Chapter III. TERRITORIAL REORGANIZATION FOR SECURITY AND DEVELOPMENT

1. See Fort William (Foreign-Secret) Progs. nos. 114–23 of 26 April 1850, pp. 559–90.
2. Ibid., p. 562.
3. Confidential Minute by the Gov. Gen., ibid., pp. 591–5.
4. Olaf Caroe, *The Pathans*, p. 329.
5. These reasons have been analysed by C. C. Davies in the first chapter of *The Problem of the North-West Frontier* (Cambridge Univ. Press, 1932).
6. Russian pressure in Central Asia had started building up in the wake of her reverses in the Crimean War. See Ft. William (Foreign and Political) Progs., nos. 52–67 of Jan. 1869 (printed) dealing with Russian progress in Central Asia since the Crimean War, containing Memorandum and Minutes covering Central Asia, Afghanistan and India's North-West Frontier.
7. *The Cambridge History of the British Empire*, V, 455.
8. Caroe, op. cit., p. 374.
9. *The Cambridge History of the British Empire*, V, p. 456.
10. Col. Sir T. Hungerford Holdich, *The Indian Borderland, 1880–1900*, pp. 226–7, Holdich had been a member of the Afghan Boundary Commission of 1884–5.
11. Ibid., p. 227.
12. Ibid., p. 228.
13. Ibid., p. 229.
14. See Caroe, op. cit., p. 381; see also the appendix giving details of the agreement, with subsequent reaffirmation by later Afghan rulers.
15. See Holdich, op. cit., pp. 234–5.
16. See Caroe, op. cit., chap. XXIV on Waziristan.
17. Davies, op. cit., p. 16.
18. Vol. II, p. 136.
19. The first chief commissioner of the N. W. F. P. was Lieut. Col H. A. Dean.

For the early administration of this new province see O'Dwyer, *India as I Knew It*, chap. VII.

20. A. J. Moffatt Mills, *Report on the Province of Assam* (Calcutta, 1854), App. M., pp. 151–2.
21. *Hill Tribes of the N. E. Frontier of Bengal*, pp. 7–8.
22. A. J. Moffat Mills, op. cit., p. 154. (from Capt. Gordon, Asst. Agent to Governor-General, Darmang, to Major Jenkins, Agent, Gauhati, 13 Feb. 1844).
23. Ibid.
24. Ibid., p. 158.
25. Alexander Mackenzie, op. cit., p. 495.
26. See John Bryan Neufville, 'On the Geography and Population of the N. E. Border of Assam', pp. 5–6, in *Selections from the Records of the Bengal Government*, no. XXIII (Papers relating to some frontier tribes on the North-East border of Assam).
27. Ibid., pp. 6–7.
28. Mills, op. cit., p. 143.
29. John Bryan Neufville, op. cit., pp. 5–6.
30. See Alexander Mackenzie, op. cit., pp. 77–143.
31. The report of the survey conducted by Jenkins and Pemberton is contained in the Foreign-Political Department Proceedings, nos. 112–114, dated 15 Oct. 1832.
32. Ibid., no. 114D (Letter no. 2 of 6 Feb. 1832).
33. In the absence of a lawful heir the Cachar plains were annexed by the British in 1830. The northen portion of Cachar had been seized by Tula Ram, a rebel. The British annexed the whole of Cachar in 1850 when Tula Ram died. Northern Cachar was mainly inhabited by Angami Nagas and Kookies, who made it difficult for the British to keep peace on their own frontier, the Cachar plains. (See Mills, op. cit., pp. 49–50.)
34. Mills, op. cit., paras 177–8.
35. Fort William Revenue Progs., no. 12 of 20 Oct. 1847.
36. See Mackenzie, op. cit., pp. 101–43.
37. Ibid., p. 496.
38. Philips (ed.), *Correspondence of Lord William Cavendish Bentinck*, II, 1033.
39. Ibid., p. 1033.
40. Letter to Chief Sec, Fort William, 13 Oct. 1832 in Foreign Pol. Progs. no. 114A of 15 Oct. 1832, para 27. See also paras 15 and 16 dealing with internal security problems.
41. Ibid., para 31.
42. Ibid., para 32.
43. From Major A. White, Political Agent, Upper Assam, to Capt. Jenkins and Lt. Pemberton, 15 April 1832. For.-Pol. Progs., no. 1141 of 15 Oct. 1832.
44. For.-Pol. Progs., no. 114A, 15 Oct. 1832, para 45.
45. Ibid., para 46.
46. Ibid., no. 114E of 15 Oct. 1832, para 8.
47. Ibid.

48. Ibid., p. 56.
49. Philips (ed.), op. cit., II, 1385–90.
50. Mills, op. cit., App. E.
51. Alexander Mackenzie, op. cit., p. 370.
52. Sir Robert Reid, *Years of Change in Bengal and Assam* (London, 1966), p. 108.
53. See Philips (ed.), op. cit., II, 1034.
54. See J. W. Edgar, civil officer with the Lushai Expeditionary Force, to the Sec., Govt. of Bengal, 5 June 1872, para 21 in Alexander Mackenzie, op. cit., p. 477. See also pp. 479–82 containing a reply from C. U. Aitchison, Sec. to the Govt. of India, Foreign Dept., to the Offg. Sec., Bengal Govt., no. 1883P, 4 Sept. 1872.
55. See Minute by Sir William Grey (13 March 1868), Home (Pub.) Cons. no. 150 of 28 March 1868, p. 11.
56. Ibid., p. 13.
57. Memorandum by the Governor-General, 30 Jan. 1868, Home (Public) Cons., nos. 148–63 of 28 March 1868, paras 12–15, pp. 25–7.
58. Home (Public) Cons., no. 160 of 28 March 1868, para. 5, p. 67.
59. See Home (Public) Cons., nos. 148–63 of 28 March 1868, pp. 87–90. A summary of discussions on the reconstitution of Bengal in 1868 is given by H. H. Risley, Home Secs., in his Note dated 6 Dec. 1904 on the redistribution between Bengal and Assam. See Home (Pub.) A Progs., no. 165, as part of nos. 155–67 of Feb. 1905, pp. 39–41.
60. Home (Public) Progs., nos. 205–8 of April 1873.
61. The Note is printed in Mackenzie, op. cit., pp. 495–503. It was forwarded to the Govt. of India with Sir Steuart Bayley's Secretary's letter no. 1921 of 1 Sept. 1879.
62. From C. Bernard, Offg. Sec., Govt of India to the Chief Commissioner of Assam, no. 119, dated 27 March 1880; ibid., p. 504.
63. See Home (Public) A Progs., Feb. 1905, no. 155 (nos. 155–67).
64. Ibid., Notes (Public, A, 155–67), 8 Feb. 1904, p. 7.
65. Home (Public) A Progs. no. 149–60 of Dec. 1903 (no. 158), para. 19, pp. 105–6.
66. Ibid., para, 23, p. 106.
67. See Home (Public) A Progs. Dec. 1903, nos. 149–60, pp. 49–69.
68. Ibid., pp. 5–18 for Hewett's Note, and pp. 29–44 for Ibbetson's.
69. Both Hewett's and Ibbetson's Notes reviewed the problems of territorial reorganization in an historical perspective, problems that emerged from the transfer of Berar by the Nizam in 1902. The territory so transferred formed part of the borders of Central Provinces, Orissa and Madras.
70. Ibid., p. 67.
71. Home (Pub.) A Progs., Oct. 1905, nos. 163–98, p. 53. The Resolution of 19 July 1905, which embodied and explained the change in its final form, came into effect on 16 Oct. 1905 as notified on 1 Sept.
72. Home (Pub.) A. Progs., nos. 163–98 of Oct. 1905, para, 14, p. 51.
73. Home (General) Progs., no. 156, p. 116.
74. Home (Pub.) A. Progs., no. 155–167 of Feb. 1905 (Notes), p. 2.

75. Foreign (Pol.) Cons., no. 124 of 4 Feb. 1853, para 16.
76. Under Acts 16 and 17 Vict., c. 95 enacted in 1853.
77. Foreign Cons. no. 28 of Feb. 1856.
78. See Foreign (Pol.) Progs. nos. 116–19, K.W., 14 Jan. 1859. The execution of the plan was expedited by Stanley's letter of 11 Nov. 1858.
79. See Foreign (Pol.) Progs., no. 48 of Nov. 1861.
80. See Foreign (Gen.) B Progs., nos. 179–186 of Jan. 1862 (Chief Com., C.P. to Govt. of India, Judl. 1045/5, 25 Dec. 1861). See also Misra, *Administrative History*, pp. 299–301.
81. Parl. Papers, Cd. 5979 of 1911, para 13 (Coronation Durbar Announcement at Delhi, 12 Dec. 1911).
82. Ibid., para 20.
83. See Misra, *The Central Administration of the East India Company*, pp. 142–3.
84. Philips (ed.) *Correspondence of Lord William Cavendish Bentinck*, I, 286.
85. Ibid., p. 246.
86. Ibid., p. 247 Court of Directors to the Gov.–Gen. in Council at Fort William 3 July 1829).
87. Ibid.
88. Political Letter from Court, 27 Dec. 1833, NAI, Pol. Dispatch no. 18 of 1833, para 9.
89. Ibid., para 11.
90. See Misra, *Administrative History*, pp. 257–8. See Bentinck to Metcalfe, 20 April 1834 and 24 April 1834 in Philips (ed.), op. cit., II, 1249–50 and 1259–60.
91. Metcalfe to Bentinck, Calcutta, 28 Feb. 1834 in Philips (ed.), op. cit., II, 1212–13.
92. Bentinck to Metcalfe, 15 March 1834; ibid., p. 1225.
93. Ibid., p. 1270.
94. Ibid.
95. See Misra, *District Administration and Rural Development in India*, Preface and chaps. II–III.
96. 5 & 6 Will. IV, c. 52, s. 2.
97. Under this new arrangement Metcalfe was to control within the limits of his jurisdiction Residents and political agents whose appointment was to proceed from the Govt. of India.
98. See papers relating to the creation of a Lieutenant-Governorship in the N.W. Provinces in Foreign-Pol. Cons., no. 1/10 of 28 March 1836.
99. See Minute by W. Muir, Finance Member, 15 July 1876, Home (Pub.) Progs., Jan. 1877/230–5; also containing letter to Sec. of State, 26 Jan. 1877.
100. Ibid. (Govt. of India Resolution and Order, no. 45 of 15 Jan. 1877).
101. See Home (Pub.) Progs., Jan. 1877/230–5, para 2. Kumaun and Garhwal were not included in the table.
102. See Misra, *Administrative History*, pp. 307–9.
103. See *Report on the Administration of the North-Western Provinces and Oudh ending 31 March 1893*, Allahabad, 1894.
104. Home (Pub.) A Progs., March 1902/246–52.

Chapter IV. INSTITUTIONAL AGENCIES AND POLITICAL STABILITY

1. See Misra, *The Indian Middle Classes—Their Growth in Modern Times*.
2. See Godley to Welby, 20 July 1899, Kilbracken Papers, I.O. Mss. Eur. F. 102/1, no. 501; also ibid., no. 502 (24 July 1899).
3. Philips (ed.), *Select Documents on the History of India and Pakistan*, IV, p. 13, citing from Parl. Papers, vol. 56, no. 102, col. 1515. I.O. Library Official Dispatch.
4. Ibid., p. 40 (Sir Charles Wood to Sir Bartle Frere, 19 Aug. 1861).
5. See *Report of the Indigo Commission*, 1860, and J. P. Grant's Minute, 17 Dec. 1860, cited in Home (Pub.), A Progs. no. 105 of Feb. 1905.
6. Philips (ed.), op. cit., p. 34 (Govt. of India to Sec. of State, 15 Jan. 1861).
7. Under the Indian Councils Act (1861) Madras and Bombay were allowed separate legislative councils. The North-Western Provinces and Punjab did not have any at the time.
8. See Memorandum from British India Association to Chief Sec., Govt. of Bengal in I.O. (Pub.) Progs., Nov. 1898, vol. 5414, p. 2324, para 13.
9. See Misra, *The Bureaucracy in India*, chap. I.
10. Enc. to Letter no. 3 in Northbrook Collection, I.O. Mss. Eur. C. 144, vol. V, 16 Feb. 1885.
11. Ibid., Enc. to no. 4, 17 March 1885, ff. 14–15.
12. Philips (ed.), op. cit., p. 14.
13. *Report of the Famine Commission* (1880), Part I, para 57.
14. See Misra, *District Administration and Rural Development in India*, chap. II.
15. W. C. Bonnerjee on the National Congress in Philips (ed.), op. cit., p. 139.
16. Ibid.
17. I.O. Mss. Eur. C. 144, no. 16, f. 3 (23 June 1886).
18. I.O. Mss. Eur. C. 144, no. 17, f. 2.
19. I.O. Leg. Progs. (India), 4 Jan. 1886, vol. 2771, no. 114, p. 35.
20. See Misra, *The Indian Middle Classes*, pp. 374–89.
21. See Enc. III in Dufferin to Cross, 29 June 1888, I.O. Mss. Eur. E. 243(24), pp. 12–13; Mss. Eur. E.243(25), pp. 150–65.
22. See Philips (ed.), op. cit., pp. 60–1. The Committee was appointed on 22 Sept. 1888 to examine the memorandum of Sir Anthony MacDonnel on the Congress demand for representative government.
23. Home (Pub.) Letter to Sec. of State, no. 75 of 24 Dec. 1889.
24. Cited in *Montford Report*, p. 45.
25. The provincial Councils of Madras, Bombay and Bengal were constituted under the Indian Councils Act of 1861. The Council of the North-Western Provinces and Oudh came into being in 1886. The Punjab Council was constituted in 1897.
26. Home (Pub.) Letter to Sec. of State, no. 7 of 21 March 1907, I.O. Mss. Eur. D.573 (Council Reforms), para 46.
27. *Montford Report*, p. 51.
28. See B. Shiva Rao(ed.), *The Framing of India's Constitution*, i, 19–30.

29. See Home (Pub.) Letter from Sec. of State, no. 15 of 30 June 1892 and Govt. of India letter dated 26 Oct. 1892.
30. Home (Pub.) A Progs. Aug. 1892, nos. 237–52.
31. Cited in *Montford Report*, p. 48.
32. See Govt. of India to Sec. of State in Home (Pub.) no. 21 of 1 Oct 1908, paras 20, 35–44.
33. See Home (Education) A Progs., nos. 37–41 of March 1882. The decline of Muslims was said to be more especially marked in Bengal and the North-Western Provinces and Oudh.
34. See Misra, *The Indian Middle Classes*, pp. 387–9.
35. See Home (Ests.) A Progs., June 1904, nos. 94–104.
36. See. B. Shiva Rao (ed.), op. cit., 15–16.
37. Ibid., pp. 27–8.
38. *Joint Committee on Indian Constitutional Reform* (Session 1933–4). vol. I (Part I), Report, pp. 109–10. The 'facts' referred to included the large extent of the country, its undeveloped communications, widespread illiteracy, differences of language and cultural norms.
39. Ibid., pp. 111–12.
40. Cited from Prasad Papers (Random Letters File) by Granville Austin in *The Indian Constitution: Cornerstone of a Nation*, p. 35.
41. See *Montford Report*, p. 191.
42. Ibid., p. 193.
43. Philips (ed.), *Select Documents*, p. 91.
44. The Birmingham University Library copy, available in the private collection of Sir Austen Chamberlain, I.O. MSS. Eur. AC. 22 (91) was used by the author.
45. *Memorandum*, p. 1.
46. See Misra, *The Administrative History of India*, pp. 62–74.
47. See encs. to Govt. of India Foreign and Political Dept. letter no. 2311—IA addressed to Local Governments from Simla on 25 July 1918 in Foreign and Pol. (Internal) A Progs., nos. 45–8.
48. *Report of the Indian States Committee 1928–9* (Chairman: Harcourt Butler), pp. 20–1.
49. See vol. I (Part II) Progs for Session 1933–4, p. 75, para 150.
50. The second one, which was attended by M. K. Gandhi on behalf of the Congress, had discussed the question of communal representation, including that of Scheduled Castes.
51. *Report*, vol. I, Part I, para 19 (35–40), pp. 11–12; also Govt. of India, dispatch of 30 Sep. 1930 communicating to the Sec. of State its views on the Statutory Commission's report.
52. *Report* (Joint Selection Committee), vol. I, Part I, para 26.
53. Ibid. para 27.
54. See Hailey Papers, I.O. MSS. Eur. E. 220/30 (Notes on the Constitutional revision written in Simla in July (1931) and presented after Governors' Conference), pp. 190–225.
55. *Report*, Joint Select Committee, vol. I, Part I, para. 23 (5–30).

Notes

56. V. P. Menon, *The Transfer of Power in India*, p. 56.
57. The Indian States had 89 representatives in the Constituent Assembly on 15 October 1949. B. Shiva Rao (ed.), op. cit., p. 101.
58. See Cmd. 7047 (Statement on Indian Policy, 20 Feb. 1947, announcing intention to transfer power to Indian hands by June 1948), paras 5–6.
59. See Cmd. 7136. Out of 296 members of the united Constituent Assembly set up in pursuance of the Cabinet Mission's scheme, the Congress had 208 seats, and the Muslim League 73. See B. Shiva Rao (ed.), op. cit., p. 97. Under the Radcliff Commission award the allocation of seats after Partition was: Governors' Provinces, 226; Indian States, 89; Ajmer-Merwara, Delhi and Coorg, 3: total 318. Ibid. p. 99.
60. See Misra, *The Judicial Administration of the East India Company in Bengal*; *The Central Administration of the East India Company*; *The Administrative History of India*; *The Bureaucracy in India*, and *District Administration and Rural Development in India*.
61. See Buckland, *Bengal Under the Lieutenant-Governors*, vol. I, pp. 26, 219.
62. Ibid., pp. 24–5.
63. Home (Pub.) A. Progs., Dec. 1903, nos. 149–160, p. 39, para 46.
64. Philips (ed.), *The Correspondence of Lord William Bentinck*, I, 773.
65. See Home (Pub.) A Progs., Dec. 1903, nos. 149–60, pp. 39–40.
66. *Cambridge History of the British Empire*, V, 35.
67. Home (Pub.) A Progs., Feb. 1905, nos. 155–67, pp. 38–9, para 3.
68. See Home (Pub.) A Progs. Feb. 1905, nos. 155–67, p. 39, paras. 4–5. See also Home (Pub.) Dispatch to Sec. of State, no. 3 of 2 Feb. 1905, para. 3.
69. Home (Pub.) A Progs., Dec. 1903, nos. 149–60, p. 3.
70. *Selections from Revenue Records, N. W. Provinces*, III, 63.
71. See Parl. Papers (House of Lords), no. 445 of 1853 (*Report From the Select Committee on the Affairs of the East India Company* ... Communicated from the Commons to the Lords, 21 June 1833).
72. Ibid., p. 136, para 6.
73. Ibid., para 4.
74. Ibid., p. 135, para. 3.
75. Ibid., p. 137, para 10.
76. Ibid., p. 153, para 67.
77. Ibid., pp. 153–4, paras. 67–8.
78. See Philips (ed.), *The Correspondence of William Cavendish Bentinck*, I, 707–14 (Bentinck's minute on the training of the civil service).
79. For the security of the rights of civil servants see sections 96B(1), 67A and 72D, 96B(2) of the Govt. of India Act (1919); Lloyd George to Austen Chamberlain, 11 Aug. 1922, I.O. Mss. Eur. AC. 18(63), p. 3; *Memoranda Submitted to the Indian Statutory Commission by the Government of India* (1930), V, 1626; and sections 240–9 of the Govt. of India Act (1935).
80. Vol. I, Part I (*Report*), p. 3.
81. Ibid., pp. 4–5.
82. Philips (ed.), op. cit., II, 1441.
83. Ibid., p. 1449.

84. Ibid.
85. See *The Army in India and its Evolution* (Govt. of India, Calcutta, 1924), pp. 16–17.
86. See *Organization of the Indian Army Report* (1859), C. 2515, pp. xiv–xv. The Commission was presided over by Major-General Jonathan Peel.
87. Philips (ed.), *Select Documents*, IV, 515.
88. *The Army in India and Its Evolution*, p. 27.
89. *Montford Report*, para. 329.
90. See *Report of the Committee appointed to inquire into the Administration and Organization of the Army in India*, Cmd. 943 (1920), pp. 88–92.
91. *The Army in India and its Evolution*, p. 57.
92. See *Legislative Assembly Debates*, vol. I, Part II (1921), pp. 1753–4.
93. *The Army in India and its Evolution*, p. 38.
94. Philips (ed.), op. cit., p. 530 (*Indian Statutory Commission Report*, Cmd. 3568 (1930), vol. I, pp. 96–8).
95. Ibid., pp. 5631–2.
96. 6th Rep. Com. Sec. H. C. 1772–73, para 301a.
97. Revenue Dispatch to India, no. 2 of 31 Dec. 1858, para 8.
98. Ibid., para 7.
99. I.O. Coll. to Rev. Disp., no. 14 of 9 July 1862, para 47.
100. Ibid., para 48.
101. Bernier, pp. 224–5.
102. See Philips (ed.), *The Correspondence of Lord William Cavendish Bentinck*, II, 1413.
103. Ibid., pp. 1403–12.
104. Ibid., p. 1412.
105. See Misra, *The Indian Middle Classes*, chap. VIII.
106. See Misra, *District Administration and Rural Development in India*, chap. II.
107. See Misra, op. cit., chap. V, dealing with the conditions leading to Community Development and National Extension Service under the recommendations of the Grow More Food Inquiry Committee (1952).
108. Rep. Com. H.C. on Petitions, 1781, p. 37b.
109. Bernier, p. 243.
110. Weber, *The Religion of India* (tr.), p. 112.
111. I.O. Disp. to Madras (Pol.), vol. 40, pp. 65–6.
112. Parl. Papers 34(664), 1845.
113. Parl. Papers 52(31), 1860.
114. See 10 July 1833, *Hansard*, 3rd ser., vol. 9, col. 536.
115. *On the Education of the People of India* (1838), pp. 135–6.
116. Ibid., p. 48.
117. *The Pioneer*, 6/7 September 1901.
118. See *Papers Relating to Technical Education*, 1886–1904, pp. 251–3.
119. J. Raleigh (ed.), *Lord Curzon in India: A selection of his speeches, 1898–1905*, pp. 584–5.
120. I.O. Austen Chamberlain Papers, AC, 22/91, para 3 (2 May 1917).
121. See Misra, *The Indian Political Parties*, chap. V.

122. Ibid., pp. 149–52, 168–9.
123. See Misra, *The Indian Political Parties*, for a detailed analysis of the process lending to partition.

Chapter V. DIVISIVE TENDENCIES IN A PLURAL SOCIETY

1. *Report*, p. 1.
2. See Misra, *The Indian Political Parties*, chap. 3.
3. 9th Rep. Sel. Com. H.C., 1782–3, p. 56a.
4. See 6th Rep. Sel. Com. H.C., 1782–3, app. 15, p. 943.
5. Rep. Sel. Com. HL, 1852, Q. 1148.
6. See Rep. Sel. Com. H.C., 1852, 10(553), app. 14.
7. See Home (Education) A Progs., 19 Aug. 1871, no. 2–8 (Memorandum by James O'Kinealy).
8. 'A Chapter from Muhammadan History', *Calcutta Review*, April 1871, p. 322.
9. *Selections from the Records of Govt.*, N.W. Provinces, second series, vol. III, no. III, pp. 216–17.
10. Ibid.
11. Ibid., p. 219.
12. Ibid., p. 220.
13. See *Selections from the Records of Govt.*, N.W. Provinces, second series, vol. III, no. IV, pp. 353–458.
14. Philips (ed.), *Selections*, IV, pp. 183–4.
15. See R. Nathan, *Progress of Education in India*, p. 119.
16. Indian Public Service Commission, 1886–7, *Report*, p. 31.
17. See *Census of India*, 1901, vol. 1, para. 366.
18. Ibid.
19. See Home (Ests.) A Progs., June 1904, nos. 94–104 (appendices).
20. See App. V A (ii and iii), p. 497. The all-India percentage of posts held by Brahmins was 32 against 16 for Kayasths and 17 for Muslims. See *Report*, p. 32.
21. See App. V for Bengal, p. 502.
22. See Philips (ed.), *Selections*, p. 191.
23. Ibid.
24. See *Montford Report*, p. 53.
25. Ibid., p. 54.
26. Ibid., p. 56.
27. Ibid., pp. 61–2.
28. Including five Muslims, namely, Mazharul Haq, M.A. Jinnah, Ibrahim Rahimtoola, Mir Asad, Mohammed Ali Mohamed. See B. Shiva Rao (ed.), *The Framing of India's Constitution*, I, 19.
29. *Report*, paras. 228–30, pp. 148–9.
30. Ibid.
31. See B. Shiva Rao (ed.), op. cit., I, 25–30.

32. See Philips (ed.), *Selections*, pp. 204–5.
33. See Savarkar's Presidential address at the 21st session of the Akhil Bharatiya Hindu Mahasabha, 1931, *Hindu Rashtra Darshan*, pp. 140–3.
34. See Foreign-Political (External) B Progs., nos. 138–42, Oct. 1921 (B. 361).
35. See App. I to Home (Pol.) B. Progs., Jan. 1920, K.W. nos. 283–92.
36. See Home (Pol.) B. Progs., April 1920, no. 189.
37. Viceroy to Sec. of State, 25 Oct. 1920, Home (Pol.) B(Print) Progs, Nov. 1920, no. 281.
38. See Misra, *The Indian Political Parties*, pp. 175–7.
39. *The Collected Works of Mahatma Gandhi*, vol. XVIII gives 1 August 1920 as the date on which Gandhi wrote to the Viceroy giving notice of his decision on non-cooperation. 1 August 1920 was in fact the date on which the Central Khilafat Committee met in Bombay with Gandhi as President and decided to renounce titles and surrender medals.
40. Home (Pol.) Deposit (Print) Progs., May 1921, no. 43.
41. See The *Collected Works of Mahatma Gandhi*, XVIII, 370–2.
42. Ibid., pp. 459–60.
43. See *Gazette of India*, 6 Nov. 1920, Res. no. 4484.
44. See Misra, *The Indian Political Parties*, pp. 191–212.
45. See Govt. of India to Under Sec. of State Home Dept. Conf. no. 8 of 12 Jan. 1928, para 6.
46. See K.W. to Home (Pol.) File 31 of 1928.
47. I.O. Chamberlain Papers, AC. 14(39), Annexe 1.
48. Conclusions of a Conference of Ministers held in Curzon's Room, Foreign Office, 20 Dec. 1921 at 4 p.m. (Secret, Final Copy no. 5), AC. 14(39).
49. See telegram from Viceroy to Sec. of State, 22 Dec. 1921 (Part II), AC. 14.
50. See *Views of the Local Governments on the Working of the Reforms* (1927), p. 251; also *Reports of the Local Governments on the Working of the Reformed Constitution* (1924), I.O. Cmd. 2362 of 1925, p. 196.
51. *Joint Committee on Indian Constitutional Reform*, vol. I, Part I, para 26 (25–30).
52. Ibid., para. 25 (5–10).
53. Ibid., para. 168.
54. The Left parties and groups, more especially the communists and revolutionary terrorists, and their influence on the Congress have been discussed in chap. V of *The Indian Political Parties*.
55. Sec. of State to Viceroy, 5 March 1930 (Enclosure), (Halifax Papers, I.O. Mss. Eur., C. 152/V, pp. 57–8.
56. Viceroy to Sec. of State, 6 Feb. 1930, ibid.
57. Sec. of State to Viceroy, 22 April 1930, Halifax Papers, I.O. Mss., Eur., C. 152/V, pp. 93–4.
58. See Report from Whitehall on the first R.T.C., 14 Nov. 1930, Halifax Papers (typed), I.O. Mss. Eur. C. 152/V.
59. See Philips (ed.), *Select Documents*, p. 236.
60. Halifax Papers, I.O. Mss. Eur. C. 152/V, p. 240.
61. Ibid., pp. 241–4.

Notes 393

62. Sec. of State to Viceroy, 20 June 1930, I.O. Mss. Eur. C. 152/V, p. 125.
63. Ibid., pp. 124–5.
64. Sec. of State to Viceroy, 29 May 1930, ibid., p. 118.
65. See Halifax Papers, I.O. Mss. Eur. C. 152/V.
66. Viceroy to Sec. of State, 6 Feb. 1930, Halifax Papers, I.O. Mss. Eur. C. 152/V, p. 29.
67. Halifax Papers (typed report of the opening speeches on 14 Nov. 1930), I.O. Mss. Eur. C. 152/V.
68. I.O. Records, R.T.C. report of 27 Nov. 1930.
69. R.T.C. report of 9 Dec. 1930.
70. India Office to Irwin, 24 Dec. 1930, Halifax Papers (typed matter).
71. Ibid.
72. Ibid.
73. I.O. Records, Conference Report, 31 Dec. 1930.
74. The Muslim seats finally sanctioned for the Punjab were 84 out of 175, while in Bengal they were to have 117 out of a total of 250. They could, however, contest for general seats also. See the Govt. of India Act, 1935, Third Schedule, p. 245. The Sikhs had 31 reserved seats in the Punjab.
75. Viceroy to Sec. of State, 2 Feb. 1931. Ibid., p. 384.
76. Ibid., pp. 378–82.
77. Ibid., p. 392.
78. Ibid., p. 378.
79. See Philips (ed.), *Select Documents*, pp. 241–2.
80. See Cmd. 4014 (statement by U.P. Govt., 14 Dec. 1931).
81. See *Proceedings of the Indian Round Table Conference* (Second Session), Cmd. 3997 (1932).
82. The original Award issued on 16 Aug. 1932 and presented to Parliament as Cmd. 4147, was concerned only with Provincial Legislatures. The Poona Pact (25 Sept. 1932) which applied to the Depressed Classes, increased their seats recommended by the Communal Award but rejected the principal of separate electorate in their case. See S. of S.'s Memorandum on Communal Award in Joint Sel. Com. Report, Conf. A3, 26.5.33, pp. 112–16.

Chapter VI. THE POLITICS OF POWER AND PARTY RULE, 1937–47

1. See Coupland, *The Constitutional Problem in India*, Part II, p. 111.
2. *Report of the Sapru Committee*, p. 151.
3. See Philips (ed.), *Select Documents*, p. 347. See also Haig Papers, I.O. Mss. Eur. 115/12–13 (District Admn. in U.P. 1937–8).
4. Coupland, *Cripps Mission*, p. 15.
5. Philips (ed.), op. cit., p. 351.
6. Ibid., p. 350.
7. *Report of the Sapru Committee*, p. 45.

8. *Report of the Joint Select Committee* (1934), vol. I, Part I, p. 86, para 154.
9. See Hailey Papers, I.O. Mss. Eur. E.220/30, pp. 1–8.
10. Ibid., p. 8.
11. Hailey Papers, I.O. Mss. Eur. E.220/30, pp. 128–31.
12. Ibid., p. 140.
13. Ibid., p. 144.
14. See Notes on Constitutional Revision written in Simla in July (1931) after the Governors' Conference. Hailey Papers, I.O. Mss. Eur. E.220/30, pp. 190–225. According to Simon, the central executive could include leaders of Indian political parties. But it was to be irremovable, not responsible to the legislature.
15. See Hailey Papers, I.O. Mss. Eur. E.220/32, Confidential, no. 13, pp. 13–22.
16. Ibid., pp. 16–17.
17. Ibid., pp. 48–9.
18. Ibid., pp. 23–7.
19. Zetland to Linlithgow, 8 Feb. 1937, Zetland Papers, I.O. Mss. Eur. D. 609/8, no. 4.
20. Zetland to Linlithgow, 6 Dec. 1937, Zetland Papers, I.O. Mss. Eur. D. 609/8, no. 51.
21. Zetland to Linlithgow, 13 Dec. 1937, I.O. Mss. Eur. D. 609/8, no. 52.
22. Linlithgow to Zetland, 3 Jan. 1939, I.O. Mss. Eur. D. 609/17, para 19.
23. Linlithgow to Zetland, 21 Feb. 1939, I.O. Mss. Eur. D. 609/17, para 10, p. 76.
24. Ibid.
25. Linlithgow to Zetland, 7 March 1939, I.O. Mss. Eur. D. 609/17, para 14, p. 112.
26. Linlithgow to Zetland, 5 Sept. 1939, I.O. Mss. Eur. D. 609/18, Enc. 2, p. 103.
27. Para 20 of enc. to Linlithgow's personal and private letter to Zetland, 25 Sept. 1939, I.O. Mss. Eur. D. 609/18, p. 160. See also Haig Papers, I.O. Mss. Eur. F. 115/2 (Left pressure on Congress to bring out its ministries as a temporary expedient; in Linlithgow to Haig, 1 Dec. 1939, para. 4).
28. Linlithgow to Zetland, 25 Sept. 1939.
29. Ibid., p. 229.
30. Ibid., pp. 200–1.
31. Ibid., p. 230.
32. Ibid., p. 89.
33. Coupland, op. cit., Part II, chap. XXI.
34. See Cmd. 6350 of 1942 (Lord Privy Seal's Mission): Statement and Draft Declaration by H.M's Govt. with Correspondence and Resolutions connected therewith. For the statement and Draft Declaration, see pp. 3–5.
35. Ibid., p. 17.
36. Ibid., pp. 7–15.
37. See B. Shiva Rao to T. B. Sapru, Allahabad, 9 April 1942 in Home (Pol.) File no. 221/42—Poll (I) of 1942, p. 31.
38. Cmd. 6350 of 1942, pp. 6–7.
39. See Home (Pol.) File no. 221/42—Poll (I) of 1942, p. 11 (21 March 1942).

The assessment was correct, for Jinnah had not yet acquired any control over the politics of Punjab and Bengal, two major Muslim-majority provinces.
40. See Misra, *The Indian Political Parties*, pp. 386–95, 538–40, 547; also *Statement published by the Government of India on the Congress Party's responsibility for the Disturbances in India 1942–43*; *Correspondence with Mr. Gandhi* (1944) published with the authority of the Govt.; Mansergh (ed.), *The Transfer of Power*, vol. II.
41. *India Today*, pp. 521–2.
42. See B. R. Ambedkar, *Federation Versus Freedom*, pp. 132–44.
43. Cited in the *Report of the Sapru Committee*, pp. 138–9.
44. See Philips (ed.), *The Partition of India: Policies and Perspectives 1935–47*, p. 465.
45. *Indian Annual Register*, 1945, I, 239.
46. Ibid., pp. 200–1; also ibid., II, pp. 142–4.
47. See Jayakar Papers, File no. 826, item 30 (19 July 1945).
48. From Chief Sec., Assam, to Home Sec., Govt. of India, Confidential Branch, no. C. 141/42/225, dated 6 Feb. 1944, under Home Dept. Dy. no. 1590/44—Poll (I), S. no. 1, pp. 3, 6–7.
49. Ibid., pp. 2–4. See also Jayakar Papers, File no. 224 (*Times of India*, 13 June 1944), p. 127.
50. These figures were supplied by Sir John Anderson, Chancellor of the Exchequer, who added that in the same period the sterling public debt of India had been reduced from £350,000,000 to £26,000,000. See Jayakar Papers, File no. 224 (*Times of India*, 14 June 1944), p. 131.
51. See Home (Pol.) File 146/42—Poll (I) of 1942 (Secret App. to Notes on the Tata strike) pp. 9ff.
52. See Home (Pol.) File no. 7/5/44—Poll (I) of 1944 and K.W., no. 1, pp. i–vi after p. 41 of Notes. See also Home (Pub.) File 7/1/45—Poll (I), pp. 2–5.
53. See Home (Pol.) File 7/5/44—Poll (I), para, 3, p. 26 (S. J. L. Oliver's Note of 2 Aug. 1944).
54. Home (Pol.) File 7/1/45 and K.W. (App. III to Notes, Secret Communist Survey no. 4, 30 June 1945, para. 5).
55. The election figures are cited in V. P. Menon, *The Transfer of Power in India*, p. 228.
56. See Misra, *The Indian Political Parties*, pp. 555–8.
57. Home (Pol.) File 18/3/46—Poll (I).
58. The Congress was represented by Jawaharlal Nehru, Abul Kalam Azad, Vallabhbhai Patel and Abdul Ghaffar Khan, and the League by Jinnah, Mohammed Ismail Khan, Liaqat Ali Khan and Abdur Rab Nishtar.
59. See Cmd. 6821.
60. See Misra, *The Indian Political Parties*, pp. 592–8.
61. Home (Pol.) File no. 18/12/46-Poll (I), Confidential Report from Bengal for the second half of Dec. 1946.
62. Terence Creach Coen, *The Indian Political Service*, pp. 122–3.
63. Ibid., p. 123. Coen quotes *The Third Killer* by Guy Wint (p. 156) to suggest that it was Panikkar, the Chief Minister of Bikaner, who induced the Princes to

acquiesce in their own liquidation.
64. John W. Wheeler-Bennett, *King George VI*, pp. 709–10.
65. Auchinleck Papers, MUL Hem 1152 (11 May 1946).
66. Ibid.
67. See Misra, *The Indian Political Parties*, pp. 625–7.
68. Ibid., f.n. 87, p. 622.
69. John Connell, *Auchinleck*, p. 876.
70. Cited in V. P. Menon, *The Transfer of Power*, p. 332.
71. See Hodson, *The Great Divide*, chap. 17; also Hugh Tinker, *Experiment with Freedom* (1967), chap. 6–7.
72. See Coen, op. cit., pp. 125–8. See also V. P. Menon, *The Story of the Integration of the Indian States*, pp. 97–8.

SELECT BIBLIOGRAPHY

MANUSCRIPTS

Birmingham University Library
  Private Papers of Sir Austen Chamberlain, AC

India [Commonwealth Relations] Office Library
  Home Miscellaneous Series [relevant volumes]
  Legislative Proceedings
  Private Papers
    Haig (MSS Eur.F.115)
    Hailey (MSS Eur.E.220)
    Halifax (MSS Eur.F.78)
    Kilbracken (MSS Eur.F.102)
    Northbrooke (MSS Eur.C.144)
    Zetland (MSS Eur.D.609)

Manchester University Library
  Private Papers of Field-Marshal Auchinleck

National Archives of India, New Delhi
  Private Papers of Jayakar

  Proceedings of the Government of India in Home, Establishment, Judicial, Political, Public and Special Departments

  Proceedings of the Legislative Councils of India

REPORTS AND PRINTED DOCUMENTS

Ninth Report of the Select Committee, House of Commons, 1782–3

Papers Relating to the Constitution of Indian Governments. India Office Records, 19 [1085]

Second Report of the Select Committee on Indian Territories, etc. Parliamentary Papers [HL], No. 445 of 1853

Report of a Commission on the Organization of the Indian Army, 1859

Report of the Indigo Commission, 1860

Parliamentary Papers, LVI. 1878–9

Report of the Famine Commission, 1880

Report of the Public Service [Aitchison] Commission, 1886–7

Report on the Administration of the N. W. Provinces and Oudh ending 31 March 1893, [1894]

Montford Report, 1917

Report of the Committee appointed to inquire into the Administration and Organization of the Army in India, 1920

Papers Relating to Technical Education

Sharp [ed.], Selections from Educational Records, Calcutta, 1920

Report of the Local Governments on the Working of Reforms, 1924

Views of the Local Governments on the Working of Reforms, 1927

Report of the Indian States' Committee, 1928–9

Memorandum Submitted to the Indian Statutory Commission by the Government of India, 1930

Report of the Indian Statutory Commission, 1930

Joint Committee on Indian Constitutional Reforms, 1933–4

Correspondence with Mr Gandhi, 1942–4, [1944]

Report of the Sapru Committee, 1945

Philips, C. H. [ed.], *The Evolution of India and Pakistan, 1858–1947*, Select Documents, vol. IV, 1962

———, *The Correspondence of Lord William Cavendish Bentinck*, 2 vols., Oxford, 1977

Hansard, 3rd Series

SECONDARY SOURCES

Abul Fazl, *Ain-i-Akbari* [trans. Jarrett], Calcutta, 1876

Aitchison, C. A., *A Collection of Treaties, Engagements and Sunruds* [14 Vols.], Calcutta, 1929–31

Anderson and Subadar, *The Development of an Indian Policy, 1818–58*

Select Bibliography

———, *The Expansion of British India, 1818–58*

Auber, P., *Rise and Progress of British Power in the East*, London, 1837

Austin, G., *The Indian Constitution: Cornerstone of a Nation*, Oxford, 1966

Badenock, *The State of the Indian Army*, London, 1926

Bernier, F., *Travels in the Mogul Empire* [ed. Smith], Oxford, 1916

Black, C. E. D., *Memoirs of the Indian Surveys*, London, 1891.

Buckland, C. E., *Bengal Under the Lieutenant Governors*, 2 vols., Calcutta, 1901

Caroe, Olaf, *The Pathans*, London, 1958

Compton, H., *European Military Adeventurers of Hindustan from 1784 to 1803*, Oxford, 1976

Creagh Coen, T., *The Indian Political Service*, London, 1971

Connell, J., *Auchinleck*, London, 1959

Cotton, C. W. E., *Memorandum on the Revenue History of Chittagong*, Calcutta, 1919

Coupland, R., *Report on the Constitutional Problem in India*, London, 1943

———, *The Cripps Mission*, London, 1942

Das Tarak Nath, *British Expansion in Tibet*

Davies, C. C., *The Problems of the North-Western Frontier*, Cambridge, 1932

Dodwell, H. H. [ed.], *The Indian Empire 1858–1919*, Cambridge History of the British Empire, Vol. 5, 1932

Fairley, Jean, *The Lion River: The Indus*, New York, 1975

Fawcett, Charles, *The First Century of British Justice in India*, Oxford, 1934

Firminger, A. [ed.], *The 5th Report on East India Affairs*, Calcutta, 1917

Forrest, *The Life of Lord Clive*, London, 1918

Government of India, *The Evolution of the Indian Army*, 1924

———, *Gazette of India*, 6 Nov., 1920

Gleig, G. R., *Memoirs of Warren Hastings*, 3 vols., London, 1841

Hamilton, F., Buchanan, *An Account of the Kingdom of Nepal*, Edinburgh, 1819

Hodson, H. V., *The Great Divide: Britain-India-Pakistan*, London, 1965

Holdich, T. H. *The Indian Borderland, 1880–1900*, London, 1902

Ibn Hasan, *Central Structure of the Mughal Empire*, London, 1936

Ilbert, C., *The Government of India*, Oxford, 1898

Irvine, W., *Army of the Indian Mughals*, London, 1903

Keith, A. B., *Speeches and Documents on Indian Policy, 1750–1921*, London, 1922

Lamb, A., *McMahon Line*, London, 1966

Long, H., *Selections from Unpublished Reccords of Government for 1748–67*, Calcutta, 1869

Ludlow, *Thoughts on the Policy of the Crown Towards India*

Mackenzie, A., *History of the Relations of the Government with the Hill Tribes of the North-East Frontier of Bengal*, Calcutta, 1884

Malcolm, C. R., *Memoirs of the Indian Surveys*, London, 1871

Mansergh, N. [ed.], *The Transfer of Power*, Documents 1942–7, 8 vols., London, 1970–9

Mehra, P., *The North-East Frontier*, Delhi, 1979

Menon, V. P., *The Transfer of Power in India*, Bombay, 1957

———, *The Story of the Integration of the Indian States*, Delhi, 1969

Mills, A. J. M., *Report on the Province of Assam*, Calcutta, 1854

Misra, B. B., *The Administrative History of India, 1834–1947*, Bombay, 1970

———, *The Indian Middle Classes: Their Growth in Modern Times*, London, 1961

———, *The Indian Political Parties ... up to 1947*, Delhi, 1977

———, *The Bureaucracy in India ... up to 1947*, Delhi, 1977

———, *District Administration and Rural Development in India*

———, *The Central Administration of the East India Company, 1773–1834*, Manchester, 1959

Mitra, H. N., *Indian Annual Register*, Calcutta, 1919–47

Muir, R., *The Making of British India, 1756–1858*, London, 1923

Murray, H., *Discoveries and Travels in Asia*, Edinburgh, 1820

Nathan, R., *Progress of Education in India, 1897–8 to 1901–2*, Calcutta, 1904

Neufville, J. B., 'On the Geography and Population of the Northeast Border of Assam', in *Selections from the Records of the Bengal Government*, no. XXIII

O'Dwyer, M., *India As I Knew It*, London, 1925

Orme, R., *Historical Fragments of the Mogul Empire*, London, 1805

Pelsaert (trans. Moreland), *Jahangir's India*, Cambridge, 1925

Philips, C. H. and Wainwright, M. D., *The Partition of India, Policies and Perspectives 1935–47*, London, 1947

Raleigh (ed.), *Lord Curzon in India: A Selection of His Speeches, 1898–1905*

Reid, R., *Years of Change in Bengal and Assam*, London, 1966

Ruthnaswamy, M., *Some Influences that Made British Administration in India*, London, 1939

Sardesai, G. S., *The Main Currents of Maratha History*, Bombay

Sarkar, J. N., *Mughal Administration*, 3rd ed., Calcutta, 1935

Scrafton, L., *Reflections on the Government of Indostan*, Calcutta, 1770

Shiva Rao, B., *The Framing of India's Constitution*, Delhi, 1968

Strachey, J., *India: Its Administration and Progress*, London, 1885

Tinker, H., *Experiment with Freedom*, London, 1967

Trevelyan, C. E., *On the Education of the People of India*, London, 1938

Trevelyan, G. O., *The Life and Letters of Lord Macaulay*, London, 1989

Turner, *An Account of an Embassy to the Court of the Teshoo Lama*, London, 1888

Weber, M., *The Religion of India*, Glencoe, 1958

Wheeler-Bennett, J. W., *King George VI*, London, 1958

Willoughby, M. E., 'The Relation of Tibet to China', *Journal of the*

Central Asian Society, vol. XI, 1924

Wilson, H.H. (ed.), *Mackenzie Collection of Oriental Manuscripts*, 2 vols., Calcutta, 1840

Younghusband, F., *India and Tibet*, London, 1910

JOURNALS

The journals consulted include the *Journal of the Asiatic Society of Bengal, Asiatic Researches, Quarterly Review* and *Christian Intelligencer*.

INDEX

Abdur Rahman, Amir of Afghanistan, 72, 125–6
Accession, Instrument of, 215, 324, 325, 335, 375
Act: Age of Consent (1891), 265, 275; Amending (1786), 6, 9, 26, 31, (1781), 26, 31; Charter (1813), 27, 31, 48, (1833), 18, 27, 28, 29, 31, 33, 34, 42, 48, 176, 182 (curtails jurisdiction of Bengal Govt.), 171, 173–4, (heightens central control over Presidencies), 48, 49, (1853), 34–5, 37, 39, 184, 187; Council of India (1869), 38; Criminal Law (Amendment) (1919), 287; Defence of India (1915), 287; Government of India (1833), 10, 12–13, 18, 175; (1854), 184; (1858), 15, 37, 50, 184, 185, 216; (1869), 185–6; (1912, constitutes Bengal and Bihar and Orissa as provinces and 3 presidencies), 170; 1919, Montford Reforms *see separate entry*; 1935-Constitution Act *see separate entry*; (1919), 204, 207, 216, 292–4; establishes 'governors' próvinces, 170, 178; (introduces dyarchy), 212; introduces separate electorates, 282–3; Muddiman Committee, on working of, 295; (1935), 208; establishes Orissa a governor's province, 170; instrument of unity, 218, 220, 299–300; provisions of, 218–19; India (1784), vi, 2, 6, 7–10, 13, 14, 17, (1853) 49–50, 34–5, 235; Indian Bishops and Courts (1823), 27; Indian Councils (1861), 38, 39, 178, 184, 188, 189, 191, 192, 195, 196, (1892) 41, 190, 196 (British govt.'s note of transmission of), 200–1, 265; (1909) 168, 202–4, 205, 211; Indian Independence (1947), 375; Indian Police (1861), 30; of 1835 suspending Agra Presidency provisions in Charter Act (1833), 176; Regulating (1773), 20–1, 24, 25; Territorial Army (1923), 253–4

Acts, replace Regulations, 33
adalats, 20–1; conflict with Supreme Court, 24; Impey's Code adopted for, 24; source of authority of, 24
'Additional Members' (of Executive Councils), 34, 188
administration, 42–6, 222–38; burgeoning workload of, 228; control of, over rural economy, 222–8; personnel for, 235–8
Afghanistan, 71, 107; Amir of, arbitrates tribal disputes, 125; Anglo-Russo Boundary Commission on, 72–3, 75–6, 78, 109, 110;—Anglo War, (1838), 74–5, 122, (1878), 70, 71–2, 123, 124, 129, 244; Burnes mission to, 69; Durrand Mission to (Durrand Line), 125–6; maintained as buffer against Russia, 68–72, 133, 134, 179; reaction to 'forward policy', 125; supports Britain in World War I, 134; surveyed by occupying forces, 75, 78
Agha Khan, 205, 303; 'address' to Lord Minto, 277
Agents, British, in Indian states, 18–19
Agra, Presidency of, established, 173–4; limited powers of, 174; merged with Oudh, 177; reduced to status of Lt.-Governor's province, 176
agrarian legislation, 192–6
agrarian unrest, 192–3, 225, 226, 227
Air Force, Royal, becomes part of India's defence services, 251; Royal Indian, 253
Aitchison Commission on civil services, 236
Akalis, 359, 360
Aksai Chin, 110–11
Alexander, A. V., 361
All-Parties Conference, proposed (1942), 348
Ambedkar, B. R., 298
Ameer Ali, Syed, 194, 205
Amery, L. S., 347, 352
Amritsar, massacre at, 287

Amir Khan, 65
Anglo-Chinese Convention (1876), 89–90; (1886), 90; (1890), 93, 95, 98
Anglo-Chinese Treaty (1906), 99, 101
Anglo-Indians, 314
Anglo-Russian Treaty (1907), 99, 101, 103, 110
Anglo-Tibetan Relations, xxiv-xxvi; Agreement on McMahon Line (1914), 104–5
annexation, British policy concerning, 9; Queen's Proclamation (1858), ends policy of, 20; with reference to Indian states, 16–17, 18–20
aristocracy, landed, 10; divested of legal powers, 24
army, 58–67; Bengal, 66; Bombay, 58; Indianization of commissioned ranks of, 322; instruments of conquest, 63–4; King's/European/Indian, 59–61; officers, British and India, 61–2, 322; organizational pattern of, 59; Presidency, 58–9; prominent in survey activities, 66–7, (during operations), 75, 78; unification of, 63; work among tribal and 'backward' communities, 67
Army Commission, on defence of Assam, 157
Army, Indian, 248–9, 252; British officers in, 243–4; deficiencies, 249–50; effects of First World War on, 248–50; Esper Committee (1919), on re-organization of, 250–1; integration of, 241–3, 247–8; King's Commissions to Indians in, 249, 253; local corps in, 246–7; strength of (1914 and 1918), 248; territorial distribution of recruits for (1919), 250; university cadet corps, 253; *see also*, forces, armed
Army Organization Commission (1859), 241–3; (1879), 244–6; against reduction in strength of armed forces in British India, 245; for abolition of Presidency Armies, 245–6; *see also* Esper Committee (1919)
Army, Territorial, proposed, 250; Act (1923), 253–4
Arya Samaj, 285, 291, 319

Asaf Ali, M., 367
Asiatic Society of Bengal, 51
Assam, 134, 148, 153, 368–9; auxilliary armed forces in, 145–7, 152, 180; cantonments in, 144–5; Chief Commissionership of, 150–1, 153, 154, 155–6, 180; communications in, 142–3, 149, 152; Company, 149–50; Congress ministry in, 360; conquered, 134–5; developments of, 157–9; enlargement of, 158–63; Frontier Police, 156–7; frontier with Bhutan, 82–3; Hindu/Muslim employment in, 275; incorporated in Eastern Bengal and Assam, 161, 180–1; integration of, 142–7; lack of capable officers in, 158–9, 180–1; Light Infantry, 145, 147; Naga Hills district of, 151; 'non-regulation' areas in, 224; 'outsiders' in, 162, 180, 181; *Report on the Province of* (Mills), 140; restoration of Chief Commissionership for, proposed, 169; Supreme Govt. takes direct control of, 150–1, 153, 154; Sylhet Dist. given choice of joining East Bengal, 373; territories of, 154–5; *see also* East Bengal & Assam *and* tribal policy, British
associations, political, rise of, 196; consultations with, 200
Attlee, Prime Minister, Clement, 220; announces decision to transfer power by June 1948, 372; appoints Mountbatten, 371–2; recommends replacement of Wavell, 370; statement on Cabinet Mission, 361–2
Auchinleck, General Sir Claude, 254; opposes Wavell, 370–1
'August Offer' (1940), 220, 339
Auckland, Lord, 17, 176; annexations of Indian states under, 17,18,19
Aungier, Gerald, 3
Aurangzeb, Emperor, iii, 261; Hindu employment under, 272
Ava, *see* Burma
Ayyangar, N. Gopalaswami, 221
Azad, Maulana Abul Kalam, 303, 338, 347, 352; *India Wins Freedom*, 318;

negotiates with Cripps, 343–4; to Cripps on control of defence, 344

Baji Rao II, xvi, xx, 64
Balaji Baji Rao, xii
Baldev Singh, Sardar, 367
Baluchistan, 121–3, 130; *jirga* system in, 123–4
Banaras movement, 320
Bande Matram, 320, 337
Baroda, Maharaja of, 369
Bassein, Treaty of, xv–xvi, 64
Basu, Bhupendra Nath, 207
Bayley, Sir Steuart, on Angami Nagas, 141; on military dispositions in Assam, 156–7
Beadon, Cecil, on policy towards Nagas, 150
Bell, Charles, 101
Bengal, ii, 41; Army, 240, 245–6; (Calcutta), becomes seat of Central Govt., 170–1; (Calcutta), Chamber of Commerce, 197, 201; community-wise distribution in judicial service in, 270; Curzon moots partition of, 160–3; Eastern, attempt to set up Muslim majority unit in, 161, 180–1; Education Council, 191; extent of (1832), 171, (1868), 153; frontier with Bhutan, 83; grant to Nawab of, 269; Hardinge on community position in Legislative Council of, 168–9; Hindu/Muslim employment position in, 275, 277; Muslim League strengthens position in, 317; 'non-regulation' areas in, 224; opposition to detachment of districts from, 155, 160–1; Rent Bill (1882), 192–3; re-unification on linguistic basis proposed, 169–70; rise of middle class in, 258; shift of capital from Calcutta proposed, 171–3; Tenancy Act (1859), 192–3, (1885), 194, 196, 225, 227; trade with Tibet, 89; *see also* Chittagong Division; Bengal, partition of; Calcutta
Bengal, partition of, 155, 160–1, 202, 277; agitation against, 167–8; begins new political era, 181; Curzon's arguments for, 160–3; employment distribution in, 275, 277; officials isolated from masses in, 225
Bengalis, in Assam, 180, 181
Benn, Wedgwood, 304, 312, 329; on consideration of both majority and minorities, 385–6; on Simon Commission Report and RTC, 300–1, 305; proposes scheme for central government for India (1930), 328; to Irwin on Gandhi as buffer against 'terrorists', 301–2, 305
Bentham, Jeremy, 10, 32
Bentinck, Lord William, 11, 14, 45, 51, 232; appointed Governor-General, 173–4; educational policy of, 258; minute on N.E. frontier, 151; on communication deficiencies, 171–2; on internal security and effect of reform on, 238–9; on need to strengthen Manipur as frontier, 143, 151; on type of European colonizers needed, 18; urges shift of capital, 171–2; withholds powers from Metcalfe, 174–5
Bernier, Francois, on merchants' concealed wealth, 257
Besant, Annie, 283, 287
Bethune, Drinkwater, 29
Bhabha, Cooverji Hormusji, 367
Bhonsles, 165
Bhopal (case), 17; Nawab of 215, 369
Bhutan, 80, 82–4, 135, 155; British relations with, xxiv–xxvi, 83–4; civil war in, 84; frontier with Assam, Bengal, 82–3; invades Sikkim, 80; links with Tibet, 82–3
Bihar, 162; Congress ministry in (1946), 360; separation from Bengal, 169, 170
Bikaner, Maharaja of, 215, 369
Bird, William Wilberforce, on Muslim judicial officers and Hindu revenue and accounts staff, 270
Birla, G. D., 302
Board of Control (Parliamentary), for Indian affairs, 7, 13, 49; approves competitive examination for recruitment, 50; superseded by Secretary of

State in Council, 37, 185
Bogle, George, 79, 89; missions to Tibet, 79, 82
Bolan Pass, 122, 123
Bolshevism, *see* Communism
Bombay, 58; Army, 240, 245–6; Congress Ministry in (1945), 360; Presidency, 241
Bombay Plan, 355
Bombay Presidency Association, 197
Border Security Force, 119–20
Bose, Sarat Chandra, 367
Bose, Subhas Chandra, 336, 344
boundaries, *see* frontiers
Boundary Delimitation Commission, Anglo-Afghan-Russo, 72–3, 75–6, 78
Boundary (Radcliffe) Commission, 374–5
Brelvi, S. A., 303
Britain, immigration to India from, 10, 13; legal status in India of immigration from, 10, 13, 17
British Indian Association, 190, 197
Brodie, Col., mission to Naga areas, 140
Brodrick, St. John, on limited British interest in Tibet, 98
'buffer' territories on N.E., 179–80
Burma, 90, 112–15, 141; Anglo-Wars, xxi-xxiii, 151; annexation of Upper, 115; annexes Manipur, 151; attracts British as market, 112–13; Bombay–, Trading Corporation, 114; Chinese movements on border of, 102; French penetration of Upper, 114; measures against incursions from, 143–7
Burnes Mission to Kabul, 69
Butler (Harcourt), Committee (1929), on Indian States' paramount power relationships, 15, 213

Cabinet Mission, 220–22, 361–7, 370; finds Pakistan impracticable, 365–6; 'grouping' scheme of, 368; recommendations, 221; recommends and sets up Interim Government, 366; Simla meetings with Congress and League, 362; statement of unilateral recommendations, 362–5; strives to maintain Indian unity, 371; three-tier scheme for Union proposed by, 362, 368
Campbell-Bannerman, Sir Henry, 99
Campbell, Sir George, 195–6, 225
Canning, Lord, 40, 49, 154, 224; moderate policy of, 37, 188; proposals concerning legislative councils, 38
capital city, shift of, 171–3
caste: conflicts arising out of constitutional reform, 296; 'depressed', reservations for, 296; elimination of considerations of, in Indian Army, 241, 242; system (Weber on), 261; upper, dominance, 298
Census of India (1901), examines Muslim employment, 205
Central Government, functions of, 182; need for strong, 155; proposed structure of, 327–32; resists decentralization, 175; Simon Commission on, 327; under 1935 constitution, 324–5
centralization: of financial control, 13; legislative, 13, 33–4; maintained with growth of local bodies, 42; undertaken for reasons of economy and trade, 12–13
Central Provinces (CP), 165–6
Chamberlain, Sir Austen, 265–6
Chamber of Princes, 16, 303; negotiates with Cabinet Mission, 221
Charles II of England, 58
Chefoo Convention (1876), 89–90
Chen I-fan (Ivon Chen), 101
Chelmsford, Lord, 212
Chhatari, Nawab Ahmad Said Khan of, 303, 318
Chief Commissionership, 164, 184, 196; in Assam, 150–1, 153, 154, 155–6, 180; in C. P., 165; in Oudh, 167; in Punjab, 163–4; on N. E. frontier, 180; suits tribal areas, 179
Chief Justice (Supreme Court), 34
Chiefs' Conference, 211
China, 78; activities in Mishmi country, 102; attitude to Sikkim, 91, 93; gains from Lhasa Convention, 99; imposes severe terms on Nepal, 80–1; Joint

Index

Commission with Britain on Tibetan trade, 94–5; Burma as land route to China, 113; concerning Anglo-Tibetan agreement (1914), 104–5; suzerainty over Tibet, 79–80, 89–90, 99; *see also* Anglo-Chinese Convention(s); Simla Conference (1913)
Chittagong, 155, 158, 159, 160, 180
Chundrigar, I. I., 367
Churchill, Winston, 309, 351; statement in Commons (1942), on Cripps mission, 340
Civil Disobedience Movement, 217, 301, 319, 338
Civil Finance Committee, 14, 171
Civil Services, 46–50; all-India, 236–7; crucial composition of, 237; classification within, 224, 225, 236–7; Commissioners, 50; depletion of European element in, 355; examinations for recruitment to, 50, 236, 237; Indianization of, 292, 322; nomination of minorities to, 296–7; personnel for, 235–8; post-War shortages in, 254; 'provincializing' of, 253; training for, 48; under 1919 Act, 293; *see also* convenanted service
Clive, Robert, ii, iv, 3–4; on need for parliamentary control over growing territories, 3–4; reorganizes Indian Army, 60–1, 66
'close-border policy', 119–21, 124, 130, 179; end of, 122; Curzon urges revival of, 131; fails with Angami Nagas, 150
Code of Civil Procedure (1859), 30
Code of Criminal Procedure (1861), 29, 30
Colaba case, 18
Collector, 44, 45; constraints on, 225–6; heads district legal system, 21; key position of, 222–3, 230–5; wide discretionary powers of, 233
College at Fort William, 48
colonization, 18, 40, 41; 1833 Act opens prospects for, 10, 17, 18, 28; clash with imperial interests, 37, 183; necessitates central legislative authority, 33, 189, 191

Colvin, Sir Auckland, 200
Commander-in-Chief, 8, 34, 242, 247–8; takes over Presidency Armies, 246–7; *see also* Auchinleck, Sir Claude; Kitchner, Lord
Commissioner, Divisional, 222
Communal Award (1932), 215, 298, 314
Communalism, 266–7, 306–15; aggravated by Wavell's broadcast, 358; and 1919 reforms, 295–7; arising out of employment rivalry, 237; and provincial autonomy, 307; encouraged by classical studies, 274–5; fuelled by Jinnah's tactics (1945), 353–4; in Bengal politics, 168–9, 268–9; intensifies, 303–5; linguistic, 319–20; reflected at first RTC, 308
communal representation, 218; extended to 'depressed classes', 296; Montford Report on, 281–2
communal riots, 366–7, 368, 369–70, 372
communications and transport, 66, 187
Communism, 252, 265, 287, 313, 336; Congress attitude to 301; emergence of, in India, 291; 'Gandhi's fear of', 288, 301; Russian, encourages Khilafat Movement, 286
Community Party of India, 365–6
Company Courts, 26, 31
Congress, Indian National, 40, 181–2, 197, 214, 220, 252, 264, 266, 275;— Akali alliance, 360–1; and Left elements, 354, 355–7; assessment of Muslim League (1942), 345; in assemblies of Muslim majority provinces, 295–6; attitude to 2nd RTC, 301, 302; attitude to war effort (1942), 343–4; blind to communal issue, 307, 314;—British interests converge, 357; Calcutta (Dominion Status) Session of (1928), 300; Civil Disobedience Movement of, 312–13; contests 1937 elections, forms ministries, 219–20; decides to launch mass movement (1942), 347; elections, Bengal (1945–6), 360; elections, general (1945–6), 359–60; favours centralized constitution, 209;

GOI Act (1935), rejected by, 220, 326–7, 334; nominates nationalist Muslim to Interim Government, 366; Lahore (Independence) Session of (1929), 300–1, 303; Leftist groups in, 301, 313, 336–7; ministries (1946), 360; ministries resign, 321–2; Muslim organizations connected with, 303, 304; 1915 Bombay Session of, with League Session, 283–4; Non-cooperation Movement, 289–92; pamphlet on post-war food situation, 354; participates in 2nd RTC (1931), 302; policy vis-a-vis Communist Party, 356; post-war resurgence of, 354; reaction to Sept. '45 announcement, 358; refuses coalition with League 316, 317, 318, 367; rejects August (1940) offer, 339; rejects Cripps (1942) proposals, 343–5; rejects League offer of coalitions, 360–1; representative character of, 316–18; resolution against enforced 'grouping', 372; resolution on Indian States and federation (1938), 326–7; revivalism and radicalism in, 319–20; Swarajist wing of, 290, 295, 297; Wavell's post-election estimate of, 361; Working Committee, 339, 352, 368; Allahabad (1942) resolutions of, 347; *see also* Quit India Movement

Congress-League Scheme for Constitutional Reform (1916), 206–7, 284; *see also* Lucknow Pact

Congress Muslim Party, 303

Constituent Assembly, 221, 326, 358, 372–3; Cabinet Mission proposals concerning, 221, 363–4, 366; committee of, negotiates with princes, 221; Indian states in, 364, 369, 375; opens (1946), 368

Constitutional Reform, Indian, Report of the Joint Committee on, 237–8

Constitution, Federal (1935), 324–36; central powers under, 325–6; dual nature of, 324–5; Gandhi's personal views on, 335–6; provisions to safeguard unity in, 327–8; role of Home Department under, 331–2

Constitution, Indian (Republic), Draft (1947–8), 208–9

Cornwallis Code *see* 'non-regulation administrations'

Cornwallis, Lord, vii-viii, 9, 26, 44, 45, 47, 230, 269; enacts Permanent Settlement, 44, 223–4, 256

Council of India, 35, 40, 174, 185–6; Act (1869), 38; financial powers of, 185–6

Coupland, Sir Reginald, 320, 340

covenanted service, 7, 47, 48, 49, 235

Cranbrook, Lord, 151

Cripps Mission (1942), 220, 333, 340–5; background to, 333–40; proposals, 341–2, 351

Cripps, Sir Stafford, 340, 347, 361, 371; negotiations with Maulana Azad, 343–4; on proposed Constituent Assembly, 342–3; rejoinder to Gandhi, 347

Curzon, Lord, 70, 130, 211; considerations of, in planning expansion of Assam, 159–60; minute of, on territorial reorganization of Assam, Bengal, Bihar and Orissa, 159–60, 180–1; on India's peasant backbone, 264; significance of formation of NWFP by, 134; Tibet policy of, 96–8; views of, on frontier defence, 131–2, 179

Dalai Lama, 84, 89; flees Younghusband's mission, 97; stay in India of (1910–12), 100; 13th, asserts right to sovereign foreign policy, 90–1; 13th, contacts Tsar, 96

Dalhousie, Lord, xxiii, 38, 117, 196; accelerates change in legislative field, 34–5, 187–8; annexationist policy of, 19–20, 270; note on Punjab administration, 164; partial to *laissez-faire*, 36; promotes postal and telegraph services, and railway construction, 66; separates legislative and executive functions of Council, 35, 187–8

Dalal, Sir Ardeshir, 355

Darbhanga, Maharaja of, 191, 194

Darjeeling, 81, 82, 161

daroga-i-adalat, 21
Das, C. R., 294
Deccan Riots Commission, Report of, 196
Declaration, Montagu (1917), on Political Reform, *see under* Montagu, Edwin
Defence of India Act (1915), 287
Delhi, transfer of capital to, 169
'deliverance day' (6 Dec. 1939), 322
Deo, Acharya Narendra, 350
Deogaon, Treaty of, xvii
'Depressed Classes', 298, 314
Desai, Bhulabhai, 220, 352
Desai-Liaqat Pact (1944), 220
des Granges, Baron Otto, survey of areas between Bengal and Burma, 112–13; of Manipur, 141
'Direct Action Day', 366–7, 368, 370
division, causes of; demand for full provisional autonomy, 298–9; employment pattern, 269–78, 296–7; extension of democracy, 306–15, 316; linguistic, 274–5, 319–20; political, 284–92; representation in legislatures, 278–84, 295–7
diwani adalats, 21, 24–5
diwani of Bengal, ii, iv, 5; legal system for, 21–2; and affect on Muslims, 269; political management of, 47; reforms in administration of, 4
Dodwell, on status of local chiefs under central power, 15
Dorjiev, 96
Dufferin, Lord, 211; attitude to National Congress, 197; Committee on Provincial Councils, 199; concessions to zamindars, 194; on 'interest groups' in legislatures, 194, 198; warning to Tibet (1888), 91
Durand, Henry Mortimer, Sir, mission to Kabul (1893), 125–7; negotiates with Chinese Amban of Tibet, 92–4
Durand Line, 124, 126, 128, 132, 179; anomalies in, 126–7, 129; demarcation of, 127
Dutteah succession case, 18
'Dyarchy', 212, 217, 266, 268, 328

Eastern Bengal & Assam, new Province of, 161–2, 180; position of Assam in, 162, 168
East India Company, abolished, 2, 15, 37; acquires territories, 3, 43, 48; administrative pattern of, 1ff, 9, 42, 47–50, 222; conquers Assam, 134–5; corruption in administration of, 43–4; Court of Directors of, 7, 185; employees back Hindu businessmen against Mughals, 257–8; end of trade monopoly of, 10, 13, 48; formation of large administrative units under, 170; Macaulay favours continuation of, 32; military policy of, 2, 58–9, 66; mission to Ladakh, 107; parliamentary control of, 2, 5; opposes free trade, 47; policy towards Indian states, 15–16; powers of Board of Control over, 7; religious policy of, 262; economic development, 259–60
Eden, Sir Ashley, mission to Bhutan, 83–4
education, 263–4; Hindu, 261; Muslim, Committee on, 273–4; nurtures rise of middle class, 258–9; policy, 198, 263
Edward, Prince of Wales, visit to India, 294
elections, 203, 204–5; (Central Council), 205–9; (general, 1945–6), 317–18, 321, 354, 357, 359–60, 367–8; direct, 206, 208, 279; (1937), 316–17; restricted franchise for, 207–8, 221; territorial constituencies for, 207; under Morley-Minto reforms, 279–80; under Montford reforms, 268–9
electorates, separate, 205–6, 278–9, 298, 308, 310, 311, 314; joint demanded, 307, 310
Elias, Nay, 74
Ellenborough, Lord, 19
Elphinstone, Mountstuart, xix, xxi, 11, 12, 44
Esher (Lord) Committee (1919), 250–1
Europeans, 41, 314; at Simla Conference (1945), 352; demand repression of mutineers, 188; gain right of free entry into India, 183; in

civil appointments, 275–6; in legislatures, 280, 359; oppose Canning, 37, 188; oppose setting up of legislatures, 188–9; resist jurisdictions of *Sadar Diwani Adalats*, 29; under judicial system, 26–7; *see also* colonization

Executive Council of India, 34–5, 50, 183; as 'Interim Government' (1946), 366–7; Indians invited to join (1940), 339; Muslims in, 318; post-war plans for, 352, 353; reserved legislative powers of, 35

Executive Councillors, 186

executive councils, creation of, 211–12; Governor-General's, 318; set up in provinces, 211

'Extremists', 275

Famine Commission (1880), 192, 227, 260

famines, 195, 259–60; Bihar (1873), 195

Fazl-i-Husain, 296; on Muslim cooperation with officials, 306; presents Muslim view on Simon Report, 304–5

Fazlul Huq, 338

Federal Court, 325

Federal Legislature, 216, 299

federation, 302–3, 305, 309, 310–11; Congress resolution (1938), 326–7; delayed, 333–5; Jt. Select Committee on legal basis for, 214; legislature for, 216–17; Linlithgow-Zetland exchanges on inauguration of, 333–5; moves towards, 213–19; princes accept, 299; Simon Commission favours, 298; suspended, 335; safeguards and reservations under, 299–300, 302; *see also* Federal Legislature; constitution, federal (1935)

forces, armed, in British India: called out in support of civil administration, 252; communist influences in, 356; divided by politics, 254; Indianization of, 322; organized as three Corps under Kitchner, 248; placed under four 'commands', 246–7; racial composition of, 242, 244; *see also* Army, Indian

Fort William, 2; College at, 48; Mayor's court at, 23; supreme powers over presidencies, 8

'forward policy', 179; acquisition of tribal territory under, 124–5; adopted against Angami Nagas, 150; tribal reaction to, 128–9

France, English wars with, 59, 63–4

Franchise Committee, 207

free trade movement, 47, 48

Frere, Sir Bartle, 38, 39, 188, 236; on need for legislatures, 202

Frontier constabulary, 120

frontiers, administrative reorganization at, 116–34 (north-west); 134–42 (north-east); anti-British rising in N.W., 128–9; areas, central control over, 130–1, 153–63; Assam, survey of, 143; between Sikkim and Tibet defined (1890), 93; 'buffer' areas on N. E., 179–80; demarcation opposed by Tibet, 96; factors in delimitation of, 68–115; in Assam Himalaya, 102; Manipur as, 143; north-eastern, 134–48, 179–80; north-west, 68–76, 179; north-Himalayan, 78–112; north-western extended by Treaty of Gandamak (1879), 123; on East Naga District, 151; of Tripura defined, 141; 'scientific', 129–30; Tibetan claims on, 91; with Burma, 112–15; with western Tibet, 106–112; *see also* close-border policy; forward policy

Gaekwar of Baroda, 212

Gaffar Khan, Khan Abdul, 303, 317

Gandamak, Treaty (Anglo-Afghan) (1879), 72, 123, 124, 125

Gandhi, M. K., 181–2, 344–5, 352; accepts Rajaji formula, 350; and Rowlatt Bills agitation, 287–8; anxious to help government (1939), 336, 339; appeals to Jinnah, 322; arrest and release (1930–1), 302, (1942, 1944), 347, 349; at second RTC, 313–14; Civil Disobedience

Movement of, 217, 301, 319, 338; concern at influence of Left, 336; correspondence with Viceroy (1942), 347; differences with Nehru and Rajagopalachari, 345–6; 'fears communism', 288, 312; intervenes in Rajkot, 334;—Irwin Pact, 302, 306, 313;—Jinnah talks (1944), 349–51; League ('Pakistan') Resolution (1940), moderating influence on, 338; Non-cooperation Movement of, 289–92, 293; on Cripps offer, 345; plea for withdrawal of Allied troops, 346; rejects compromise on visit of Prince of Wales, 294–5; regrets losing opportunity to shape 1935 constitution, 335–6; represents Congress at second RTC (1931), 302; seeks interview with Viceroy, 312, 313; seeks post-war reconciliation with Britain, 356; supports Khilafat Movement, 288, 293; talks with Viceroy (1940), 338–9; urged by big business to compromise, 312; wants nationalist Muslims accepted as representative, 337; wooed by Cripps and Pethick-Lawrence, 371

Gartok, trade mart, 97, 102

Ghazanfar Ali Khan, 367

Ghose, Ram Gopal, 190–1

Gladstone, William, 200

Goddard, General, xiii

Gokhale, Gopal Krishna, 283, 289; scheme for post-war reforms (1914), 206–11; urges self-government, 204

Governor-General, assent of, required for laws, 35, 42, 192; powers of, 38, 39, 201, 217; powers under GOI Act, (1919), 292–3, (1935), 218–19, 293–300; 'Special Responsibilities' of under 1935 constitution, 300

Governor, 2; widening political involvement of, 2–3; appointment of, 6; 'assent' of, required for legislation, 42; exercise of 'Special Responsibilities' in provinces, 300; financial powers of, restricted, 13; military status of, 59; overriding powers of, 217, 218, 299, (under 1935 Act) 219; subordinated to G.G., 14; transfer of 'reserved' subjects to, 292–3

Governor-General and Council, 184, 186–91; appointment and powers of, 5–6; Bengal civilians' monopoly in membership of, 38, 50; direct responsibility for Bengal withdrawn from, 14; enhancement of personal powers of G.G., 8; 9; financial control centralized with, 13; G.G. holds ultimate power in, 194–5; growing responsibilities of, 164–5; growing powers of, 13, 14; Indians invited to join (1940), 339; legislation becomes collective responsibility of, 31; legislative powers of, 33

Grenville, Lord, 13, 236

Grey, Sir Charles, on pluralities in legal process in British India, 27

Gulab Singh Dogra, 107–9

Gurkhas, 147

Gwalior, case 17: Maharaja of, 211

Gyantse trade mart, 97, 102

Haig, Sir Harry, 328; on need to keep Congress out of office, 336–7

Haileybury College, 48, 50

Hailey, Sir Malcolm, 328, 329; scheme for central government, 330–2

Halliday, Sir Frederic, recommends Collector-Magistrate system, 225

Haque, Mazharul, 207

Hardinge, Lord, xxix, 19; border policy of, 103, 111; *Confidential Memorandum* (1915), on provincial self-government, 212; on non-official elements in councils, 168–9; policy towards princes, 211; promotes railway construction, 66

Hart, James (Irish adviser to Amban in Lhasa), 93; negotiates Tibetan trade matters with British, 94–5

Hastings, Lord, xviii, xix, xx; Himalayan policy of, 80, 82, 88; policy towards Indian States, 209, 230

Hastings, Warren, xxv-xxvi, 6, 24, 47,

65, 88; appoints Collectors, 44; fights corruption, 43; initiates Surveys, 54–5; legal reforms in diwani area, 21–2; on chaos of administration, 43–4; on need for political base for government, 4–5; policy of 'subordinate isolation', 16; promotes Sanskrit studies, 51; Tibet policy of, 79, 84
Hewett, J. P., on territorial reorganization, 160
High Courts, 30
Himalaya, Assam, 100–6; surveys of, 84–88
Hindi, 179, 275; challenges Urdu in N. W. Provinces and U.P., 272, 320;—Hindustani controversy, 321; promoted for vernacular education, 319
Hindu Mahasabha, 285, 290, 291, 307, 319, 333, 360
Hindu-Muslim comparisons— employment, education, language, 269–77
Hindus, 55, 295; acceptance of Mughal rule, 261; at first RTC, 307–12; as political workers, 319; Bengali, in legislative council, 168–9; become adept in Persian and Urdu under Mughals, 271–2; business groups among, 257–8; communalism among, fuelled by Jinnah's intransigence, 353–4; employment position of, 270, 275–7; in Bengal Partition agitation, 168; take to English education, 272
Hoare, Sir Samuel, 316
Holkar, xiv–xv, 64
'Home Charges', 185
Home Rule League, 287, 289
Home Secretary, 39
Hume, A. O., 197
Hyderabad, Nizam of, 15; delays accession to Indian Union, 375

Ibbetson, Denzil, 229; on need to transfer Chittagong division to Assam, 158; on territorial reorganization, 160; on tribal fear of the administration, 226; on zamindari oppression of tribal people, 225–6

I.C.S., 183
Ilbert, Sir Courteney, on jurisprudential confusion, 27–8
immigration, British, 19–20, 40; necessitates uniform political legal system, 10, 13, 28, 34
Imperial vs. colonial interests, 37, 48
Imperial, Advisory Council, 211; Council of Ruling Princes, 211; Advisory Council, 211
Impey, Sir Elijah, 24–5
Income Tax Bill (1886), 198–9
India Act, see under Act
India Council, 14
Indian Association, 197
Indian Councils Act, (1861), 38, 39, 41; (1892), 41; (1909) enlarges councils, 168
Indian Penal Code 1860), 29, 30, 31, 262
Indian Sandhurst Committee Report, 253; on increasing number of Indian King's Commissioned Officers, 253
Indian States, 15–20; accede to Indian Union, 375; accept federation, 215, 217, 218, 299, 306, 325, 327; Agents in, 18–19, 210; alarmed by Congress resolution (1929), 303; and Constituent Assembly, 364, 369; and integration, 16–17, 209–22; annexation of, by British, 16–17, 19–20; delegation to first RTC, 215, 299; institutional agencies in, 209; Indian Independence Act not applicable to, 375; Linlithgow-Zetland exchanges on, 333–5; relations with British, 210–19, (examined by Butler Committee), 15, 213; relations with central power, 15–16; relations with, transferred from GOI to Crown representative, 219; status and powers of, 15, 210; status under 1935 constitution, 324–6; steps towards federation with British India, 213, 299; succession in, 16–17; treaties with, 15–16; Queens Proclamation (1858) on, 20
India Office, 38
Indian National Army (INA), 358
indigo: disturbances (1889), 192, 225; cultivation ended, 225

Indore case, 17, 19
Indore, Maharaja of, 211
industry, growth of, 259–60, 265, 284
integration, 41; administration as force for, 42, 183, 222–41; agencies of, 1, 183; army's role in, 63–7, 182; beginnings of, with Indian States, 213; civil service role in, 46–50, 182; collectorates contribute to, 45–6; administration's contribution to, 20, 42, 222; emergence of all-India polity, 11, 12, 28; GOI Act (1935) provides for, 218; in political and diplomatic affairs, 9; legislative, 10, 13; political, 1–20; position of Indian States in process of, 16–17, 209–22; promoted by uniformity of judicial practice, 25; provisions in 1935 Act to ensure, 327; role of English language in, 275; setbacks to, with extension of democracy, 306–15; surveys contribute to, 50–7; threatened by demand for full provincial autonomy, 298–9; through Board of Control, 7–8; through central legislative authority, 31–42; through Central Secretariat, 46; through federation, 213; through legal reform, 30–1; through socio-economic policies, 254–62; unified civil service and, 49–50, 183, 237
Interim Government, post-war, 221, 351, 368; dissension within, 369; paralysed in face of Punjab riots, 372; proposed and formed, 366–7
Iqbal, Muhammad, 303; and idea of 'Muslim India', 297
Irwin, Lord, forwards Fazl-i-Husain Note to Wedgewood Benn, 304;—Gandhi Pact, 302, 306, 313; on Jinnah, 307; statement on 'dominion status', 300
Islamic law, 22, 25
Islington Commission, 249; on distribution of posts between Muslims, Hindus and others, 277
Ismay, Lord, 372, 373

Jagat Narain Lala, 346
Jagjivan Ram, 367

Japan, 340, 345, 346
Jaipur, 334
Jayakar, M. R., 302, 309, 368
Jenkins, 139, 149; on passage through Naga country, 139; *see also* Jenkins, Francis and R. B. Pemberton
Jenkins, Francis and R. B. Pemberton, survey Assam frontiers, 143; recommend disposition of cantonments in Assam, 144–5
Jervis, Sir John, 30
Jinnah, Mohammad Ali, 207, 283, 289, 303, 322, 338, 352, 362, 369; at first RTC, 301–7, 310–11; calls for observance of 'deliverance day', 322; declines/agrees to assist formation of Interim Government, 367; exploits cultural and linguistic fears of Muslims, 320–1; 'fourteen points' formulated by, 303, (at 1st RTC), 308, 311;—Gandhi talks (1944), 349–51; meeting with Viceroy (1945), 353; on Quit India movement, 347; reaction to Congress demand for immediate independence (1940), 339; reacts to Nehru's 'arrogance', 321; reported anxiety to settle with Congress (1942), 345; tactics vis-a-vis Wavell Plan, 353–4; terms of, for wartime cooperation with Congress, 337; view on Rajaji formula, 350
Johnson, Col. Louis, 220, 344, 346
Joint Select Committee on Constitutional Reform (1919), 204, 207, 292; (1934), 207–8, 209, 214, 215, 218, 298–9
Jones, Sir William, 51
Judge-Magistrate system, 44
judges, represented in Executive Council of India, 34, 39
judicial system; Company Courts, 26; criminal justice, 25, 26; European supervisors in, 22; executive acts subject to, 232–3; jury, trial by, 25–6, 271; justices of the peace, 27; King's Courts established in Calcutta, Madras and Bombay, 22–3, 26; Mayor's court (Fort William), 23; Mughal, 21; Muslims lose

employment opportunities in, 270–1; on acquisition of diwani, 21–2; pluralities in, 26–8; prior to 1834, 20–8; Sir Courteney Ilbert on, 27–8; supreme courts (Madras and Bombay), 27; Supreme Court of Judicature (Fort William), 20–4; *see also* law

Kalat, 122–3
Kashipur, Raja Sheoraj of, advocates Hindi *vs* Urdu, 273
Kashmir, 107–9, 375
Khilafat Movement, 285–6, 287, 290, 295; communist infiltration of, 291; Gandhi's support to, 288, 293
Khaliq-uz-Zaman, Chaudhury, 318
Khudai Khidmatgars, 303
Khyber Pass/Rifles, 124
Kidwai, Rafi Ahmed, 319
Kimberly, Lord, 200
King's Courts, 22–3, 26, 31
Kishen Singh, Pandit, 85, 88, 109
Kitchner, Lord (later Field Marshall), on principles guiding defence policy, 247; takes steps to unify army in India, 247–8
Kol rebellion, 226
Kotach case, 17, 19

Ladakh, 106–9, 110–11
laissez-faire, 9–10; abandoned, 259–60; antithesis of welfare measures, 195; European legislators promote, 36; inadequate for tackling agrarian problems, 192–3
Lajpat Rai, 285
Lake, Gerald (Lord), 64
Landsdowne, Lord, 92, 98, 123
language, politics of, 167–70, 175, 179, 181; in Bengal Partition agitation, 167–8; proposed reorganization of provinces on the basis of, 181, 298
'lapse', *see* annexation
law, 20–31; civil, 30; Member 34, 183; replaces custom in revenue administration, 222; under Collectors, 45–6; *see also* judicial system
Law Commission: First (1834), 28, 29; Second (1853), 29–31

Lawrence, Sir Henry, xxx, 163–4
Lawrence, Sir John, xxx, 71, 163; appointed Chief Commissioner, Punjab, 164; appointed Lt. Governor, Punjab, 165; opposes council form of government for Bengal, 154; urges constitution of Assam as a separate Chief Commissionership, 154
Lee Commission, 253
legislation: by councils, 9; centralization of, 13; consultation with public opinion on, 40; for social reform, 262; *see also* Acts
Legislative Assembly, Central: discusses Esher Committee Report on Indian Army, 251; results of 1945 elections to, 359; *see also* Legislative Council of India
Legislative Council of India, 30, 33, 35, 40, 165, 192–209; composition of (1892), 205–6; Dalhousie establishes procedures for, 35, 39, 196; discusses Esher Committee Report on army reorganization, 251; elections to, 206–9; elective principle indirectly recognized for, 201; exclusive European control of, 36; expansion of, 203; GG's power to override, 194–5; memorandum of Indian members of (1916), 204, 207; Ordinary (legislative) member of, 34; powers reduced, 38; restricted elected representation in, 279; zamindars dominant non-officials in, 194
legislative council(s), 31–42, 187–91, 185–209; Cannings' proposals concerning, 38–9; Dufferin Committee's recommendations concerning, 199–200; enlargement of, 168–9; established for N. W. Provinces, 178; European domination of, 39; expansion of, 202–3; Indians enter, 180; nominated members of, 39; opposed by Europeans, 188–9; preponderence of lawyers in, 279–80; representation of Muslims, 277–84; Swarajists contest elections for, 295
Lhasa Convention (1904), 97–8, 99, 110

Index 415

Liaqat Ali Khan, 352, 367
Liberals and Liberalism, 10, 28; growing influence of, 10–11; Indian, 55, 275, 283, 290, 309, 314–5; Indian policy, 45; Tibet policy of, 99, 310
linguistic provinces/states, 181
linguistic rivalries, 319–20
Linlithgow, Lord, 322, 326, 335–6, 337, 338; 'August (1940) Offer' of, 220, 339; on Indian States, 333–5; on minorities, 323; local bodies, 235
Lonchen Shatra, 101
Lothian, Lord, proposed scheme for central government, 328–30
Lowe, Robert, 30
Lucknow Pact (1916), 282; Fazl-i-Husain's critique of, 304–5
Lytton, Lord, 71, 130, 151, 211, 244, 274

Macaulay, Lord, 40, 187, 262–3; appointed '4th Ordinary Member' of Legislative Council, 34, 173; frames criminal code, 28–9; Minute on education, 198, 258; on central executive and legislature, 32; on civil service recruitment, 49–50, 236; 'Standing Orders' of, 183
MacDonald, Ramsay; backs federal scheme, 217–18; policy statement on India (1931), 309–10, 311, 312; proposes scheme for central Indian government (1930), 328
Mackenzie, Alexander, 67, 135, 139
Mackenzie, Holt, 45, 256; opposes Judge-Magistrate system, 230; on fragility of sepoy 'loyalty', 240; urges Collector-Magistrate system, 231–5
Madras and Bombay Armies Act (1893), 246
Madras Mahajan Sabha, 197
mahalwari system, 45, 222
Mahmudabad, Raja of, 318, 320
Mahomed Reza Khan, 22
Malaviya, Madan Mohan, 207, 294–5, 312, 360
Malcolm, Sir John, xvii, xix, xx, 11, 12, 69, 239
Mandal, Jogendra Nath, 367
Mandavi case, 19

Mandlik, Rao Sahib V. N., 194, 198
Manipur, 151–2; as frontier, 143; people of, as soldiers, 147
mapping, 52–4, 56–7
Marathas, vii, xii-xxi, 64, 165
market, control of, 195–6
Markham, on Boyle's mission to Tibet, 79
Marris, Sir William, on distribution of civil appointments, 276; opposes linguistic provinces, 181
Mathai, Dr John, 367
Mayo, Lord, 71; encourages Muslim education, 274–5; separates Assam from Bengal, 155–6
Mazumdar, Rai Jadunath, 293–4
McMahon, Sir Henry, 101, 103, 104; Commission, 101; Line, 102, 103, 104, 110, 111; on Simla Conference (1913), 101; talks with Tibetan representative, 102, 180
Meherally, Yusuf, 303
Menon, V. P., appointed Secretary, States Department, 375–6; and the transfer of power, 373–4; and on the Act of 1935, 220
Mesopotamia Commission, 186
Meston, Sir James, 284–5
Metcalfe, Charles, Lord, xxi, 12, 18, 44, 45; accepts Governorship of N.W. Provinces, 176; acts as Governor-General, 175; appointed Governor of Agra Presidency, 173–4; letter to Bentinck concerning powers of, 174–5; on the Kol tribe, 226
middle classes, 183, 184; divisions among, 201; fear of communism, 288; growth of, 189–90, 255–57, 257–9; growing British dependence on, 265–6; non-English-speaking, 319
military policy and organization, 238–54; all-Indian unification of, 241–8
Mindon, King of Burma, 112, 113
Minto, Lord, 69, 99, 205, 211; concedes separate electorates for Muslims, 205, 278–9; expands representative element in legislatures, 202–4; Muslim 'address' to, 277; proposes 'buffer' between Indian and Chinese

territory, 102–3; *see also* Morley-Minto Reforms
Mir Jafar, 64
missions, Christian, 37, 226, 262
modernization, *see* westernization
Mohammad Ali, 285–6, 289, 294–5; leads delegation to Europe, 285–6
Mohammad Shafi, 303, 310, 311
Montagu-Chelmsford Report/Reforms, 16, 204, 252, 280–1; establishes 'responsible government', 268; Fazl-i-Husain on Muslims under, 304; inquiry into by (Muddiman) Committee, 295; on communal representation, 281–2; position of Muslims under, 318–19; concerning Indian States, 212–13; rejected by nationalists, 286, 289; rejects linguistic provinces, 181; on wartime record of councils, 280–1; *see also* Act, G.O.I. (1919)
Montagu, Edwin, 207; aftermath of resignation of, 295; Declaration on Political Reform (1917), 266, 268, 280–1, 300; on parliamentary control and Indian affairs, 292–3
Montford Report *see* Montagu-Chelmsford Report
Moonje, B. S., 307, 308, 309, 310, 311, 333
Moorcraft, William, 107, 108, 109
Moplah rebellion (1921), 252
Morley, John (Lord), on attitude of 'gentlemen' towards elections, 205; on his reforms, 203–4; pro-Chinese attitude of, 99; proposes system for Muslim representation, 278
Morley-Minto Reforms (1909), 202–4, 206, 268; councils, 278–9, 283
Mountbatten, Lord Louis, 372–4; accepts V. P. Menon's plan, 374; holds back Radcliffe Boundary Commission Report, 374–5; persuades princes to accede, 375; sees inevitability of partition, 372, 373; time-table for transfer of power, 374
Muddiman, Sir Alexander (Reforms Inquiry Committee), 295, 296
Mughal(s), ii, iii, xii, 13, 25, 178; fall of, effect on Muslims, 271; Emperor, powers of, 9 (n. 15); grant diwani to Company, ii, iv, 4, 5; Hindus prominent in service of, 271; Hindu traders etc. under, 269; judicial system of, 21; mansabdari system of, 46; replaced by British, 14; revenue system of, 254–6; rule, basis of, 261
Muhammadan Anglo-Oriental College, 274, 276
Muhammadan Association, National, 197; Central, 205
Muir, Sir William, 274
Mukerji, Babu Joy Kissan, 192
Mukerji, Raja Pearey Mohan, 192, 194, 198
Mundas, 226
Munro, Sir Thomas, xxi, 11, 12, 44, 256; on dangers from sepoys, 239–40; on role of Collector, 230
Muslim(s), 55, 205–6; 'address' to Lord Minto, 277–8, 284; adversely affected by replacement of Persian, 271; alarmed at Congress strength, 313–14; alienated by Nehru's mass contact programme, 319; and Simon Report, 304–5, and trade, 269; as judicial officers, 270; cooperation with official block in legislatures, 306; divisions among, 303, 307; education, committee on (1872), 273–4; employment among, 205, 270, 273, 275–7; fall of Mughals and, 271; interest groups among, 304; in Viceroy's Executive Council, 318; land grants to, 271; reaction to Congress demands, 199; literacy among, 275; majority in E. Bengal & Assam, 161–2, 168, 180–1; nationalist and leftist, 303–4, 359, 360; pan-Islamism among, 285–6; representation in legislatures, 205, 206, 277–84; separate electorates for, 205–8, 277–8; suspicious of Labour Govt., 308; under 1919 constitution, 295–7
Muslim Conference, All-India (1929), 297, 303
Muslim League, All-India, 205, 220,

Index

221, 278, 286, 297; alienated by Nehru's 'mass contact' programme, 319; attitude to Nehru Report, 298;— Congress relations in U.P., 318–19; failure in 1937 elections, 316–17; forms ministry in Punjab and Bengal, 360; impracticable form of Pakistan demand, 365–6; in Assemblies, 295–7; Iqbal's 'Pakistan' address to (1930), 297–8; 1915 session of (Bombay), 283–4; observes 'Direct Action Day', 366–7; offers coalitions to Congress, 360–1; opposition to federation, 333; pact with Congress (1916), 282, 284; Pakistan resolution (1940), 322–3, 338, 350; Quit India movement, effect on, 345, 348; reaction to 'August (1940) Offer', 339; refuses to name executive councillors at Simla Conference (1945), 353; regarded as ally by Wavell, 371; rejects Constituent Assembly, 368–9, 372, Cripps Proposals, 342–3, Interim Govt. offer, 366; revival of, 290–1; rise of, 318–24; setback in collapse of States' Negotiating Committee, 369; success in 1945–6 elections, 317–18, 355, 359–61; two factions in, 303; vetos constitutional advance, 339, 354; welcomes September (1945) announcement, 358; welcomes Viceroy's assurance, 323; withdraws from Constituent Assembly, 221

Muslim Nationalist Party, All-India, 303

Mutiny, 30, 33, 36, 37, 57, 67, 82, 108, 188, 241

'Nabobs', 4
Nana Fadnis, xii, xiii, xiv
naibs, 22
Naidu, Sarojini (Mrs), 356
Nain Singh, Pandit, 89, 109; trans-Himalayan surveys by, 85–8, 180
Nandkumar, Maharaja, 262
Naoroji, Dadabhai, 197
Napier, Gen. Sir Charles, occupies Sind, 121, 241; on internal security, 117

Narayan, Jayaprakash, 313
Nationalist Party, 352, 360
nationality and nationalism, Indian, emergence of, 238
naval revolt, 358
Nehru, Jawaharlal, 291, 322, 344, 346, 347, 356, 372; attitude to Jinnah and Muslim League, 321; concedes monarchical system in Indian States, 369; differences with Gandhi, 345–6; invited to form Interim Government, 367; 'mass contact' programme of, 319; moves objectives resolution in Constituent Assembly, 368; opposes Mountbatten's plan for freedom for multiple successor states, 373; persuaded by V. P. Menon to accept partition plan, 373–4
Nehru, Motilal, 302, 303, 312
Nehru (Motilal) Committee, Report of (1928), 207, 297, 298
Nepal, 80–2; Anglo-War, 80, 81, 82; conflict with Tibet, 80–1; invades Sikkim, 80–1; policy towards Russia, 96
Neufville, Capt., survey of Sadiya Tract by, 138
Nizam of Hyderabad, vii, viii, xvii, 213, (army of), 240
nomination, of minorities to services, 296–7; to legislative bodies, 203, 206
Non-cooperation Movement, 289–92, 293–4, 295, 319; govt resolution on, 289–90; proposed as form of resistance to Japanese invasion, 346
'non-regulation' administrations, 224; in C.P., 165–6; difficulties of, in mixed areas, 226; in eastern India, 224; in Punjab, 163–4; in Santhal Parganas, 227
Northbrook, Lord, 71
North-Western Provinces, 41, 45, 166, 177, 189; constituted with Metcalfe as Lt. Gov., 176; Hindu and Muslim employment in, 275; merged with Oudh and Agra to form United Provinces, 177; need for separation from Bengal, 171; Urdu-Hindi

confrontation in, 272
North-West Frontier Province, 70, 117–21, 128, 130, 132–4; 178; administered by Political Service, 118; Agency System in, 128, 132; boundary of, 117–18; Congress forms ministry in (1946), 360; constituents of, 132; defence of, by political management, 118; extended under 'forward policy', 124–7, 130; legal system in, 120–1; problems of maintaining law and order in, 133; referendum in, 372; trans-border territories of, 119, 126; detached from Punjab, 165; why constituted, 133–4

Ochterlony, General, xxi
Orissa, attached to Bihar, 170; Congress Ministry in (1946), 360; Famine Commission, 195–6; language politics in, 170; 'non-regulation' areas in, 224; separation from Bengal proposed, 169, 229
Oudh, 167, 171, 270, 275; merged with Agra to form UP, 177–8
Outram, General James, 167

Palme Dutt, R., 348
panchayats, 208; 209; Holt Mackenzie on role of, 234
Panchen Lama, 99
pan-Islamism, 285–6, 287, 288; communist exploitation of, 291
paramountcy, 16, 28, 31, 42, 63–4, 210, 325; over tribes, 148
Partition, 221–2, 266–7, 375–6; and Indian States, 375–6; and 1945–6 elections, 359–60; cause of, 316, 318; long-term responsibility for, 375–6; Mountbatten sees inevitability of, 372
Patel, Vallabhbhai, 347, 360, 367; heads States Department 375–6; persuaded by V. P. Menon to accept partition, 373
Pathans, 119, 120–1, 221, 124
Patiala, Maharaja of, 369
patronage, 6–7, 47, 48, 49, 235; obstacle to development of all-India service, 48–9; conflict over, 175

Paul, A. W., 92; negotiates Tibetan trade matters with Chinese, 94–5
Peacock, Sir Barnes, 29
Pemberton, R. B., Captain, mission to Bhutan, 83; see also Jenkins, Francis and R. B. Pemberton
Permanent Settlement, 44–5, 223, 256–7, 260; Hindus gain from, 271; obstructs contact between officials and people, 225–6
Perron, General, 64
Perry, Sir Erskine, on anomalies in judicial system, 27
Persia, attacks Afghanistan, 70
Persian, 51; Hindu mastery of, 271–2; in universities, 274; replaced by English in courts, 271
personnel, administrative, 235–8
Pethick Lawrence, Lord, 220, 357, 361, 371
Phillip, William, 346–7
Pindaris, xix, 64–5
Pitt, William, 4
planters, 191, 192, 259–60; allied with zamindars, 191, 225; Association of, 206; oppose Tenancy Act (1885), 225; oppression by, 192, 224
Plassey, Battle of (1757), 64
Police Act, Indian (1861), 30
Police Commission (1860), 30
Political Agencies, 132
Political Service, India, 32–5; administers NWFP, 118, 128; wound up, 375; see also Agencies, Political
political system (Brit. India), 184–209
Poona Pact (1932), 215, 298, 314
Poona Sarvajanik Sabha, 197
Portuguese, 69
Prasad, Dr Rajendra, 347, 367, 368; on restricting direct elections to village panchayats, 209
Presidencies, 2, 47, 48, 59, 170–1, 184; armies of, 240–1 (reorganized), 241–3 (abolished), 245–6; subordinated to Fort William, 8
press, 41, 196, 197, 202
Princes (Chamber of), 213, (Council of), 211, 212; see Indian States
Privy Council, 211

Index

Property Bill, 193
provinces, autonomy and full democracy for, 217, 219, 298–9, 309, 324–5; establishment of, 170; Jinnah's 14 points for reorganization of, 303; Joint Select Committee urges strong executive for, 218; proposals for redistribution of, 297–8, 305, 307, 314; results of 1945 elections in, 360
Provincial Council, 21
public opinion, 167
Public Service Commission, (1886–7), 205
Punjab, 41, 69, 189; administration of tribal areas in, 163–5; anomalous position of N. W. frontier areas in, 130–1; Commission, 118; constituted Lt.-Governorship, 165; Land Alienation Act (1901), 265, 296; League ministry in, 360; League gains strength in, 317, 360; Pathan tract retained in, 132; Unionist Party in, 296
Punjab Irregular Force, 119–20
Purandhar, Treaty of, xiii

qazis, 21, 25
Quetta, 122, 123, 130
Quinton, J. W., 152
Quit India movement, 254, 345, 347–8, 354

Radcliffe, Sir Cyril, Boundary Commission, 374–5
radicalism, 198, 206
Raghunath Rao, xii, xiii
railways, 66
raiyats and raiyatwari system, 44, 45, 222; protected under Central Rent Act, 192; protection to, against tribals, 148
Rajagopalachari, C., 322, 348, 367; constitutional proposals for solution of deadlock, 349; formula, basis for Gandhi-Jinnah talks (1944), 349–51; urges Congress to concede right to federating units to secede, 346
Rajkot crisis, 334
Rajput states, 19

Ranjit Singh, xxix, 107
Rau, B. N., 209
Rawlinson, General, C-in-C, 252–3
Reading, Lord, 294, 309
'Red Shirts', 303
Reform Bill, 1832 (England), 10–11
Reforms Inquiry (Muddiman) Committee, 295
regionalism, 175, 181–2
Regulating Act (1773), 2, 5–6, 20–21, 24, 25, 44
Regulations, 262; concerning Collectors, (1772), 44, (of 1822) 230; concerning elections (1893), 205, 206 (1909), 206; concerning rent or revenue civil suits (of 1831), 232–3
'regulation system', 181
religion, British policy on, 262
Rent Act, Central, 192
'reservations', 217, 218
'responsible government', British government declaration on (1917), 204; under 1919 Act, 293–4
Retrenchment Committee (Inchcape) (1922), 251–2
revenue, Boards of, 222; Committee, 21; system, 222–8; (assessments subject to judicial review), 232–3; (British Indian), 256–7; (Mughal), 254–6
revivalism, 199, 206, 264–5, 260
revolutionary movements, 168
riots, communal, 290
Ripon, Lord, 205
Risley, H. H., on administrative workload, 228, 229; on need for larger Assam, 159
Romilly, Sir John (later Lord), 30
Roosevelt, Franklin D., 344, 346, 351
Rose, Archibald, 101
Round Table Conference, First (1930), 214–15, 306–9, (official report on), 308–12, (Minority Sub-Committee at), 310, 311, (Federation Committee), 310–11, 324, 326; Second (1931), 301, 302, 313–14; Third (1932), 215, 314–15
Rowlatt, Justice, (Committee), 287, (Bills/Acts), 287

Roy, M. N., 338
Roy, Raja Rammohun, 11, 55
Russia: activities in Central Asia, 70–1, 111–12, 121, 127, 130; activities on Afghanistan's northern border, 72–3; Afghanistan maintaned as buffer against, 68–72, 133, 134; Anglo-Afghan Boundary Commission, 72–3, 75–6; influence of, excluded from Tibet, 99; Lhasa Convention and, 98; movements towards Ladakh, 107, 110; 19th century British policy in relation to, 109–112; presence of, in Tibet, 96; relations with Afghanistan strengthened, 71; support to pan-Islamism, 286–7; war with Japan (1904–5), 202
Ryan, Sir Edward, 30

Saadat Ali, viii
sadar diwani adalats, 21, 24–5; jurisdiction extended over Europeans, 29
sadar faujdari adalat, 21
Sadr or *Sadr Amins*, 29
'safeguards', 216, 217, 218
Sandeman, Sir Robert, 123; frontier defence policy of, 179; presses agreement with Khan of Kalat, 122; tribal policy of, 121–2
Sandhurst, 253
Sanskrit studies, 51–2, 274
Santhal rising, 227
Sapru, Tej Bahadur, 207, 215, (Committee), 220, 308, 310; seeks reconciliation (post-1942), 348
sati, 262, 265
Satyagraha Sabha, 287
Savarkar, V. D., 285
Scott, David, 147, 149
Secretariat, Government, 46, 228–30, 235
Secretaries to government, 228–9
Secretary of State for India, 7, 184–7; powers under GOI Act (1858), 37–8, 39, (and 1869), 185–6; role under 1919 Act, 292–3; under Act of 1935, 324; takes over as appointing authority for civil services, 50

Select Committee (1832), 32, 35, 40, 49
Shafaat Ahmad Khan, 303, 321, 367
Shafi Daudi, 303
Shah Alam, ii, xii
Salbai, Treaty of, xiii
Shaukat Ali, 289, 294–5
Sheng Tai, Amban at Lhasa, 94; negotiates Tibetan trade matters with British, 94–5
Sher Ali, Afghan Amir, 71, 123
Shradhanand, Swami, 285
Sikandar Hayat Khan, Sir, 337–8
Sikh(s), kingdom, xxviii–xxx, 69, 107–8, (frontier administration of), 117–18; representation in councils, 282–3
Sikkim, 80–2; relations with British and Tibet, 91, 96; Anglo-Chinese Convention on (1890), 92–3, 95; attacked by Tibet, 91; Britain extends control over, 80, 81, 82; invaded by Nepal and Bhutan, 80–1
Simla Conference (1945), 351–4, 355
Simla Conference and Convention (1913), 101–5, 111; China disapproves McMahon Line at, 104; direct Anglo-Tibetan talks at, 102, 104, 180; Russo-Chinese declaration on Mongolia at, 104–5
Simon Commission, 181–2, 207, 214, 217, 252, 298, 300–1, 304–5, 307, 327–8
Sind, 69; conquest of, 241
Sindia, xii, xiii, xiv, xvi, xvii, xx 64, 65
Sinha, Dr Sachchidanand, 368
Siraj-ud-daula, 64
social reform, 260–7
Southborough Committee (1919), 207
Srinivasa Sastri, V. S., 207, 308
Stalin, Joseph, 351
Stanley, Lord, endorses Collector-Magistrate system, 225; Revenue Dispatch of, 256
States Department (India), 375
Stephen, Sir James, 40
Stephen, Fitzjames, 29
sterling balances, 355
Strachey, Sir John, i
strikes, 284–5, 286
Subsidiary Alliance, x
Suez Canal, 38

Index

Supreme Council 23–4
Supreme Court (Fort William) of judicature, 20–1, 23–4, 31; revenue agents excluded from jurisdiction of, 26; empowered to frame its own rules, 25–6; conflict with Supreme Councils and *adalats*, 23–4; represented in Executive Council of India, 34; source of authority of, 24
Supreme Courts, Madras and Bombay, 27
Suhrawardy, H. S., 360, 366; campaigns for an independent Bengal, 373
surveys, 50–7, 66–7, 68, 77–80, 196; by des Granges of Manipur, 141; geological, 57; in Afghanistan and Persia, 73; Indians trained in techniques of, 74; in foreign territory, 74–5, 78; in Santhal areas, 227–8; land, collectors entrusted with, 230–1; marine, 76–7; of Assam frontier, 143; of Brahamaputra Valley, 138–9; of Ladakh and Central Asia, 108–9; revenue, 55–7, 265; statistical, 54–5; topographical, 52–4, 56
Survey of India, 77–8
Sutherland, Colonel, 19
Swadeshi Movement, 168
Swaraj Party, 290, 295, 297
Swaraj Sabha, 289
Syed Ahmed Khan, Sir, 273, 320
Sydenham, Captain, on the Pindaris, 65
Sylhet, 373

Tagores, 190
Tara Singh, Master, 352
Tashi Lama, 79, 85
Tata, Sir Dorab, 285
Tawang, 102–4, 135–6
tea, 161, 163, 181
Temple, Sir Richard, 84
territorial reorganization, 116–82; along land frontiers, 179; motivation for, 116, 155
Thakurdas, Sir Purshotam, 307
Thibaw, King of Burma, 114–15
Tibet, 78–112; and McMahon Line, 104; Anglo-Chinese Convention concerning (1890), 93; attacks Sikkim, 91; attitude to Britain, 90–1; boundary and trade disputes with, 88–100; Britain's limited interest in, 98; British missions to, 79–80, 90, 91; British ultimatum to (1887) and operations against (1888), 91–2; Chinese forces enter, 100, 102, 110, 149; defies Chinese authority, 90–1, 95–6; direct relations with British, 100–1; gains full internal autonomy, 102; Hastings' policy towards, 79–80; Indo–, Trade Regulations, 95, 102; 'inner' and 'outer', 104–6, 111; western, and Indian boundary, 106–12; Joint Anglo-Chinese Commission on trade of, 94–5; marts in, for trade with India, (Yatung) 94, 95, (Gartok and Gyantse) 97; opposition to free trade and boundary demarcation, 95–6; status under Anglo-Russian Treaty (1907), 99; Younghusband Mission to and Lhasa Convention, 97–8; Zorawar Singh's invasion of, 107; *see also* Anglo-Tibetan Agreement; Simla Conference (1913)
Tilak, Bal Gangadhar, 264–5, 283, 288
Tilsit, Treaty of (1807), 69
Times, The (London), 356; on desirability of annexing Indian territories, 17; report on Leftist presence at Lahore session (1929) of Congress, 301
Tipu Sultan, vii, x–xi
trade, 259–60
Trade Union Congress, All-India, 285
tradition and change, 11–12
Trevelyan, Charles, on education, 263; on need for legislative council for each presidency, 38–9
tribal areas, N.-eastern, defence of, 149; tea-gardens in, 149–50
tribes, 224–7
tribal area (N. West), administration of, 163–5; British troops withdrawn from, 132; *jirgas* in, 121, 133; militia raised in, 132–3; post-war turbulence in (1919), 252
tribes, N. W. Frontier, 118; agreements with, 120; anti-British turbulence

among (1897–8), 128–9, (post World War I), 252; disunity among, 125; hostilities against, 120, 128–9, 131; react to Turkey's entry into World War I, 134; territory of, 119, (dual authority in), 131, (British 'forward policy' in), 124–5, 128–9
tribal policy, British, 148–153, 224
tribes, north-eastern: Abors, 137, 148, 180; Akas, 136–7, 148; as auxiliary armed forces, 146; Bhutias, 135; Daflas, 137, 148; British relations with, 135; Garo, 135, 142; Jaintas, 142; Khamtis, 137–8, 146; Khasis, 143, 153; Kookies, 141, 151, 152; Lushais, 141, 152–3, 155; Mirs, 24; Mishmis, 137; Mooamarias (Morans), 138, 146; Nagas, 139–41, 151, (Angami), 141, 150–1, 180; Singphos, 137–8, 146
Turkey, 285–6, 287, 290
Turner, Capt. Samuel, mission to Tibet, 79

Unionist Party, 296, 317, 360
United Provinces of Agra and Oudh, 177–9; Congress forms ministry in (1946), 360; Congress refuses to form coalition in (1937), 316, 317, 318
urban areas, 257–9

Vellore Mutiny, 261–2
Victoria, Queen, Proclamation of (1858), 20, 30, 241, 262; follows Canning's line, 37

Wade, Sir Thomas, 90
Warburton, Robert, 124
Wardha Scheme, 321
Wavell, Lord, 352, 355, 371; announces elections, 357; broadcast, 357–8; consultations in London, 357; Plan, 351–4; post-election estimate of League and Congress, 361; replaced, 370; sets up Interim Govt., 366–7; surrenders to Jinnah, 353–4
Waziristan operations, 127, 128–9

Weber, Max, on caste system, 261
Welby Commission (Royal Commission on Expenditure), 185
Wellesley, Arthur (Duke of Wellington), 64, 245
Wellesley, Marquis, vi, x, xv, 15, 64, 256; builds central secretariat, 46; urges subordination to Crown, 47–8
westernization, 12, 18, 31
White Paper (1933) on constitutional reforms, 207, 215, 324
Willingdon, Lord, 316
Wood, Sir Charles, 39, 49, 170, 188–9, 193, 256–7, 258, 260; Dispatch on education (1854), 259, 263
Workers' and Peasants' Party, 301; *see also* Communism
working class, 284–5
World War, First: effect on military history of India, 248–50; raised nationalist demands during, 251; uncovers deficiencies in equipment of army, 249–50; weakens imperial system, 265–6

Yakub Khan, Amir of Afghanistan, 72, 123
Yalta Conference, 351
Yatung, 94, 95, 103
Younghusband, Francis Edward, 88, 98; expedition to N. W. frontiers areas, 109–10; Mission to Lhasa, 97–8, 110
Yule, Andrew, 40
Yule, George, presides over Congress session (1888), 197

Zaheer, Syed Ali, 367
zamindars, 44; form group in legislatures, 193–4; in alliance with planters, 191, 225; oppose Tenancy Act, 225; oppression of tribals, 225–6
zamindari system, 222, 254–6
Zetland, Lord, 333; policy on States, 333–35
Zorawar Singh, 107